PROCEEDINGS OF THE ASIAN ACCELERATOR SCHOOL

Physics and Engineering of High-Performance Electron Storage Rings

and

Application of Superconducting Technology

Physics and Engineering of High-Performance Electron Storage Rings

and

Application of Superconducting Technology

Huairou and Beijing, China

22 November – 4 December 1999

Editors

Shin-ichi Kurokawa
Kenji Hosoyama
Yoko Hayashi

KEK, Tsukuba, Japan

Zhang Chuang
Guo Zhiyuan
Hou Rucheng

IHEP, Beijing, China

World Scientific
New Jersey • London • Singapore • Hong Kong

Published by

World Scientific Publishing Co. Pte. Ltd.

P O Box 128, Farrer Road, Singapore 912805

USA office: Suite 1B, 1060 Main Street, River Edge, NJ 07661

UK office: 57 Shelton Street, Covent Garden, London WC2H 9HE

British Library Cataloguing-in-Publication Data
A catalogue record for this book is available from the British Library.

ISBN 981-02-4716-8

Printed in Singapore by World Scientific Printers (S) Pte Ltd

AAS, THE ASIAN ACCELERATOR SCHOOL

Physics and Engineering of High-Performance Electron Storage Rings

Members of the Organizing Institutions

Japan:
Shin-ichi Kurokawa (KEK)
Ken Kikuchi (JSPS)
Yukihide Kamiya (Univ. of Tokyo)

China:
Xu Zhihong (CAS)
Zhang Chuang (IHEP)
Chen Shenyu (SSRC)

Local Executive Committee

Shin-ichi Kurokawa (KEK)
Guo Zhiyuan (IHEP)
Kenji Hosoyama (KEK)
Hou Rucheng (IHEP)

Editors of the Proceedings

Margaret Dienes, Guo Zhiyuan, Yoko Hayashi, Kenji Hosoyama,
Hou Rucheng, Shin-ichi Kurokawa, Zhang Chuang

Sponsors

Japan Society for the Promotion of Science (JSPS)
Chinese Academy of Sciences (CAS)
High Energy Accelerator Research Organization (KEK)
Institute of High Energy Physics (IHEP)

PREFACE

The first Asian Accelerator School (AAS) was held at Huairou and Beijing, China, November 22 – December 4, 1999, as one of the Asian Science Seminars, which is a program of the Japan Society for Promotion of Science (JSPS). The school was organized jointly by JSPS, the Chinese Academy of Sciences (CAS), the High Energy Accelerator Research Organization (KEK), and the Institute of High Energy Physics (IHEP). The main topic of the school was "Physics of Electron Storage Rings and Application of Superconducting Technology to These Machines". Forty students from 9 Asian countries participated in the school: 1 from Bangladesh, 20 from China, 4 from India, 1 from Indonesia, 15 from Japan, 2 from Korea, 1 from Nepal, 1 from the Philippines, and 1 from Vietnam.

The rapid development of accelerator sciences based on electron storage rings in Asia was one of the strongest incentives for us to organize the AAS. At present seven electron-positron colliders are operational in the world, and two of them are located in Asia: KEKB (the KEK B-Factory) at KEK, and BEPC at IHEP. It is also notable that one-third of the operating synchrotron light sources are Asian machines: Among them we find the largest light source machine in the world, SPring-8, in Harima, and one of the most active machines, KEK-PF, in Tsukuba. Beijing Synchrotron Radiation Facility, BSRF, and Hefei Light Source have been in operation for more than 10 years, and young light sources, 1.5-GeV Taiwan Light Source at SRRC in Hsinchu and 2.5-GeV Pohang Light Source of PAL in Pohang have been operating steadily and the numbers of users have been increasing. A 450-MeV light source, INDUS-I, has been commissioned at CAT in Indore. The Siam Photon Source is now being assembled in Nakhon Ratchasima. In Shanghai a 3.5-GeV light source is being planned at SSRF. Interest in constructing light sources are growing in other regions in Asia where no such machines exist at present.

To further improve the performance of electron storage rings, the use of superconducting magnets and cavities is of vital importance; therefore, the curriculum of the school was arranged not only to teach the basic physics of storage rings but also to give students a basic knowledge of superconducting technology. The first week of the school was held in a hotel at Huairou located 100 km northeast of Beijing, and the whole week was devoted to lectures on storage rings and superconducting technology. Then the site of the school was moved to IHEP in Beijing, where the students were given a one-week long hands-on training course on superconducting technology. In this course, we set up five teams: four teams for the training on superconducting magnets, and one team for training on superconducting cavities. In each of the first four groups, the students were asked to wind a superconducting solenoid, cool it down with liquid helium, and measure its performances. All four of the magnets they wound reached 7 Tesla magnetic field. Training of superconducting cavities was done by cooling down an L-band, single-cell superconducting cavity with liquid helium and measuring parameters, such as Q-value, at low temperature.

The success of the school owed much to the lecturers, who gave enthusiastic lectures in the first week, and to the trainers, who worked hard every day from early morning to late evening in the second week. Also we should not forget the selfless devotion of many people who worked behind the scenes to support the school. In this respect we especially thank Mr. Hou Rucheng, head of the foreign affairs office of IHEP.

Many lecturers of the school are not native speakers of English; Ms. Margaret Dienes checked their manuscripts in detail to make them much more readable. We owed much to her work.

I am happy to note that we organized the first AAS in the country where the international school was invented more than 2500 years ago (and also the school with hands-on training). Indeed the first and second sentences in Confucius' "Analects" refer to the international school:

子曰學而時學之不亦説乎
有朋自遠方來不亦樂乎

Confucius said, "Is it not a pleasure after all to practice in due time what one has learnt? Is it not a delight after all to have friends come from afar?"[*]

In organizing the AAS, I became fully confident of the correctness of these words of Confucius.

The enthusiasm of the students shown during the school had allowed me to catch a glimpse of our bright future, in the field of accelerators in Asia. Lastly, as my conclusion to the school, let me quote an excerpt from a famous poem of Mao Zedong, "Changsha (長沙)":

---

[*] Translation from "Analects of Confucius", Sinolingua, 1994.

恰同學少年
風華正茂
書生意氣
揮斥方遒

Young we were, schoolmates,
At life's full flowering;
Filled with students enthusiasm
Boldly we cast all restraints aside.[†]

Shin-ichi Kurokawa
Chairperson of AAS

[†]Translation from "Mao Zhedong Poems", Foreign Language Press, 1998

CONTENTS

TRANSVERSE MOTION OF CHARGED PARTICLES

QING QIN

Institute of High Energy Physics, Beijing, 100039, P.R. China
E-mail: qinq@mail.ihep.ac.cn

The transverse motion of charged particles in a synchrotron facility is our main subject, along with descriptions of basic magnets, solutions of Hill's equation, Liouville's theorem, elementary Hamiltonian treatment, etc.

1 Introduction

In this fundamental course on accelerators, we focus on the storage ring of a synchrotron. In such a machine, dipoles, or bending magnets, keep the revolution of the charged particle beams within a vacuum chamber. This confines them in a closed orbit. The strong focusing principle needs quadrupoles to make beams oscillate around the closed orbit, i.e., the betatron motion. Such an arrangement of dipoles and quadrupoles for a certain purpose is called the accelerator "lattice".

Charged particle beams are stored in high energy rings with transverse and longitudinal motions. The transverse particle motion can be divided into two parts: a closed orbit and a small-amplitude oscillation around the closed orbit.

In the following sections, we will introduce the basic knowledge of the transverse motion of charged beams in circular accelerators.

2 Coordinate System

The bending field produced by dipoles is usually vertically directed, which makes the charged particle follow a curved path in the horizontal plane. Fig 1 simply shows a curvilinear coordinate system for particle motion in a synchrotron.

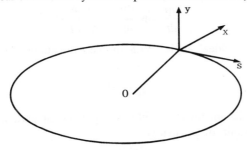

Figure 1 Curvilinear coordinate system for charged particles

Under such a coordinate system, the force acting on the charged particle in the horizontal direction is

$$\mathbf{F} = e\,\mathbf{v}\times\mathbf{B} \ , \tag{1}$$

where $\mathbf{v}$ is the velocity of the charged particle in the direction tangential to its path, $\mathbf{B}$ the magnetic guide field, and e the charge of the electron.

3 Displacement and Divergence

In the above curvilinear coordinate system, the transverse displacements for a charged particle in an accelerator are defined as x and y, in the horizontal and vertical planes respectively, and the divergence angles, x' and y', are

$$x' = dx/ds, \qquad y' = dy/ds \ . \tag{2}$$

Particles will leave the ideal orbit when divergences exist. In the meantime, quadrupole magnets will provide the restoring fields in the lattice to make particles oscillate about the ideal orbit.

4 Dipoles and Magnetic Rigidity

We assume a particle traveling perpendicularly to a guide field $\mathbf{B}$ with a relativistic momentum $\mathbf{p}$. Following a curved path of radius ρ and length ds after a duration dt, the particle's momentum becomes $p + dp$. From Eq. (1), we have

$$e\,\mathbf{v}\times\mathbf{B} = \frac{d\mathbf{p}}{dt} \ . \tag{3}$$

Since the magnitude of the force can be written as

$$e|\mathbf{v}\times\mathbf{B}| = e|\mathbf{B}|\frac{ds}{dt} \ , \tag{4}$$

where s is the longitudinal displacement and

$$\frac{d\mathbf{p}}{dt} = |\mathbf{p}|\frac{d\theta}{dt}\hat{s} = \frac{|\mathbf{p}|}{\rho}\frac{ds}{dt}\hat{s} \ , \tag{5}$$

in which θ is the azimuthal angle. With simple evaluations, we find the quality magnetic rigidity:

$$(B\rho) = \frac{p}{e} \ . \tag{6}$$

Conveniently, the magnetic rigidity is expressed in particle dynamics with electron-volts as the units of pc, that is

$$(B\rho)[\mathrm{T\cdot m}] = \frac{pc\,[\mathrm{eV}]}{c\,[\mathrm{m/s}]} = 3.3356\,(pc) \ . \tag{7}$$

5   Quadrupoles

Quadrupoles are used in modern accelerators for focusing beams. The poles of quadrupoles are truncated rectangular hyperbolae and alternate in polarity. Fig 2 shows the shape of quadrupole poles. Force analysis tells us that the strength increases linearly with the distance from an axis, while remaining zero on the axes.

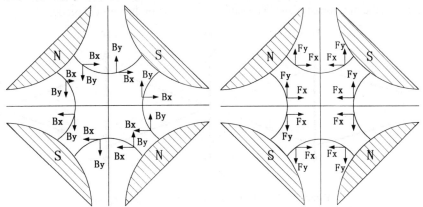

Figure 2 Components and force in a quadrupole. (Negatively charged particles enter the plane of the paper with paths perpendicular.)

In fig 2, the quadrupole would focus negative particles going into the plane of the paper, i.e., the horizontal plane. It is easy to see that, in the meantime, the negative particles are defocused in the vertical direction.

The strength of a quadrupole can be expressed by its gradient dB_y/dx normalized with respect to magnetic rigidity:

$$k = \frac{1}{(B\rho)} \frac{dB_y}{dx} \quad . \tag{8}$$

When a particle goes through a short quadrupole of length l and strength k, at a distance x the angular deflection is

$$\Delta x' = lkx \quad . \tag{9}$$

Like a focusing lens, a focusing quadrupole has the focal length f, so that

$$\Delta x' = -x / f \quad . \tag{10}$$

Thus,
$$f = -1/(kl) \quad . \tag{11}$$

6   Alternating Gradient Focusing

The most important milestone in modern accelerator development is the discovery of alternating gradient focusing. Under this theorem, quadrupoles can be used to focus beams in one plane though their defocusing property in another plane, obtaining a much stronger focusing system.

The principle of the alternating gradient system is depicted as an optical system in fig 3, where each lens is concave in one plane and convex in the other plane. If the spacing of the lenses is kept at a certain value, even with lenses of equal strength, it is still possible to find a ray always on axis at the defocusing lenses in the horizontal plane. The same case is true in the vertical plane.

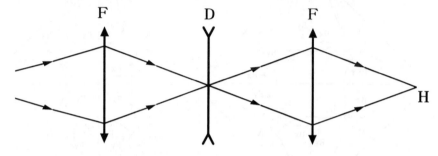

Figure 3 Alternating gradient system in horizontal plane.

For quadrupole lenses, the idea of alternating gradient works when the F lenses and D lenses are arranged along the beam line alternately. The particle trajectories in fig 4 show that the trajectories tend to be closer to the axis in D lenses than in F lenses.

The envelope of all such trajectories is described by the betatron function $\beta(s)$, which expresses the biggest excursion of a beam. With suitable strengths and spacing of lenses, $\beta(s)$ can be made periodic and kept large at all F quadrupoles and small at all D's. The same procedure can be done in the other plane.

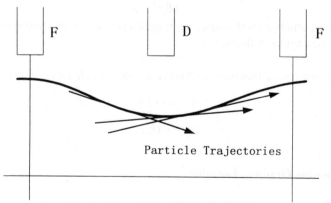

Figure 4   Particle trajectories and the envelope of betatron motion in a FODO lattice

7 Equation of Transverse Motion

Under the coordinate system shown in fig 1, which gives the unit vectors of $\hat{x}, \hat{y}, \hat{s}$, for **x, y, s** directions, we can derive the equation of transverse motion from the very beginning.

Suppose a charged particle moves along the ideal orbit in a storage ring. The radius of the orbit is $\rho(s)$ and its inverse, the curvature, is expressed as $G(s)$,

$$G(s) = \frac{1}{\rho(s)} \ . \tag{12}$$

The displacement from the origin to the charged particle, **r**, and the magnetic field, **B**, can be written as

$$\mathbf{r} = \mathbf{r_0} + x\hat{x} + y\hat{y} \tag{13}$$

and

$$\mathbf{B} = B_x\hat{x} + B_y\hat{y} + B_s\hat{s} \ . \tag{14}$$

From Eqs. (3) and (13) we have

$$\frac{d}{dt}(m\frac{d\mathbf{r}}{dt}) = e \cdot [\frac{d\mathbf{r}}{dt} \times \mathbf{B(r)}] \tag{15}$$

and

$$\frac{d\mathbf{r}}{d\theta} = \frac{d\mathbf{r}}{dt}\frac{dt}{d\theta} = \frac{e}{G_0}[(1+Gx)\hat{x} + G_0x'\,\hat{y} + G_0z'\,\hat{s}] \ . \tag{16}$$

Since $B_x = B_s = 0$, after some manipulations, we get

$$\begin{cases} \dfrac{d^2x}{ds^2} + (\dfrac{1}{\rho^2(s)} + \dfrac{1}{B\rho}\dfrac{\partial B_y}{\partial x})x = \dfrac{1}{\rho(s)}\cdot\dfrac{\Delta p}{p_0} \\[2mm] \dfrac{d^2y}{ds^2} - \dfrac{1}{B\rho}\dfrac{\partial B_y}{\partial x}y = 0 \ . \end{cases} \tag{17}$$

Here $\Delta p = p - p_0$ is the energy deviation of the charged particle. After defining

$$\begin{cases} K_x(s) = \dfrac{1}{\rho^2(s)} + \dfrac{1}{B\rho}\cdot\dfrac{\partial B_y}{\partial x} \\[2mm] K_y(s) = -\dfrac{1}{B\rho}\cdot\dfrac{\partial B_y}{\partial x} \end{cases}, \tag{18}$$

we can finally get the equation of transverse motion as

$$\begin{cases} \dfrac{d^2x}{ds^2} + K_x(s)x = \dfrac{1}{\rho(s)}\cdot\dfrac{\Delta p}{p_0} \\[2mm] \dfrac{d^2y}{ds^2} + K_y(s)y = 0. \end{cases} \tag{19}$$

8 Solution of Hill's Equation

When $\Delta p/p_0 = 0$, Eq. (19) becomes Hill's equation,

$$\frac{d^2 z}{ds^2} + K(s)z = 0 \ , \tag{20}$$

where z stands for both x and y. $K(s)$ is periodic in the circumference, C, i.e.,

$$K(s) = K(s+C) . \tag{21}$$

Equation (20) is reminiscent of simple harmonic motion but with a restoring constant $K(s)$ which varies around the storage ring. The solution of Hill's equation can be assumed to have a form similar to the simple harmonic motion:

$$z(s) = a\omega(s)e^{\pm i\psi(s)} . \tag{22}$$

Here, a is a constant, but $\omega(s)$ is a function of s, different from the amplitude of simple harmonic oscillation. Note that the phase advance in Eq. (22) is also a function of s, not of time t.

Furthermore, we have

$$z'(s) = a[\omega'(s)e^{\pm i\psi(s)} \pm i\omega(s)\psi'(s)e^{\pm i\psi(s)}] . \tag{23}$$

and

$$z''(s) = a[\omega''(s) \pm 2i\omega'(s)\psi'(s) \pm i\omega(s)\psi''(s) \mp \omega(s)\psi'^2(s)]e^{\pm i\psi(s)} . \tag{24}$$

With some manipulations, we can get

$$\begin{cases} \omega'' + K(s)\omega \mp \omega\psi'^2 = 0 \\ 2\omega'\psi' + \omega\psi'' = 0 . \end{cases} \tag{25}$$

After defining $\mu = \psi'$, we have

$$\psi' = \mu = \frac{1}{\omega^2} \quad \text{or} \quad \psi = \int \frac{1}{\omega^2} ds . \tag{26}$$

Then, defining $\beta = \omega^2$, we finally get the famous formula

$$2\beta\beta'' - \beta'^2 + 4K\beta^2 - 4 = 0 . \tag{27}$$

Now we have to define the so-called Twiss parameters, $\alpha(s)$, $\beta(s)$ and $\gamma(s)$ as

$$\begin{cases} \alpha(s) = -\omega(s)\omega'(s) = -\dfrac{\beta'(s)}{2} \\[2mm] \beta(s) = \omega^2(s) \\[2mm] \gamma(s) = \dfrac{1 + [\omega(s)\omega'(s)]^2}{\omega^2(s)} = \dfrac{1 + \alpha^2(s)}{\beta(s)} \end{cases} \tag{28}$$

so that the solution of Hill's equation can be written as

$$\begin{cases} z(s) = a\sqrt{\beta(s)}\cos[\psi(s)+\psi_0] \\ z'(s) = -a\sqrt{\beta(s)}\alpha(s)\cos[\psi(s)+\psi_0] - \dfrac{a}{\sqrt{\beta(s)}}\sin[\psi(s)+\psi_0] \end{cases} \tag{29}$$

with ψ_0 the initial phase advance.

9 Matrix Description

The solution of Hill's equation can also be expressed with matrices, which represent the transportation of the beam in storage rings. From one point, s_1, to another point, s_2, the 2×2 transport matrix can be expressed as

$$\begin{pmatrix} z(s_2) \\ z'(s_2) \end{pmatrix} = \begin{pmatrix} a & b \\ c & d \end{pmatrix}\begin{pmatrix} z(s_1) \\ z'(s_1) \end{pmatrix} = M_{12}\begin{pmatrix} z(s_1) \\ z'(s_1) \end{pmatrix}. \tag{30}$$

As the displacement and divergence of a particle should have the form of Eq. (29), we first suppose $\psi_0 = 0$, then choose $\psi(s_1) = 0$ and $\psi(s_2) = \psi$. In this way, we obtain four simultaneous equations which can be solved for a, b, c, d in terms of ω_1, ω_2 and ψ. The result is the most general form of the transport matrix:

$$M_{12} = \begin{pmatrix} \dfrac{\omega_2}{\omega_1}\cos\psi - \omega_2\omega_1'\sin\psi, & \omega_1\omega_2\sin\psi \\[2mm] -\dfrac{1+\omega_1\omega_1'\omega_2\omega_2'}{\omega_1\omega_2}\sin\psi - (\dfrac{\omega_1}{\omega_2}-\dfrac{\omega_2}{\omega_1})\cos\psi, & \dfrac{\omega_1}{\omega_2}\cos\psi + \omega_1\omega_2'\sin\psi \end{pmatrix}. \tag{31}$$

To simplify the above matrix, we can restrict M_{12} to be between two identical points in successive turns or cells of a periodic structure. This causes $\omega_1 = \omega_2$, $\omega_1' = \omega_2'$ and ψ becomes the phase advance per cell, μ. Thus

$$M = \begin{pmatrix} \cos\mu - \omega\omega'\sin\mu, & \omega^2\sin\mu \\[2mm] -\dfrac{1+(\omega\omega')^2}{\omega^2}\sin\mu, & \cos\mu + \omega\omega'\sin\mu \end{pmatrix}. \tag{32}$$

Combining this with Eq. (28), the matrix now becomes even simpler — the Twiss matrix:

$$M = \begin{pmatrix} a & b \\ c & d \end{pmatrix} = \begin{pmatrix} \cos\mu + \alpha\sin\mu, & \beta\sin\mu \\ -\gamma\sin\mu, & \cos\mu - \alpha\sin\mu \end{pmatrix}. \tag{33}$$

Suppose

$$M(s) = \begin{pmatrix} m_{11} & m_{12} \\ m_{21} & m_{22} \end{pmatrix}, \tag{34}$$

the Twiss parameters can have other expressions

$$\begin{cases} \alpha = \dfrac{m_{11} - m_{22}}{2 \sin \mu}, \quad \beta = \dfrac{m_{12}}{\sin \mu}, \quad \gamma = -\dfrac{m_{21}}{\sin \mu} \\[2mm] \cos \mu = \dfrac{m_{11} + m_{22}}{2} \end{cases} \tag{35}$$

where μ is the phase advance per cell,

$$\mu = \oint \frac{ds}{\beta(s)} \ . \tag{36}$$

10 Stability of Transverse Motion

When particles move in a storage ring without any energy deviation, say, $\Delta p = 0$, Eq. (19) has periodic solutions, which are also the solutions for an ideal machine. Thus, the particles will go around the ring on an Equilibrium Orbit.

For a real machine, due to errors of K_x and K_y, and the energy deviation, particles with different energies have different closed orbits.

In the alternating gradient focusing lattice, particles would remain focused only if they were close to the axis when passing through defocusing quadrupoles. With the matrix description, the stability condition can be achieved when the product $\prod_{Nk}\{M(s)\}^{Nk}$ does not diverge after N periods, which consist of one turn, and k turns, which define the real stable life of beam. The conditions for this to be so are (1) automatically satisfied by a linear transport matrix, det $|M| = 1$, and (2) the matrix has real non-vanishing eigenvalues.

If (z, z') is depicted as the vector $\mathbf{Z}$, this implies

$$MZ = \lambda Z \tag{37}$$

and

$$\det|M - \lambda I| = 0 \ . \tag{38}$$

With the Eq. (33), which is the form of M for a period, we have

$$\cos \mu = \frac{a + d}{2} = \frac{1}{2} \operatorname{Tr} M \ . \tag{39}$$

Solving for λ, we can get the eigenvalues

$$\lambda = \cos \mu \pm i \sin \mu = e^{\pm i\mu} \tag{40}$$

so that the stability condition for real values of λ is obtained provided that

$$|\cos \mu| = \frac{|a + d|}{2} = \frac{1}{2}|\operatorname{Tr} M| < 1 \ . \tag{41}$$

Then we can easily see that M has the characteristics of an exponential:

$$M^K = (I \cos \mu + J \sin \mu)^K = I \cos K\mu + J \sin K\mu \ . \tag{42}$$

Equation (42) can be compared with

$$(e^{i\mu})^K = e^{iK\mu} \ . \tag{43}$$

Thus, M^K will remain stable only if μ is real.

By decomposing M into two matrices, one can have another demonstration of the stability condition:

$$M = \begin{pmatrix} \cos\mu + \alpha\sin\mu, & \beta\sin\mu \\ -\gamma\sin\mu, & \cos\mu - \alpha\sin\mu \end{pmatrix} = I\cos\mu + J\sin\mu \tag{44}$$

where I is the unit matrix and J is described by

$$J = \begin{pmatrix} \alpha, & \beta \\ -\gamma, & \alpha \end{pmatrix} . \tag{45}$$

11 Betatron Tunes and Envelope Functions

In the above section one of the most important parameters in a circular machine was shown to be the Betatron Tune, i.e., the betatron wave number. If, in a constant gradient machine, a particle with the largest amplitude in the beam, $a\beta^{1/2}$, starts off with phase ψ_0, after one turn its phase has increased by

$$\Delta\psi = \oint ds/\beta = 2\pi R/\beta \tag{46}$$

where R is the radius of the ring. The quantity v is defined as the number of betatron oscillation per turn:

$$v = \frac{\Delta\psi}{2\pi} = \frac{R}{\beta} \quad \text{or} \quad \beta = \frac{R}{v} \ . \tag{47}$$

In a normal machine, this is also approximately true. It is often written as

$$\bar{\beta} = R/v \ . \tag{48}$$

It is important that v must not be a simple integer or vulgar fraction. Otherwise, particles will be in the resonance of $nv = p$ (n, p are integers) and lost with this dangerous condition.

In a regular storage ring, the betatron tunes, v_x and v_y in the horizontal and vertical planes respectively, are also called the working points.

Another important parameter in machine design and commissioning is the β function, which describes the envelopes of betatron oscillations made by particles and is therefore also called the Envelope Function.

In the equation $2\beta\beta'' - \beta'^2 + 4K\beta^2 - 4 = 0$, if $K(s) = 0$, the solution will be

$$\beta(s) = \beta_0 + \frac{(s - s_0)^2}{\beta_0} \ , \tag{49}$$

where β_0 and s_0 are the initial values with $K(s) = 0$. For example, at the interaction point of a collider, $K(s) = 0$, we can calculate the beta function of the location of the first quadrupole from Eq. (49). The β function and other lattice functions define the size of the aperture the stored beam needs, while the tunes of a machine determine the stability.

The Twiss functions, α, β, γ, and the tune, μ, are local and apply to the point chosen in the period as a starting and finishing point. Each individual element, such as drift space, dipole, quadrupole, sextupole in the ring has its own transport matrix. One can first choose the starting point, s_0, where the Twiss parameters are assumed. Starting from that point and multiplying the element matrices for one turn, one can find a, b, c and d in the matrices numerically and then apply the Eq. (33) to find the Twiss matrix.

12 Transport Matrices for Individual Components in a Ring

Drift spaces, dipoles, and quatrupoles comprise most of the elements of a storage ring. The simplest among these is the drift space. The divergence of the trajectory and the angle of the ray have the following relation:

$$\theta = \arctan^{-1}(x') . \tag{50}$$

It is easy to see that in phase space, a horizontal drift length is a translation from (x, x') to $(x+lx', x')$. Thus the matrix for a drift space can be written as

$$\begin{pmatrix} x_2 \\ x'_2 \end{pmatrix} = \begin{pmatrix} 1 & l \\ 0 & 1 \end{pmatrix} \begin{pmatrix} x_1 \\ x'_1 \end{pmatrix} . \tag{51}$$

A thin quadrupole, with a very small length but a finite integrated gradient of

$$lk = \frac{l}{B\rho} \cdot \frac{\partial B_y}{\partial x} , \tag{52}$$

could be compared with a focusing lens. A ray diverging from the focal point passes through the lens with a displacement x, becoming parallel by a deflection

$$\theta = \frac{1}{f} \cdot x . \tag{53}$$

Thus, the behavior of the ray could be described by a simple matrix:

$$\begin{pmatrix} x_2 \\ x'_2 \end{pmatrix} = \begin{pmatrix} 1, & 0 \\ -1/f, & 1 \end{pmatrix} \begin{pmatrix} x_1 \\ x'_1 \end{pmatrix} . \tag{54}$$

For a quadrupole, moving a displacement x, a particle sees an integrated field:

$$\Delta(Bl) = l \cdot \frac{\partial B_y}{\partial x} \cdot x . \tag{55}$$

Then the small deflection θ is

$$\theta = \frac{\Delta(Bl)}{(B\rho)} = \frac{l}{(B\rho)} \cdot \frac{\partial B_y x}{\partial x} = lkx . \tag{56}$$

One can see at once that $lk = 1/f$. Thus the matrix of a thin quadrupole is

$$\begin{pmatrix} x_2 \\ x_2' \end{pmatrix} = \begin{pmatrix} 1, & 0 \\ -kl, & 1 \end{pmatrix} \begin{pmatrix} x_1 \\ x_1' \end{pmatrix}. \tag{57}$$

Normally, the quadrupoles in a synchrotron are not short compared to their focal length. Then the matrices for long quadrupoles are

$$M_F = \begin{pmatrix} \cos(l\sqrt{k}), & \dfrac{1}{\sqrt{k}}\sin(l\sqrt{k}) \\ -\sqrt{k}\sin(l\sqrt{k}), & \cos(l\sqrt{k}) \end{pmatrix} \tag{58}$$

and

$$M_D = \begin{pmatrix} \cosh(l\sqrt{k}), & \dfrac{1}{\sqrt{k}}\sinh(l\sqrt{k}) \\ -\sqrt{k}\sinh(l\sqrt{k}), & \cosh(l\sqrt{k}) \end{pmatrix}. \tag{59}$$

The dipole magnets, which bend the beam in the accelerator, can be thought of as drift spaces as a first approximation. But in a real bending magnet, one must consider the focusing effect of their ends. In a pure sector dipole, the particles that pass at a displacement x away from the center of curvature will have longer trajectories in the magnets. This is just like a quadrupole, focusing horizontally but not vertically. The matrices for a sector dipole are

$$M_x = \begin{pmatrix} \cos\theta, & \rho\sin\theta \\ -\dfrac{1}{\rho}\sin\theta, & \cos\theta \end{pmatrix}, \qquad M_y = \begin{pmatrix} 1, & \rho\theta \\ 0, & 1 \end{pmatrix}. \tag{60}$$

Most dipoles are not sector magnets, but still have parallel end faces because of easier lamination in manufacturing.

13 Regular FODO Lattice

Arranging the quadrupoles as shown in fig 5, we get a regular FODO lattice. The matrix for one period between mid-planes of focusing quadrupoles is

$$\begin{aligned} M &= \begin{pmatrix} 1 & 0 \\ \mp 1/(2f) & 1 \end{pmatrix}\begin{pmatrix} 1 & L \\ 0 & 1 \end{pmatrix}\begin{pmatrix} 1 & 0 \\ \pm 1/f & 1 \end{pmatrix}\begin{pmatrix} 1 & L \\ 0 & 1 \end{pmatrix}\begin{pmatrix} 1 & 0 \\ \mp 1/(2f) & 1 \end{pmatrix} \\ &= \begin{pmatrix} 1 - L^2/2f^2 & 2L(1\pm L/2f) \\ -L/2f^2(1\mp L/2f) & 1 - L^2/2f^2 \end{pmatrix} \\ &= \begin{pmatrix} \cos\mu + \alpha\sin\mu & \beta\sin\mu \\ -\gamma\sin\mu & \cos\mu - \alpha\sin\mu \end{pmatrix}. \end{aligned} \tag{61}$$

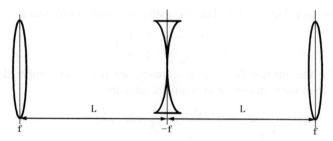

Figure 5   Regular FODO cell composed of thin lenses

Multiplying out the product of M and using Eq. (33), we obtain

$$\begin{cases} \cos \mu = 1 - \dfrac{L^2}{2f^2} \\ \alpha_{x,y} = 0 \\ \beta = 2L \dfrac{1 \pm \sin(\mu/2)}{\sin \mu} \end{cases} \qquad (62)$$

where the plus sign indicates the matrix between mid-planes of focusing quadrupoles and the minus sign that of defocusing quadrupoles. The α, which means the slope of the beta function, is zero at the planes of symmetry. The ratio of the maximum and minimum β functions in one plane will be

$$\frac{\hat{\beta}}{\bar{\beta}} = \frac{1 + \sin(\mu/2)}{1 - \sin(\mu/2)} \ . \qquad (63)$$

If, more generally, we solve in the horizontal plane the matrix product for an F lens with a focal length f_1 and a D lens with a focal length f_2, we can draw a graph for the stable area of f_1 and f_2 when $\sin \mu < 1$ (plotted in fig 6).

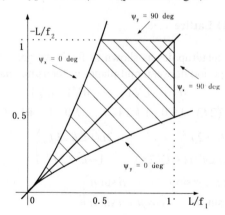

Figure 6   Stability diagram (cross hatched) for a FODO lattice. ψ is the phase advance per cell

14 Liouville's Theorem and Emittance

Liouville's theorem is a conservation law of phase space obeyed by particle dynamics. It is stated as (also depicted as fig 7):

"In the vicinity of a particle, the particle density in phase space is constant if the particles move in an external magnetic field or in a general field in which the forces do not depend upon velocity."

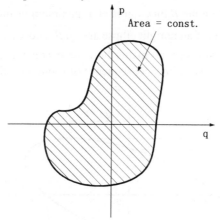

Figure 7   Illustration of Liouville's theorem

As a direct consequence of Liouville's Theorem, it is shown that, though the beam's cross section may have different shapes over the ring, the area of its phase space will not change (fig 8).

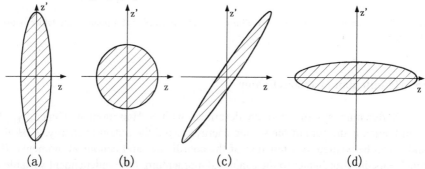

Figure 8 Development of a constant emittance beam in phase space. (a) narrow waist; (b) broad waist or maximum in beta; (c) diverging beam; (d) broad maximum at the center of an F lens

In regions where the β function is not a maximum or minimum, the ellipse of the beam phase space diagram will be tilted, as in fig 8 (c). In fig 9, we see that the sizes of the ellipse are related to the Twiss parameters. From Eq. (29), we can get the equation of the ellipse,

$$\beta z'^2 + 2\alpha z z' + \gamma z^2 = a^2 = \varepsilon ,\tag{64}$$

which is often called the Courant-Snyder invariant.

Usually, a beam of particles can be represented as a cloud of points within a closed contour, normally an ellipse in the phase space diagram. The area within the ellipse is proportional to the "emittance" of the beam. When energy is fixed, this area can be expressed as $\varepsilon = \int z' dz$ in units of $\pi \cdot \text{mm} \cdot \text{mrad}$. Figure 9 shows the contour at the place where the β function is at a maximum or minimum and where the major and minor axes of an upright ellipse are $\sqrt{\varepsilon \beta}$ and $\sqrt{\varepsilon / \beta}$.

The emittance is conserved, no matter how the bending and focusing magnets operate on the beam. This is a direct consequence of Liouville's theorem.

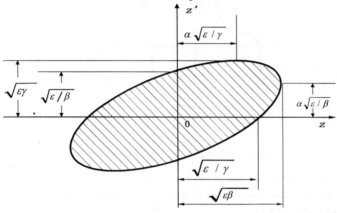

Figure 9   Parameters of a phase-space ellipse with an emittance ε at a point in the lattice between quadrupoles.

15   Hamiltonian in an accelerator

A dynamic system is often described with a Hamiltonian. Generally, the Hamiltonian is the sum of the kinetic energy, T, and the potential energy, V. Both T and V can be written as a function of the coordinates and canonical momenta. If q represents the coordinates, p the canonical momentum, t the independent variable or time, and $H(q, p, t)$ the Hamiltonian, the equations of motions now are

$$\frac{dq_i}{dt} = \frac{\partial H}{\partial p_i}, \quad \frac{dp_i}{dt} = -\frac{\partial H}{\partial q_i} .\tag{65}$$

With the Hamilton method applied in beam dynamics, or a relativistic charged particle in an electromagnetic field, the Hamiltonian is given by

$$H = c[m^2c^2 + (\mathbf{p} - 2\mathbf{A})^2] + e\varphi \tag{66}$$

where m is the rest mass of a charged particle, $\mathbf{p}$ the momentum, c the velocity of light, e the electron charge, φ the electromagnetic potential, and $\mathbf{A}$ the magnetic vector potential.

In our curvilinear coordinate system, the Hamiltonian becomes

$$H = c[m^2c^2 + (p_s - eA_s)^2(1 + \frac{x}{\rho})^{-2} + (p_x - eA_x)^2 + (p_y - eA_y)^2]^{1/2} + e\varphi \tag{67}$$

where p_x and p_y are the projections of $\mathbf{p}$ in the x and y directions and

$$p_s = (\mathbf{p} \cdot \mathbf{k})(1 + \frac{x}{\rho}) . \tag{68}$$

Like these canonical momenta, the canonical vector potentials A_x, A_y and A_s are also defined analogously, especially

$$A_s = (\mathbf{A} \cdot \mathbf{k})(1 + \frac{x}{\rho}) . \tag{69}$$

As H and t are conjugate variables, a new Hamiltonian can be obtained by exchanging s and t as independent variables:

$$H = -eA_s \mp (1 + \frac{x}{p})[(\frac{p_0 + e\varphi}{c})^2 - m^2c^2 - (p_x - eA_x)^2 - (p_y - eA_y)^2]^{1/2} \tag{70}$$

where $p_0 = \sqrt{\mathbf{p}^2 - m^2c^4}$.

After some simplification and manipulation, we have (in the vertical plane)

$$H = -\frac{eA_s}{p} - (1 + p_y^2)^{1/2} \approx -\frac{eA_s}{p_0} + \frac{p_y^2}{2} = \frac{p_y^2}{2} + \frac{k(s)y^2}{2} , \tag{71}$$

where $k(s)$ is the normalized gradient in a quadrupole.

Applying Hamilton's equations

$$\frac{dy}{ds} = \frac{\partial H}{\partial p_y}, \qquad \frac{dp_y}{ds} = -\frac{\partial H}{\partial y} , \tag{72}$$

we get

$$y' = p_y , \qquad p_y' = -k(s)y . \tag{73}$$

Finally,

$$y'' + k(s)y = 0 , \tag{74}$$

which is the Hill's equation in the vertical plane.

References

1. M. Sands, SLAC-121, 1977.
2. E.J.N. Wilson, AIP Conference Proceedings 153, p.3, 1983.

IMPERFECTIONS, CHROMATICITY AND LINEAR COUPLING

QING QIN

Institute of High Energy Physics, Beijing, 100039, P.R. China
E-mail: qinq@mail.ihep.ac.cn

In a real machine, we have to deal with the errors due to the magnets' manufacture and alignment, which influence the beam performance in a storage ring. Energy differences among charged particles cause chromatic effects, which need to be corrected in the operation of a synchrotron. Extra apparatuses, such as detectors, bring in coupling in a collider and lead to luminosity degradation. In this paper, we discuss mainly the effects due to imperfections in the magnets in a storage ring, chromaticity and linear coupling.

1 Errors from Dipoles

In an ideal machine, the dipoles around the ring bend the beam exactly 2π in one turn. Such a beam path is called the central (or synchrotron) momentum closed orbit. But in reality, since even the best synchrotron magnets can not be manufactured to be absolutely identical and to have perfect fields, the central orbit of the beam is distorted by the imperfections of the bending magnets. Therefore, the ideal or design orbit is not possible, but an orbit with a small distortion is possible and is called the distorted orbit. Table 1 lists the main imperfections in a synchrotron that cause closed-orbit distortion.

Table 1. Main sources of closed-orbit distortion

Element	Kick source	RMS value	$\langle \Delta Bl/(B\rho)\rangle_{rms}$	Direction
Drift space ($l = d$)	Stray field	$\langle \Delta B_s \rangle$	$d\,\langle \Delta B_s \rangle/(B\rho)$	x, y
Dipole (angle $= \theta$)	Field error	$\langle \Delta B/B \rangle$	$\theta\,\langle \Delta B/B \rangle$	x
Dipole	Tilt	$\langle \Delta \rangle$	$\theta\langle \Delta \rangle$	y
Quadrupole (Kl)	Displacement	$\langle \Delta x, y \rangle$	$K\,l\langle \Delta x, y \rangle$	x, y

In addition, other machine imperfections such as survey errors, resolution and ripple of power supplies, movement or oscillation of the ground, etc., also influence the central beam orbit; all of these are assumed to be randomly distributed around the ring.

Next, we will find the orbit distortion with the transverse motion equation. Suppose the perturbation is $f(s)$, which can be added on the right side of Hill's equation as

$$\frac{d^2 z}{ds^2} + K(s)z = f(s) \ . \tag{1}$$

To get the orbit distortion, the variables (z, s) are first converted to (η, θ). In Hill's equation, Eq. (1), we assume

$$z = \sqrt{\beta}\,\eta(\theta) \tag{2}$$

so that Hill's equation can be rewritten as

$$\frac{d^2 z}{ds^2} + K(s)z = \frac{1}{v^2 \beta^{3/2}}[\frac{d^2\eta}{d\theta^2} + v^2\eta] = f(\theta) \ . \tag{3}$$

Suppose the solution of Eq. (3) has the form

$$\eta(\theta) = C_1 \cos v\theta + C_2 \sin v\theta \tag{4}$$

where C_1 and C_2 are functions of v. Then, with the method of Lagrange's variation of constants, we have

$$\begin{cases} \dfrac{dC_1}{d\theta}\cos v\theta + \dfrac{dC_2}{d\theta}\sin v\theta = 0 \\[2mm] -\dfrac{dC_1}{d\theta}\sin v\theta + \dfrac{dC_2}{d\theta}\cos v\theta = v\beta^{3/2} f(\theta) \ . \end{cases} \tag{5}$$

Solving this equation, we obtain

$$\begin{cases} \dfrac{dC_1}{d\theta} = -v\beta^{3/2} f(\theta)\sin v\theta \\[2mm] \dfrac{dC_2}{d\theta} = v\beta^{3/2} f(\theta)\cos v\theta \end{cases} \tag{6}$$

with

$$\begin{cases} C_1(\theta) = -\int v\beta^{3/2} f(\theta)\sin v\theta d\theta + A \\[2mm] C_2(\theta) = \int v\beta^{3/2} f(\theta)\cos v\theta d\theta + B \end{cases} \tag{7}$$

and

$$\begin{cases} \eta(\theta) = A\cos v\theta + B\sin v\theta + \int v\beta^{3/2}(\varphi) f(\varphi)\sin(v\theta - v\varphi)d\varphi \\[2mm] \dfrac{d\eta(\theta)}{d\theta} = -Av\sin v\theta + Bv\cos v\theta + \int v^2\beta^{3/2}(\varphi) f(\varphi)\cos(v\theta - v\varphi)d\varphi \ . \end{cases} \tag{8}$$

As the condition of periodicity leads to $\eta(\theta) = \eta(\theta+2\pi)$, $\eta'(\theta) = \eta'(\theta+2\pi)$, $\beta^{3/2}(\varphi) = \beta^{3/2}(\varphi+2\pi)$, and $f(\varphi) = f(\varphi+2\pi)$, A and B have the following form:

$$\begin{cases} A = \dfrac{1}{4\sin^2 \pi v}\int_0^{2\pi} v\beta^{3/2}(\varphi) f(\varphi)[\sin(2\pi v - v\varphi) + \sin v\varphi]d\varphi \\[2mm] B = \dfrac{1}{4\sin^2 \pi v}\int_0^{2\pi} v\beta^{3/2}(\varphi) f(\varphi)[\cos(2\pi v - v\varphi) - \cos v\varphi]d\varphi \ . \end{cases} \tag{9}$$

Thus, $\eta(\theta)$ and $z(\theta)$ will be

$$\begin{cases} \eta(\theta) = \dfrac{1}{2\sin\pi v}\displaystyle\int_0^{2\pi} v\beta^{3/2}(\varphi)f(\varphi)\cos(\pi v + v\theta - v\varphi)d\varphi \\[4mm] z(\theta) = \dfrac{\sqrt{\beta(\theta)}}{2\sin\pi v}\displaystyle\int_0^{2\pi} v\beta^{3/2}(\varphi)f(\varphi)\cos(\pi v + v\theta - v\varphi)d\varphi \ . \end{cases} \tag{10}$$

Replacing θ with s, we can finally get the orbit distortion as

$$z(s) = \frac{\sqrt{\beta(s)}}{2\sin\pi v}\oint\sqrt{\beta(\bar{s})}f(\bar{s})\cos[\psi(s)-\psi(\bar{s})-\pi v]d\bar{s} \tag{11}$$

where $\psi(s) = \int_0^s \dfrac{ds}{v\beta}$. From Eq. (11), we see that the tune v must avoid integers, i.e.,

$v \neq n,\ n \in J$.

2 Correction of Closed Orbit Distortion

As mentioned above, the closed-orbit distortions are due mainly to the dipole errors around a synchrotron. If there is a dipole kick ΔBl at s_1, with Eq. (11) the orbit distortion at s_2 can be calculated as

$$z(s_2) = \frac{1}{2\sin\pi v}\sqrt{\beta(s_1)\beta(s_2)}\cdot\frac{\Delta Bl}{B\rho}\cdot\cos[v\pi - |\psi(s_2)-\psi(s_1)|] \ . \tag{12}$$

For beams perform well, orbit correction is indispensable. In most cases, beam position monitors (BPMs) are used to measure the closed-orbit distortion, and dipole correctors are applied to correct this distortion around the storage ring. Suppose there are M BPMs along a ring, whose coordinates are $s_1, s_2, \ldots, s_m$, and the measured differences between the distorted orbit and the original orbit on each BPM are u_{01}, $u_{02}, \ldots, u_{0m}$. There are N correctors around the ring with strengths of $\theta_1, \theta_2, \ldots, \theta_n$, respectively. With these correctors, the orbit difference at each BPM is Δu_m, and the displacement of the orbit at s_m is $u_m = u_{0m} + \Delta u_m$ ($m=1, 2, \ldots, M$). If $u_m = \Delta u_m$, then $u_{0m} = 0$, i.e., the distortion orbit has been corrected to the design orbit. But normally, $u_{0m} \neq 0$ and we can only ask for

$$F = \sum_{m=1}^{M} u_m^2 = \sum_{m=1}^{M}(u_{0m} + \Delta u_m)^2 \tag{13}$$

to have a minimum, where the u_{0m} is calculated with Eq. (11).

Since the Δu_m are generated by N correctors with strengths θ_n, i.e., $\Delta u_m = \sum_{n=1}^{N} T_{mn}\theta_n$, where T_{mn} is an $m\times n$ matrix, $F = \sum_{m=1}^{M}(u_{0m} + \sum_{n=1}^{N} T_{mn}\theta_n)^2$ will have a minimum when $\dfrac{\partial F}{\partial \theta_n} = 0$. Then, we can get θ_n , and from $\theta_n = \dfrac{\Delta B_n}{B\rho}\Delta s$, the ΔB_n for correctors will be finally obtained.

Generally, the correctors used in operation are less than installed in a storage ring. Getting the same correction, the less correctors, the smaller their effect, and the better the beam performance. Such a kind of calculation and iteration of orbit distortion is called MICADO (Minimisation des Carres des Distoution d'Orbite) by a computer code, and is widely used on orbit correction in accelerators.

Another method of correcting orbit distortion is to apply beam bumps deliberately at one part of the circumference without affecting the orbit elsewhere. Figures 1 and 2 give the schematics of 3-bump and 4-bump corrections.

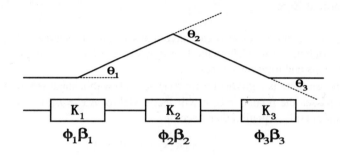

Figure 1   Schematic of 3-bump correction

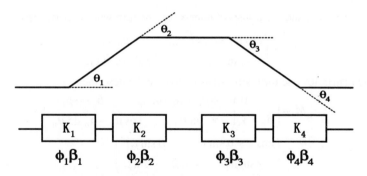

Figure 2   Schematic of 4-bump correction

With the orbit distortion equation, we can get the relations of 3 or 4 kickers' strengths (θ), phase advances (ϕ), and beta functions as follows:

$$\frac{\theta_1\sqrt{\beta_1}}{\sin(\phi_2 - \phi_3)} = \frac{\theta_2\sqrt{\beta_2}}{\sin(\phi_3 - \phi_1)} = \frac{\theta_3\sqrt{\beta_3}}{\sin(\phi_1 - \phi_2)}, \quad \text{(3-bump)} \qquad (14)$$

$$\begin{cases} \sqrt{\beta_1}\,\theta_1 \sin\phi_{41} + \sqrt{\beta_2}\,\theta_2 \sin\phi_{42} + \sqrt{\beta_3}\,\theta_3 \sin\phi_{43} = 0 \\ \sqrt{\beta_1}\,\theta_1 \sin\phi_{13} + \sqrt{\beta_2}\,\theta_2 \sin\phi_{23} + \sqrt{\beta_4}\,\theta_4 \sin\phi_{43} = 0 \ . \end{cases} \quad \text{(4-bump)} \quad (15)$$

The orbit correction around the whole ring can be found by summing the local bumps, which correct parts of the orbit in different places. The local bump method of orbit correction is often used in synchrotron light sources, e.g. at the front end, where the synchrotron light is extracted and the orbit must be corrected locally.

3   Gradient Errors

Like dipoles, quadrupoles also can have many errors in manufacturing, surveying, etc. The gradient errors change the focusing strength $K(s)$, causing the variation of beta functions and transverse tunes.

Suppose there is a gradient error $k(s)$ at $s = 0$ with a small length Δs. When an electron passes the point $s = 0$, it will feel a kick of $\Delta x' = k(s)\cdot \Delta s\cdot x$. With the perturbation, Hill's equation then becomes

$$\frac{d^2z}{ds^2} + [K(s) + k(s)]z = 0 \ . \tag{16}$$

If the matrix form of the perturbation is written as $\begin{pmatrix} 1 & 0 \\ \delta & 1 \end{pmatrix}$ where $\delta = |k(s)\Delta s| \ll 1$, and the transport matrix for one turn with no perturbation is

$$M_0 = \begin{pmatrix} \cos\mu_0 + \alpha_0 \sin\mu_0 & \beta_0 \sin\mu_0 \\ -\gamma_0 \sin\mu_0 & \cos\mu_0 - \alpha_0 \sin\mu_0 \end{pmatrix}, \tag{17}$$

then the matrix for one turn with perturbation can be described as

$$\begin{aligned} M &= \begin{pmatrix} 1 & 0 \\ \delta & 1 \end{pmatrix}\begin{pmatrix} \cos\mu_0 + \alpha_0 \sin\mu_0 & \beta_0 \sin\mu_0 \\ -\gamma_0 \sin\mu_0 & \cos\mu_0 - \alpha_0 \sin\mu_0 \end{pmatrix} \\ &= \begin{pmatrix} \cos\mu + \alpha \sin\mu & \beta \sin\mu \\ -\gamma \sin\mu & \cos\mu - \alpha \sin\mu \end{pmatrix}. \end{aligned} \tag{18}$$

Thus we can get

$$\frac{1}{2}\text{Tr}(M) = \cos\mu = \frac{1}{2}\beta_0\delta \sin\mu_0 + \cos\mu_0 \ . \tag{19}$$

As $\cos\mu = \cos(\mu_0 + \Delta\mu) \approx \cos\mu_0 - \Delta\mu \sin\mu_0$, and $\Delta\mu = 2\pi\Delta v$, we have

$$\Delta v = -\frac{1}{4\pi}\beta_0\delta = -\frac{1}{4\pi}\beta_0 k(s)\Delta s \ . \tag{20}$$

Integrating for one turn, the tune shift due to gradient errors can be expressed as

$$\Delta v = -\frac{1}{4\pi}\oint \beta(s)k(s)ds \ . \tag{21}$$

Next, we will consider the variation of beta function caused by the gradient error $k(s)$. If a particle starts from s_2, and after passing s_1, it returns to s_2 for a revolution, then the whole path, which the particle covers, can be seen as $s_2 \rightarrow s_1$ (B) and $s_1 \rightarrow s_2$ (A). If there is no error, the matrix is

$$M_0 = AB = \begin{pmatrix} a_{11} & a_{12} \\ a_{21} & a_{22} \end{pmatrix}\begin{pmatrix} b_{11} & b_{12} \\ b_{21} & b_{22} \end{pmatrix} = \begin{pmatrix} a_{11}b_{11} + a_{12}b_{21} & a_{11}b_{12} + a_{12}b_{22} \\ a_{21}b_{11} + a_{22}b_{21} & a_{21}b_{12} + a_{22}b_{22} \end{pmatrix}. \quad (22)$$

If there is a perturbation δ whose matrix is $M_\delta = \begin{pmatrix} 1 & 0 \\ \delta & 1 \end{pmatrix}$ with $\delta \ll 1$, then the

matrix for the whole ring will be

$$\begin{aligned} M &= \begin{pmatrix} m_{11} & m_{12} \\ m_{21} & m_{22} \end{pmatrix} = \begin{pmatrix} a_{11} & a_{12} \\ a_{21} & a_{22} \end{pmatrix}\begin{pmatrix} 1 & 0 \\ \delta & 1 \end{pmatrix}\begin{pmatrix} b_{11} & b_{12} \\ b_{21} & b_{22} \end{pmatrix} \\ &= \begin{pmatrix} a_{11}b_{11} + a_{12}b_{11}\delta + a_{12}b_{21} & a_{11}b_{12} + a_{12}b_{12}\delta + a_{12}b_{22} \\ a_{21}b_{11} + a_{22}b_{11}\delta + a_{22}b_{21} & a_{21}b_{12} + a_{22}b_{12}\delta + a_{22}b_{22} \end{pmatrix} \\ &= \begin{pmatrix} m_{110} + a_{12}b_{11}\delta & m_{120} + a_{12}b_{12}\delta \\ m_{210} + a_{22}b_{11}\delta & m_{220} + a_{22}b_{12}\delta \end{pmatrix}, \end{aligned} \quad (23)$$

if $M_0 = \begin{pmatrix} m_{110} & m_{120} \\ m_{210} & m_{220} \end{pmatrix}$ and $\begin{cases} m_{11} = m_{110} + \Delta m_{11}, & m_{12} = m_{120} + \Delta m_{12} \\ m_{21} = m_{210} + \Delta m_{21}, & m_{22} = m_{220} + \Delta m_{22} \end{cases}$.

Thus we get

$$\Delta m_{12} = m_{12} - m_{120} = a_{12}b_{12}\delta . \quad (24)$$

As the matrix from s_1 to s_2 without any perturbation can be described as

$$M = \begin{pmatrix} \sqrt{\beta_2/\beta_1}\,[\cos(\mu_2 - \mu_1) + \alpha_1 \sin(\mu_2 - \mu_1)] & \sqrt{\beta_1\beta_2}\,\sin(\mu_2 - \mu_1) \\ \dfrac{-1}{\sqrt{\beta_1\beta_2}}[(1 + \alpha_1\alpha_2)\sin(\mu_2 - \mu_1) + (\alpha_2 - \alpha_1)\cos(\mu_2 - \mu_1)] & \sqrt{\dfrac{\beta_1}{\beta_2}}[\cos(\mu_2 - \mu_1) - \alpha_2 \sin(\mu_2 - \mu_1)] \end{pmatrix}, \quad (25)$$

we can get

$$\begin{cases} a_{12} = \sqrt{\beta_1\beta_2}\,\sin(\mu_2 - \mu_1) \\ b_{12} = \sqrt{\beta_1\beta_2}\,\sin[\mu_0 - (\mu_2 - \mu_1)] \end{cases} \quad (26)$$

where $\mu_0 = 2\pi\nu$ is the phase advance for one revolution. Combining Eqs. (24) and (25), we have

$$\Delta m_{12} = \delta\beta_1\beta_2 \sin(\mu_2 - \mu_1)\sin[\mu_0 - (\mu_2 - \mu_1)] . \quad (27)$$

Since $m_{12} = \beta_2\sin\mu_0$, the differentiation of m_{12} will be

$$\Delta m_{12} = \Delta(\beta_2 \sin \mu_0) = \Delta\beta_2 \sin \mu_0 + \beta_2 \cos \mu_0 \cdot \Delta\mu_0 . \quad (28)$$

From the equation of tune shift due to gradient errors, Eq. (21), and equating Eq. (27) to Eq. (28), we get

$$\Delta\beta_2 = \frac{1}{2\sin 2\pi\nu} \cdot \delta\beta_1\beta_2 \cos[2\pi\nu - 2(\mu_2 - \mu_1)] . \quad (29)$$

Integrating $\Delta\beta$ along the whole ring, the final expression of the β function variation is

$$\Delta\beta(s) = \frac{\beta(s)}{2\sin 2\pi v} \oint \beta(\bar{s})k(\bar{s})\cos[2\pi v - 2(\mu(s) - \mu(\bar{s}))]d\bar{s} . \tag{30}$$

From Eq. (30), we know that the transverse tunes must avoid half integers, otherwise $\Delta\beta$ will approach infinity. This implies the resonance lines

$$\left.\begin{array}{ll} 2v_x = p, & 2v_y = p \\ v_x = v_y = p, & v_x + v_y = p \end{array}\right\} . \tag{31}$$

4   Working Diagram and Multipole Field

Up to now, we know that the perturbation due to dipoles is independent of the transverse displacement, and that the tunes should not cross the integer lines, which would cause resonances. The gradient errors, which have fields proportional to the displacement, will enhance the resonance when the tunes approach half integers. Thus we can deduce that the sextupole errors will excite third-integer resonances, and the transverse tunes should avoid 1/3 integers, which have the following forms:

$$\left.\begin{array}{ll} 3v_x = p, & 3v_y = p \\ 2v_x + v_y = p, & v_x + 2v_y = p \\ 2v_x - v_y = p, & v_x - 2v_y = p \end{array}\right\}, \quad p \in J . \tag{32}$$

We may further think that the fourth-order resonance is driven by octupoles, and so on. Then, a diagram with v_x and v_y as its axes can be drawn, on which the mesh of lines marks danger zones for the particles. Such a diagram is called a "working diagram" and the transverse tunes are "working points." Generally, the working points should satisfy

$$lv_x + mv_y \neq p \tag{33}$$

where l, m, p are integers. If Eq. (33) is not true, the resonance of $|l| + |m|$ order will be driven, while p is the azimuthal frequency. Figure 3 shows the working diagram for the BEPC storage ring up to fifth-order lines.

Summarizing the field errors, which drive the resonances, we can get the magnetic vector potential of a magnet with $2n$ poles in Cartesian coordinates as

$$A = \sum_n A_n f_n(x, y) \tag{34}$$

where f_n is a homogeneous function in x and y of order n. Table 2 lists $f_n(x, y)$ for

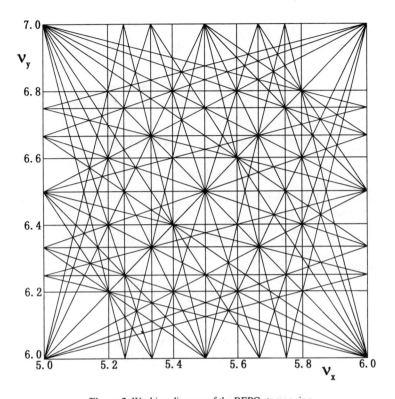

Figure 3 Working diagram of the BEPC storage ring

Low-order multipoles with both regular and skew terms. Figure 4 shows the difference between regular and skew multipoles.

Table 2 Solutions of magnetic vector potential in Cartesian coordinates.

Multipole	n	Regular f_n	Skew f_n
Quadrupole	2	$x^2 - y^2$	$2xy$
Sextupole	3	$x^3 - 3xy^2$	$3x^2y - y^3$
Octupole	4	$x^4 - 6x^2y^2 + y^4$	$4x^3y - 4xy^3$
Decapole	5	$x^5 - 10x^3y^2 + 5xy^4$	$5x^4y - 10x^2y^3 + y^5$
Dodecapole	6	$x^6 - 15x^4y^2 + 15x^2y^4 + y^6$	$6x^5y + 20x^3y^3 + 6xy^5$

The function $f_n(x, y)$ can be expanded in the form of

$$f_n(x, y) = (x + iy)^n \qquad (35)$$

where the real terms stand for regular multipoles, and the imaginary parts correspond to skew ones.

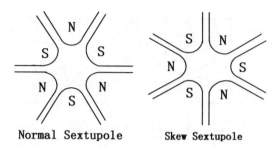

Figure 4 Normal and skew sextupoles

Furthermore, the scalar potential $\phi(r, \theta)$ in Cartesian and polar coordinates obeys the Laplace equation

$$\frac{\partial^2 \phi}{\partial x^2} + \frac{\partial^2 \phi}{\partial y^2} = 0 \quad \text{and} \quad \frac{1}{r}\frac{\partial}{\partial r}\left(r\frac{\partial \phi}{\partial r} \right) + \frac{1}{r^2}\frac{\partial^2 \phi}{\partial \theta^2} = 0 \tag{36}$$

which has the solution

$$\phi = \sum_{n=1}^{\infty} \phi_n r^n \sin n\theta \tag{37}$$

in polar coordinates.

As the fields in polar coordinates have the expressions

$$B_r = \frac{\partial \phi}{\partial r}, \qquad\qquad B_\theta = \frac{1}{r}\frac{\partial \phi}{\partial \theta} \tag{38}$$

or $\quad B_r = \phi_n n r^{n-1} \sin n\theta, \qquad B_\theta = \phi_n n r^{n-1} \cos n\theta$,

we can get the vertical field (when $y = 0$)

$$B_y = B_r \sin\theta + B_\theta \cos\theta$$

$$= \phi_n n r^{n-1}[\cos\theta \cos n\theta + \sin\theta \sin n\theta] \tag{39}$$

$$= \phi_n n r^{n-1} \cos(n-1)\theta = \phi_n n x^{n-1} .$$

The Taylor series of multipoles then can be expanded as

$$B_y = \phi_1 + \phi_2 \cdot 2x + \phi_3 \cdot 3x^2 + \phi_4 \cdot 4x^3 + \cdots$$

$$= B_0 + \frac{1}{1!}\frac{\partial B_y}{\partial x}x + \frac{1}{2!}\frac{\partial^2 B_y}{\partial x^2}x^2 + \frac{1}{3!}\frac{\partial^3 B_y}{\partial x^3}x^3 + \cdots . \tag{40}$$

The first term of Eq. (40) represents dipole, the second quadrupole, the third sextupole, the fourth octupole, etc. Some multipole field shapes are drawn in Fig 5.

In general, multipoles can drive resonances of lower order. For instance, octupoles drive fourth- and second-order, sextupoles third- and first-order, etc. The nonlinear resonances are those of third-order and above, driven by nonlinear multipoles, which can reduce the dynamic aperture severely, especially in a superconducting machine.

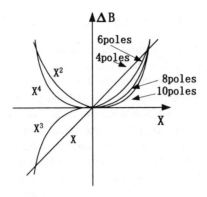

Figure 5 Multipole field shapes

5 Chromaticity

In an ideal machine, particles move along the ring with identical energy, as described in Hill's equation. But normally there are momentum differences among particles, which will cause tune spread for the beam. As the tune spread due to momentum is equivalent to the chromatic aberration in an optical lens, it is called "chromaticity," which is expressed as

$$\xi = \frac{\Delta v / v}{\Delta p / p} \quad \text{or} \quad \xi = \frac{\Delta v}{\Delta p / p} \ . \tag{41}$$

Other imperfections, such as the non-uniform guide field, the sextupole component in magnetic field, $dB_y/dt \neq 0$, which causes the ripple effect in the vacuum chamber, and the saturation of bending magnets, will also introduce chromaticity into the lattice. The chromaticity due to $\Delta p/p$ is called natural chromaticity.

When $\Delta p/p \neq 0$, Hill's equation becomes

$$z'' + Kz + \delta \, kz = 0 \tag{42}$$

where $K = \dfrac{1}{B\rho} \left(\dfrac{\partial B_y}{\partial x} \right)_0$ and $\delta \, k = \dfrac{\Delta p}{p} K$. Applying the equation derived above in

the section on gradient errors, the result of tune spread is

$$\Delta v = -\frac{1}{4\pi} \oint \beta(s) \delta k(s) ds = -\frac{1}{4\pi} \oint \beta(s) K(s) ds \cdot \frac{\Delta p}{p} \ . \tag{43}$$

Thus the natural chromaticity is

$$\xi = -\frac{1}{4\pi v}\oint \beta(s)K(s)ds \quad \text{or} \quad \xi = -\frac{1}{4\pi}\oint \beta(s)K(s)ds \ . \tag{44}$$

From the definition of natural chromaticity, Eq. (44), we find that the natural chromaticity remains negative and it comprises most of the chromaticity in a storage ring without any chromaticity correction. For the BEPC, the natural chromaticities in the horizontal and vertical planes are $\xi_x \sim -10$ and $\xi_y \sim -20$, respectively.

In some other cases, such as injection, the momentum deviation can be as high as $\pm 2\times10^{-3}$, which may cause a large tune spread and make it difficult to the beam inject. Another effect of negative chromaticity is head-tail instability, which will cause the beam loss. Thus the chromaticity must be corrected to be positive.

One way to correct the chromaticity is to introduce extra focusing which is stronger for high momentum orbits far from the central axes. Such a magnet is a sextupole. Sextupoles placed in the nonzero dispersion locations will introduce a focusing

$$\delta k = -\frac{B''D}{B\rho}\cdot\frac{\Delta p}{p} \tag{45}$$

where D is the dispersion function. This perturbation can cause a tune shift of

$$\Delta v = \frac{1}{4\pi}\oint \beta(s)\cdot\frac{B''(s)D(s)}{B\rho}ds\cdot\frac{\Delta p}{p} \ . \tag{46}$$

Thus, the chromaticity introduced by sextupoles will be

$$\xi = \frac{1}{4\pi v}\oint \beta(s)\cdot\frac{B''(s)D(s)}{B\rho}ds \ . \tag{47}$$

We can simply correct chromaticity with Eq. (47) to make it balance the natural chromaticity Eq. (44) and other chromaticities.

Sextupoles used for chromaticity correction are called chromatic sextupoles. We generally need two sets of sextupoles to correct horizontal and vertical chromaticities. One is near focusing quadrupoles where β_x is large, to correct the horizontal chromaticity, while the other is near defocusing quadrupoles where β_y is large to correct the vertical chromaticity.

The introduction of sextupoles will cause some effects on stored beams, such as damping some instabilities, generating linear coupling, exciting third order resonance, bringing about non-linearity, etc.

6   Linear Coupling

Coupling, i.e. periodic energy exchange between two oscillators, exists widely in physics. Figure 6 shows an amplitude-coupled pendulum system, in which a pair of pendulums is linked by a weightless spring.

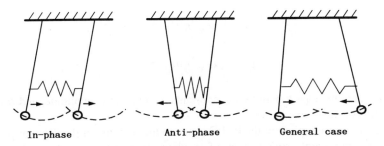

Figure 6 Example of coupling: amplitude-coupled pendulums

In a real storage ring, the betatron motions are also coupled. The sources of coupling are mainly skew quadrupole fields and solenoids. The skew quadrupole fields arise from quadrupole rolls, fringe field of the Lambertson magnet for injection, vertical closed-orbit in sextupoles, horizontal closed-orbit in skew quadrupoles, and feed-downs from higher-order multipoles. Solenoids are used mainly in the high energy detector at the interaction point of a collider, and in electron cooling storage rings.

In the operation of synchrotrons, linear betatron coupling can have both good and bad effects. On one hand, the dilation of the vertical emittance in electron rings will increase the Touschek lifetime limitation, but on the other hand, the dynamic aperture of particle motion may be reduced.

In the following section, we will briefly introduce linear coupling combined with coupling compensation in the Beijing Electron Positron Collider (BEPC).

Generally, when linear coupling exists, the transverse motion equation for particles in a storage ring can be approximated to first order as

$$\begin{cases} x'' + K_x x = -(K + \dfrac{M'}{2})y - My' \\ y'' + K_y y = -(K - \dfrac{M'}{2})x + Mx' \end{cases} \tag{48}$$

where $K_x = \dfrac{1}{\rho^2} + \dfrac{1}{B\rho}\left(\dfrac{\partial B_y}{\partial x}\right)_0$, $K_y = -\dfrac{1}{B\rho}\left(\dfrac{\partial B_x}{\partial y}\right)_0$, $K = -\dfrac{1}{2(B\rho)}\left(\dfrac{\partial B_x}{\partial x} - \dfrac{\partial B_y}{\partial y}\right)_0$,

and $M = -\dfrac{1}{B\rho}B_s$ with B_s the solenoid field and $B\rho$ the rigidity of the magnet. If we apply the method of Lagrange's variation of constants, the solutions will be

$$\begin{cases} x = \left(\int_0^s \dfrac{i}{2}F_x ds + a_{x0}\right)\beta_x^{1/2}e^{i\phi_x} + cc. \\ y = \left(\int_0^s \dfrac{i}{2}F_y ds + a_{y0}\right)\beta_y^{1/2}e^{i\phi_y} + cc. \end{cases} \tag{49}$$

where

$$\left\{ \begin{array}{l} F_x = \left(Ky - \dfrac{1}{2} M'y \right) \beta_x^{1/2} e^{-i\phi_x} - My(\beta_x^{1/2} e^{-i\phi_x})' \\[3mm] F_y = \left(Kx + \dfrac{1}{2} M'x \right) \beta_y^{1/2} e^{-i\phi_y} + Mx(\beta_y^{1/2} e^{-i\phi_y})' \end{array} \right. \tag{50}$$

and $\beta_{x,y}$ are envelope functions, $a_{x,y}$ constants of transverse amplitude, and $\phi_{x,y}$ phase advances of transverse motion.

The linear coupling of the BEPC is due to the solenoid of the detector placed at the interaction point. Other sources of coupling are minor and can be neglected. The method of compensating the linear coupling is to install skew quadrupoles at appropriate places around the ring. With the extra coupling provided by the skew quadrupoles, we can compensate the effect of the solenoid field.

From Eq. (49), we know that if $\int_0^s \dfrac{i}{2} F_{x,y} ds = 0$, the linear coupling can be compensated, and Eq. (49) becomes the solution of transverse motion without any coupling. When $F_{x,y} \neq 0$, it is impossible to satisfy the above condition on any point around the ring. But we can realize the compensation globally, that is, the effects of elements can cancel each other with no amplitude growth. In this way, we can conclude that if there are n solenoids and N pairs of skew quadrupoles, the equations for coupling compensation can be deduced as

$$\left\{ \begin{array}{l} \displaystyle\sum_{j=1}^{N} \left[Kl\sqrt{\beta_x \beta_y} \, \sin(\phi_x - \phi_y) \right]_{ji} - \left[\dfrac{Ml}{4} \left(\sqrt{\dfrac{\beta_y}{\beta_x}} + \sqrt{\dfrac{\beta_x}{\beta_y}} \right) \right]_i = 0 \\[5mm] \displaystyle\sum_{j=1}^{N} \left[Kl\sqrt{\beta_x \beta_y} \, \cos(\phi_x - \phi_y) \right]_{ji} = 0 \\[5mm] \displaystyle\sum_{j=1}^{N} \left[Kl\sqrt{\beta_x \beta_y} \, \sin(\phi_x + \phi_y) \right]_{ji} - \left[\dfrac{Ml}{4} \left(\sqrt{\dfrac{\beta_y}{\beta_x}} - \sqrt{\dfrac{\beta_x}{\beta_y}} \right) \right]_i = 0 \\[5mm] \displaystyle\sum_{j=1}^{N} \left[Kl\sqrt{\beta_x \beta_y} \, \cos(\phi_x + \phi_y) \right]_{ji} = 0 \end{array} \right. \tag{51}$$

$$I = 1, 2, 3, ..., m, \quad j = 1, 2, 3, ..., N ,$$

where M_i is the strength of the i-th solenoid and l_i its length, β_{xi}, β_{yi} are the beta functions in the mid-solenoid, ϕ_{xi}, ϕ_{yi} the phase advances, $\pm (Kl)_{ji}$ the strengths of the j-th pair of skew quadrupoles, β_{xji}, β_{yji} the beta functions of the skew quadrupoles, and $\phi_{xi} \pm \phi_{xji}$, $\phi_{zi} \pm \phi_{zji}$ the phase advances of skew quadrupoles. There is no any consideration of dispersion coupling and compensation.

With Eq. (51), we can choose the phase advance of the skew quadrupoles and get the corresponding strengths. In the light of BEPC's boundary conditions, we can find that when $|\phi_x - \phi_y| = \pi/2$, $\phi_x + \phi_y = 2K\pi + \pi/2$, $K \in J$, the second and fourth equations of Eq. (51) would be satisfied. By choosing proper strengths of skew quadrupoles, the first or third equation of Eq. (51) can be satisfied too, while the left equation is approximately satisfied. Thus, the linear coupling is compensated.

The calculation in the BEPC tells us that, installing skew quadrupoles in the injection region, or ± 64.507 m away from the IP, can cancel the effect of solenoid. The constraint to tunes caused by compensation is $|v_x-v_y| = 1$. There are three pairs of skew quadrupoles in the BEPC storage ring. Two pairs near the IP are used to compensate the coupling when beams are injected, and another pair located in the injection region is for compensation when beams collide.

The calculation of coupling shows that without skew quadrupoles the coupling reaches more than 20%, but with skew quadrupoles it drops to less than 1%. This result is consistent with the measured linear coupling of the BEPC.

By changing the strength of a certain quadrupole, and measuring the corresponding tunes, we can get the curves shown in Fig 7. The coupling can be found approximately from the following formula:

$$\kappa = \frac{(\Delta v)^2}{(v_x - v_y)^2 + (\Delta v)^2} \tag{52}$$

where v_x, v_y are the transverse tunes of the machine, and Δv the distance between the peak of the lower curve and the valley of the upper curve in Fig 7.

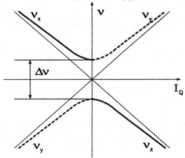

Figure 7 Method of tune splitting for coupling measurement

References

1. M. Sands, SLAC-121, 1977.
2. E.J.N. Wilson, AIP Conference Proceedings 153, p. 3, 1983.
3. S.X. Fang, N. Huang, and C. Zhang, 'The Compensation of solenoid effect on BEPC', IHEP internal report, 1990.
4. P.J. Bryant, CERN 94-01, p. 207, 1994.

LONGITUDINAL MOTION OF A PARTICLE IN A CIRCULAR ACCELERATOR

GUO ZHIYUAN

Institute of High Energy Physics, Chinese Academy of Sciences
Beijing, China
E-mail: guozy@mail.ihep.ac.cn

1 Basic equation of a particle in a circular accelerator

In the discussion of the transverse motion of a particle, the energy of the particle was assumed to be constant. In fact, it would change with synchrotron radiation or if the beam is accelerated. To keep the beam orbit constant at the ideal orbit, energy should be added from the RF accelerator system. The following equation needs to be satisfied in a circular accelerator:

$$\frac{d}{dt}(m\frac{d\mathbf{r}}{dt}) = \frac{e}{c}[\frac{d\mathbf{r}}{dt}\times\mathbf{B}(\mathbf{r},t)] + e\varepsilon(\mathbf{r},t),\tag{1}$$

where $\mathbf{B}$ is the magnetic field; $\varepsilon(\mathbf{r},t)$, the electric field; m, the mass of the particle; e, the charge of the particle; and c, the speed of light.

At the orbit coordinates, the energy E of the particle is

$$\frac{dE}{dt} = e\varepsilon(\mathbf{r},t)\cdot\frac{d\mathbf{s}}{dt}.\tag{2}$$

One can see from Eq. (2) that only the electric field contributes to the energy gain in the motion of a particle in an accelerator; there is no contribution from the magnetic field.

2 Synchrotron status

To keep the beam orbit constant in a circular accelerator, the particle should satisfy the following relation:

$$Pc = eB\rho,\tag{3}$$

where P is the momentum of the particle, and ρ is the curvature of the orbit. The physics meaning of Eq. (3) is that the energy, the speed or the revolution frequency, and the magnetic field must be synchronized in this relation to keep the orbit of particle motion constant in an accelerator. Eq. (3) can be also written as

$$\omega_s = \frac{-ec <B(t)>_s}{E_s(t)} , \qquad (4)$$

where the subscript s refers to the synchrotron condition, and $<B(t)>_s$ is the average magnetic field in a revolution. We call a particle revolving in an accelerator with frequency ω_s, as in Eq. (4), the synchrotron particle; and we call Eq. (4) the synchrotron condition.

Normally, a standard RF cavity is used in a circular accelerator to increase the energy of a particle. The frequency of the electric field in the RF cavity must be a multiple of the revolution frequency:

$$\omega_{rf} = h\omega_0 , \qquad (5)$$

where h is the harmonic number. The electric field seen on the RF cavity is

$$V_{rf} = V_0 \sin \omega_{rf} t , \qquad (6)$$

where V_0 is the amplitude of the electric field. The electric field seen by the particle is

$$E(Z,t) = E_1(t)E_2(Z) , \qquad (7)$$

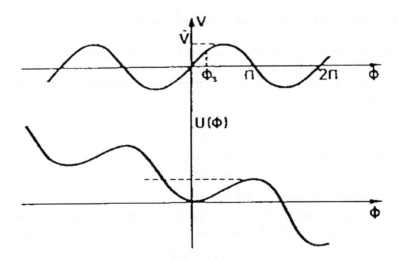

Figure 1: The electric field seen by a particle in time domain

In time domain, $E_1(t)$ can be expressed by following equation, as shown in Figure 1:

$$E_1(t) = E_0 \sin(\int_{t_0}^{t} \omega_{rf} \, dt + \phi_0) \,. \tag{8}$$

where ω_{rf} is a slowly varying function of t, or constant. In space domain, $E_2(Z)$ is as shown in Figure 2.

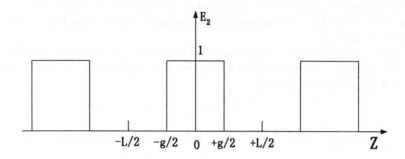

Figure 2: The electric field seen by a particle in space domain

The energy gain in one revolution is

$$\Delta E = eV \sin \phi_s \,. \tag{9}$$

3 Non-synchrotron particle and momentum compaction

It is obvious that only a particle that satisfies Eq. (4), the synchrotron particle, can be accelerated in a circular accelerator. But in fact, we can not use only one particle in the beam. For particles that deviate from the ideal synchrotron particle, the parameters are:

energy $\qquad\qquad E = E_s + \Delta E$,

momentum $\qquad\quad P = P_s + \Delta P$,

revolution frequency $\quad \omega = \omega_0 + \Delta\omega$,

RF phase at the cavity $\quad \phi = \phi_s + \Delta\phi$,

angle at particle coordinate $\quad \theta = \theta_s + \Delta\theta$,

where θ is the azumuthal angle of the particle in the accelerator, measured as positive in the direction of motion. Then we have

$$\Delta\phi = -h\Delta\theta \,. \tag{10}$$

That is, a particle that arrives at the gap of the RF cavity before the synchrotron particle, $\Delta t > 0$, $\Delta\phi > 0$, has the negative coordinate $\Delta\theta < 0$:

That is, a particle that arrives at the gap of the RF cavity before the synchrotron particle, $\Delta t > 0$, $\Delta\phi > 0$, has the negative coordinate $\Delta\theta < 0$:

$$\Delta\omega = \frac{d}{dt}(\Delta\theta) = -\frac{1}{h}\frac{d}{dt}(\Delta\phi) = -\frac{1}{h}\frac{d\phi}{dt} . \tag{11}$$

The dispersion in revolution frequency is defined as

$$\eta = -\frac{P}{\omega}\frac{d\omega}{dP} . \tag{12}$$

The momentum compaction factor is defined as

$$\alpha_p = \frac{P}{L}\frac{dL}{dP} = \frac{P}{R}\frac{dR}{dP} , \tag{13}$$

where P, L, and R are the particle momentum, the circumference of the accelerator, and the average radius ($L = 2\pi R$).

Since $\omega = c\beta/R$, we have

$$\frac{d\omega}{\omega} = \frac{d\beta}{\beta} - \frac{dR}{R} = (1-\beta^2)\frac{dP}{P} - \alpha_p\frac{dP}{P} , \tag{14}$$

$$\Delta P = \frac{1}{\beta^2}\frac{P}{E}\Delta E = \frac{1}{\omega R}\Delta E . \tag{15}$$

Then we have

$$\frac{d\omega}{\omega} = (\frac{1}{\gamma^2}-1)\frac{dP}{P} , \tag{16}$$

$$\eta = \alpha_p - \frac{1}{\gamma^2} . \tag{17}$$

For an electron machine, $1/\gamma^2 \ll \alpha_p$, we have $\eta = \alpha_p$. \hfill (18)

4   Longitudinal equation of particle motion

From Eq. (12)

$$\Delta P = -\frac{P}{\eta\omega}\Delta\omega = \frac{P}{\eta h\omega}\frac{d\phi}{dt} , \tag{19}$$

and on each revolution a particle gians energy as

$$[\Delta E]_1 = eV\sin\phi . \tag{20}$$

$$[\Delta P]_1 = \frac{1}{\omega R}[\Delta E]_1 = \frac{eV}{\omega R}\sin\phi . \tag{21}$$

Since the revolution period $T_0 = 2\pi/\omega$, the average rate of momentum gain over one revolution is

$$\dot{P} = \langle\frac{dP}{dt}\rangle = \frac{eV}{2\pi R}\sin\phi , \tag{22}$$

or

$$R\dot{P} = \frac{eV}{2\pi}\sin\phi_s . \tag{23}$$

On the other hand, for the synchrotron particle, we have

$$R_s\dot{P}_s = \frac{eV}{2\pi}\sin\phi_s . \tag{24}$$

Thus the difference in energy gain between the synchrotron particle and a non-synchrotron particle is

$$\Delta(R\dot{P}) = R\dot{P} - R_s\dot{P}_s = \frac{eV}{2\pi}(\sin\phi - \sin\phi_s) . \tag{25}$$

Expanding Eq. (25) to first order,

$$\Delta(R\dot{P}) = R_s\Delta\dot{P} + \dot{P}_s\Delta R + \cdots$$

$$= R_s\Delta\dot{P} + \dot{P}_s[(\frac{dR}{dP})_s\Delta P + \cdots]$$

$$= R_s\Delta\dot{P} + (\frac{dR}{dt})_s\Delta P$$

$$= R_s\Delta\dot{P} + \dot{R}_s\Delta P$$

$$= \frac{d}{dt}(R_s\Delta P)$$

$$= \frac{d}{dt}(\frac{\Delta E}{\omega_s}) .$$

That is,

$$\Delta(R\dot{P}) = \frac{d}{dt}(\frac{\Delta E}{\omega_s}) , \tag{26}$$

or
$$\frac{d}{dt}(\frac{\Delta E}{\omega_s}) = \frac{eV}{2\pi}(\sin\phi - \sin\phi_s) \,. \tag{27}$$

From Eq. (15) and (19),

$$\frac{\Delta E}{\omega_s} = R_s\Delta P = \frac{R_sP_s}{h\eta\omega_s}\frac{d\phi}{dt} \,. \tag{28}$$

Then we have

$$\frac{d}{dt}[\frac{R_sP_s}{h\eta\omega_s}\frac{d\phi}{dt}] - \frac{eV}{2\pi}[\sin\phi - \sin\phi_s] = 0 \,. \tag{29}$$

Eq. (29) is the longitudinal equation of particle motion in a synchrotron, in which we use the differential to replace the difference.

5 Small amplitude oscillation

Normally, in a circular accelerator the parameter R_s is constant, and during the acceleration of a particle, the parameters P_s, ω_s and η are slowly variant. Then Eq. (29) can be written as

$$\frac{R_sP_s}{h\eta\omega_s}\frac{d^2\phi}{dt^2} - \frac{eV}{2\pi}(\sin\phi - \sin\phi_s) = 0 \,, \tag{30}$$

or
$$\frac{d^2\phi}{dt^2} + \frac{\Omega_s^2}{\cos\phi_s}(\sin\phi - \sin\phi_s) = 0 \,, \tag{31}$$

where
$$\Omega_s^2 = \frac{-eVh\eta\omega_s\cos\phi_s}{2\pi R_sP_s} \,. \tag{32}$$

If a particle is executing longitudinal oscillation with small amplitude, $\Delta\phi = \phi - \phi_s \ll 1$, then $\sin\phi = \sin\phi_s + \Delta\phi\cos\phi_s$, and Eq. (31) can be written as

$$\frac{d^2\phi}{dt^2} + \Omega_s^2\Delta\phi = 0 \,. \tag{33}$$

Normally, during the acceleration of a particle, the parameter ϕ_s is constant or slowly variant. Then Eq. (33) can be written as

$$\frac{d^2\phi}{dt^2} + \Omega_s^2\phi = 0 \,. \tag{34}$$

Equation (34) shows the harmonic oscillation of a particle that deviates from the synchrotron particle with a frequency of Ω_s as expressed in Eq. (32).

In an electron storage ring, $\eta = \alpha_p$, $\omega_s = c/R_s$, $P_s = E_s/c$, the oscillation frequency can be written as

$$\Omega_s = \frac{c}{R_s}[-\frac{heV\alpha_p \cos\phi_s}{2\pi E_s}]^{\frac{1}{2}}. \tag{35}$$

Then we have the longitudinal tune

$$\nu_s = \frac{\Omega_s}{\omega_s} = [-\frac{heV\alpha_p \cos\phi_s}{2\pi E_s}]^{\frac{1}{2}}. \tag{36}$$

As an example, for the Beijing Electron Positron Collider (BEPC), the related parameters in normal operation are about $\nu_s = 0.0208$ and $f_s = 26$ kHz.

6   Transition energy

It is obvious from Eq. (34) that for $\Omega_s^2 > 0$ the oscillation will be stable. The sign of Ω_s^2 depends only on the factor $\eta \cos\phi_s$ in Eq. (32). It is necessary that $\eta \cos\phi_s < 0$ for stable particle oscillation, and from Eq. (20), $\sin\phi_s$ must be positive for a particle to get energy from the RF cavity. If we write $1/\gamma_t^2 = \alpha_p$ in Eq. (17), we have

$$\eta = \frac{1}{\gamma_t^2} - \frac{1}{\gamma^2}. \tag{37}$$

Then, at the low energy, $\gamma < \gamma_t$, $\eta < 0$, one must have $\cos\phi_s > 0$ and $\sin\phi_s > 0$, that is the synchrotron phase ϕ_s should be in the range $0 < \phi_s < \pi/2$. At high energy, $\gamma > \gamma_t$, $\eta > 0$, we must have $\cos\phi_s < 0$ and $\sin\phi_s > 0$; that is, the synchrotron phase ϕ_s should be in the range of $\pi/2 < \phi_s < \pi$.

As the stable synchrotron phase ϕ_s of the oscillation has to be changed when the accelerated energy of the particle passes though $\gamma = \gamma_t$, this energy γ_t is called the transition energy.

7   Longitudinal acceptance and RF bucket

Multiplying both side of Eq. (31) by $d\phi/dt$ and integrating, we get

$$\int [\frac{d^2\phi}{dt^2}\frac{d\phi}{dt} + \frac{\Omega_s^2}{\cos\phi_s}(\sin\phi - \sin\phi_s)\frac{d\phi}{dt}]dt = C \qquad (38)$$

or
$$(\frac{\dot\phi}{\Omega_s})^2 - \frac{2}{\cos\phi_s}[\cos\phi + \phi\sin\phi_s] = C . \qquad (39)$$

In $\left(\phi, \dot\phi/\Omega_s\right)$ phase space, we have the plots shown in Figure 3 (a) and (b) for the energy of the particle below and above the transition energy respectively.

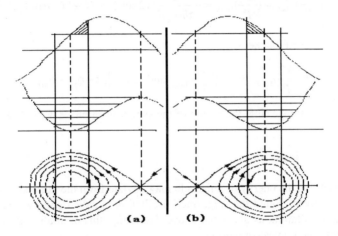

Figure 3: The energy oscillation of a particle below (a) and above (b) transition energy

We can see the trajectory of the particle in the phase-space of the stable region and the unstable region in Figure 3; these two regions are separated by a separatrix line on which the speed of the particle is infinite. The area of the stable region in the phase-space is called the RF bucket, in which particles execute stable longitudinal oscillation. The separatrix passes though the point $\phi_c = \pi - \phi_s$ and ϕ_m , which satisfies the following equation:

$$\cos\phi_m + \phi_m\sin\phi_s = (\pi - \phi_s)\sin\phi_s - \cos\phi_s . \qquad (40)$$

The bucket width is $\Delta\phi = |\phi_m - \phi_c|$.

At $d^2\phi/dt^2 = 0$, $\dot\phi = \dot\phi_{max}$, which satisfies the following equation:

$$\dot\phi_{max}^2 = 2\Omega_s^2[2 - (\pi - 2\phi_s)tan\phi_s] \qquad (41)$$

From Eq. (19), we have

$$\left(\frac{\Delta P}{P_s}\right)^2_{\max} = \left(\frac{\dot{\phi}_{\max}}{h\alpha_p\omega_s}\right)^2 . \tag{42}$$

Since $\dfrac{\Delta E}{E} = \beta^2 \dfrac{\Delta P}{P}$, and $\beta = 1$ for the electron storage ring, we have the energy acceptance for an electron storage ring as

$$\left(\frac{\Delta E}{E}\right)_{\max} = \left(\frac{\Delta P}{P}\right)_{\max} = \left[\frac{-eV}{\pi h\alpha_p E_s}(2\cos\phi_s - (\pi - 2\phi_s)\sin\phi_s)\right]^{1/2} . \tag{43}$$

The region of the RF bucket is

$$\pm\left(\frac{\Delta P}{P}\right)_{\max} \text{ or } \pm\left(\frac{\Delta E}{E}\right)_{\max} \text{ and } \Delta\phi = \left|(\pi - \phi_s) - \phi_m\right| . \tag{44}$$

For an electron storage ring, the region of the synchrotron phase is $\pi/2 < \phi_s < \pi/2$.

8   Transverse dispersion

In a real machine, the energy of a particle may deviate from the nominal value when the particle moves around the ring. Such oscillation is called energy oscillation or longitudinal oscillation, as discussed above. Normally, this oscillation is very small, affecting only the transverse motion.

Suppose the trajectory of particle is on the horizontal plane so that the vertical motion of particles is still betatron oscillation. In the horizontal plane, the displacement of a particle should be written as

$$x = x_\beta + x_c + x_D , \tag{45}$$

where x_β is the displacement of betatron oscillation, x_c the orbit distortion, and x_D the displacement due to energy oscillation. We can have

$$x_D = D(s)\frac{\Delta E}{Es} , \tag{46}$$

where $D(s)$ is defined as the dispersion function along the ring. The dispersion function means that a particle without any betatron oscillation or orbit distortion moves along a path proportional to $\Delta E/E_s$ and the ratio is $D(s)$.

Consider Hill's equation again. We add the energy deviation on the right side of the equation as (in the horizontal plane)

$$x'' + K_x x = \frac{1}{\rho} \frac{\Delta p}{p} \ . \tag{47}$$

Neglecting the orbit distortion, since $x = x_\beta + x_p$ and $x_p = D(\Delta p/p)$, we get

$$D'' + K_x D = \frac{1}{\rho} \ . \tag{48}$$

After one turn of a particle moving from any point, the matrix equation becomes

$$\begin{pmatrix} x_D \\ x'_D \\ \Delta p / p \end{pmatrix} = \begin{pmatrix} m_{11} & m_{12} & m_{13} \\ m_{21} & m_{22} & m_{23} \\ 0 & 0 & 1 \end{pmatrix} \begin{pmatrix} x_D \\ x'_D \\ \Delta p / p \end{pmatrix} \ . \tag{49}$$

From $x_p = D(\Delta p/p)$ and $x'_p = D'(\Delta p/p)$, we get

$$\begin{pmatrix} D \\ D' \\ 1 \end{pmatrix} = \begin{pmatrix} m_{11} & m_{12} & m_{13} \\ m_{21} & m_{22} & m_{23} \\ 0 & 0 & 1 \end{pmatrix} \begin{pmatrix} D \\ D' \\ 1 \end{pmatrix} \ . \tag{50}$$

It is easy to find the dispersion functions solved as

$$\begin{cases} D = \dfrac{m_{13}(1 - m_{22}) + m_{12}m_{23}}{2(1 - \cos\mu)} \\[2mm] D' = \dfrac{m_{23}(1 - m_{11}) + m_{21}m_{13}}{2(1 - \cos\mu)} \end{cases} \ , \tag{51}$$

where $\cos\mu = (m_{11} + m_{22})/2$.

Similar to the orbit distortion equation, the dispersion function $D(s)$ is

$$D(s) = \frac{\sqrt{\beta(s)}}{2 \sin \pi\nu} \oint \frac{\sqrt{\beta(s')}}{\rho(s')} \cos(|\,\psi(s') - \psi(s)\,| - \pi\nu) ds' \ . \tag{52}$$

From the above description, we know that the change in path length with momentum must depend on the dispersion function. Thus the closed orbit will have a mean radius

$$R = R_0 + \overline{D} \frac{\Delta p}{p} \ . \tag{53}$$

9 References

1. B.W. Montague, Single Particle Dynamics RF Acceleration, CERN 77-13, 1977.
2. J. Le Duff, Longitudinal Beam Dynamics in a Circular Accelerator, CERN 94-01,1994.
3. W. Pirkl, Longitudinal Beam Dynamics, CERN 95-06, 1995.
4. M. Sands, The Physics of Electron Storage Rings, SLAC Report 121, 1970.

AN INTRODUCTION TO THE DYNAMIC APERTURE

GUO ZHIYUAN

Institute of High Energy Physics, Chinese Academy of sciences
Beijing, China
E-mail: guozy@mail.ihep.ac.cn

1 What is the dynamic aperture?

As all the particles of a beam execute transverse oscillations in the magnet focusing system of an accelerator when the beam circles in a machine, we need to observe the beam in the accelerator for a enough long time for a physics experiment. Thus we need a certain space in the transverse dimension to maintain particle motion with no loss. The sufficient space refers to the aperture for particle motion.

1.1 Physics aperture

The beam circles along the orbit of the accelerator in a vacuum chamber composed of vacuum components including vacuum tube, vacuum gates, bellows, flags, injection kickers, separators, detector, diagnostic elements, RF cavities, etc. These components comprise the transverse apertures for the beam along the beam path, that is, the physics aperture of an accelerator. Obviously a larger physics aperture is better for maintaining beam motion with no loss of particles, but the size is limited by other factors such as the scale and budget of the machine, the magnet gap and electric field, etc.

1.2 Dynamic aperture

As there are nonlinear effects in an accelerator, which may be due to the nonlinear magnet elements and the higher-order field of the magnet, the amplitude of the particle oscillation may go to infinity because of the nonlinear field effects when the amplitude of the oscillation is large enough. Thus, even with a large physics aperture in an accelerator, it is still impossible to have an infinite region for particle stable motion.

We may define this limitation due to nonlinear effects on the particle as follows: the threshold amplitude beyond which the betatron oscillation becomes unbounded is called the dynamic aperture. This means that, if the initial conditions of amplitude and phase of the betatron oscillation (J_0, ϕ_0) are given, the amplitude at any time satisfies $|J(t)| < A$, where A is a finite boundary. In other words, the largest amplitude up to which the particle executes stable betatron oscillation is called the dynamic aperture. An example of the relation between dynamic aperture and physics aperture is shown in Figure 1. If we consider the real stability of the particle motion, the longitudinal oscillation effect should be taken into account; more generally, the stable region of the particle oscillation has a value in six-dimensional phase space X, P_x, Y, P_y and S, P_s. If the dynamic aperture is smaller than the physics aperture, we need to compensate and reduce the nonlinear effects that are responsible for the reduction of the dynamic aperture, for sufficient use of the physics aperture. Thus one of the major tasks for an accelerator designer is to make the dynamic aperture larger than the physics aperture.

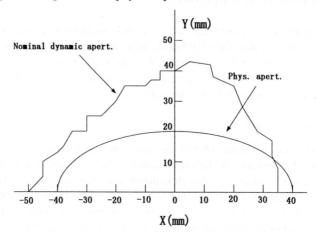

Figure 1: Physics aperture and dynamic aperture

2   What is the limitation to the dynamic aperture?

It is clear that the limitation to the dynamic aperture is the influence of the nonlinear effects on the particle motion. Normally, the nonlinear effects in an accelerator are due mainly to the nonlinear elements such as sextupoles, octupoles, injection kickers, separators, RF cavities, detectors, solenoids, etc., and to the tolerances of magnet fabrication. Nonlinear effects on the particle motion may arise also from closed-orbit distortion, which adds to the amplitude of the particle oscillation and enhances the nonlinear effects on the particle. Such effects also may arise from the oscillation frequency of the particle motion, the tune V_x, V_y and V_s; as the tune and its coupling come close to the resonant stop band, the amplitude of the particle oscillation will increase to infinity.

2.1   Nonlinear elements

Natural chromaticity is due to the frequency deviation caused by the different momenta of particles in the beam, $\xi = \Delta\nu\ /\ (\Delta p/p)$. Sextupoles are needed to compensate this chromaticity effect in the lattice of the accelerator. The sextupole is a magnet with a nonlinear magnetic field. In the new generation of synchrotron light sources, in which the focusing is extremely strong to achieve low emittance and high brightness, the chromaticity is strong.   In particle factories, electron-positron colliders and other particle colliders, a very small transverse beam size at the interaction point is required for high luminosity. In this kind of machine, a low-β insertion section is needed in the lattice, which makes the focusing very strong in the insertion quadrupole, but introduces strong chromaticity. Because these two kinds of accelerator are characterized by strong chromaticity, strong sextupoles have to be used to compensate it, and therefore strong nonlinear effects are introduced in the accelerator. Designers face a huge challenge to optimize the sextupole scheme in the lattice to get a large dynamic aperture with the chromaticity correction.

To overcome the beam instabilities, octupoles are installed in the lattice of some circular accelerators to provide the necessary Landau damping. The octupole is a magnet with a nonlinear field. Optimum arrangement of the octupoles in the lattice is necessary to keep the dynamic aperture from being seriously decreased by the octupole effect.

Some other single components may have sextupole, octupole and high-order magnetic fields due to their structure. These also give rise to nonlinear effects on particle motion in the accelerator.

Strong nonlinear effects arise from all of the above elements. These cause dynamic aperture reduction. We need to study strategies to compensate these nonlinear effects. The important work to be done is to minimize these effects in the accelerator lattice by analytical and numerical methods.

2.2   High-order magnetic field

The focusing system for particles in an accelerator is composed of many magnets, mostly bending magnets and quadrupoles. Tolerances in the magnet fabrication introduce high-order field into the magnetic field, resulting in nonlinear effects. Therefore tolerance control is very important during fabrication and correction. The field error is especially serious in an accelerator using super-conducting magnets as the field quality of a super-conducting magnet is not easy to control.

2.3   Closed-orbit distortion and tune

The dipole field error can cause a distortion, making the closed-orbit of the beam deviate from the ideal center orbit. This distortion of the beam orbit can be seen as an addition to the amplitude of particle oscillation from the center orbit as

$$A_{\beta_1}(s) = A_\beta(s) + A_{cod}(s).$$ (1)

We must make attempts to minimize the closed-orbit distortion in the accelerator, including control of the magnetic field error and putting correctors in suitable positions in the lattice. Then the physics aperture and the dynamic aperture can be effectively used.

When the particle oscillation frequencies ν_x, ν_y and ν_s are near the resonant number integer, half integer, third integer, etc., the amplitude of the particle oscillation is increased, then the nonlinear effects are enhanced, and consequently the dynamic aperture becomes smaller.

3   How to determine the dynamic aperture?

The dynamic aperture is very important for the performance of a circular accelerator. We must investigate in detail during the design of an accelerator, such as a proton synchrotron, electron-positron collider, synchrotron light source, or particle factory, whether it uses normal magnets or super-conducting magnets.

We can numerically calculate the dynamic aperture by direct launching of a particle and tracking its motion to determine the particle stability. This tracking procedure is an important tool for determining the dynamic aperture of an accelerator.

However, for a good understanding of the relevant features driving particle motion in the presence of nonlinear effects, there is strong motivation to study the different analytical methods dealing with amplitude limitations of stable motion. These methods can not give an exact estimation of the dynamic aperture, but they provide information on the nonlinear behavior of particle oscillation, which may depend sensitively on the initial conditions, and may eventually result in unbounded motion.

Another important aspect of work on the dynamic aperture is to do beam experiments on an operating machine to see what the dynamic aperture is in a real accelerator. Not only can we check the dynamic aperture with the design estimation, but also we can try to find ways to increase the real dynamic aperture to improve the performance in an operating accelerator.

3.1 Analytical approach

The analytical approach to the study of the dynamic aperture is to try to give closed expressions for amplitude limits up to which the motion remains bounded. In this short lecture we can not review all the analytical methods recently developed but we try to give an insight into them. Most of the methods introduced here are based on the Hamiltonian formalism and related equations of particle motion.

3.1.1 Resonance analysis

If the tune of a particle in a circular accelerator is the ratio of two integers, the motion of the particle could be in resonance with the magnet focussing system of the accelerator; then the particle would be lost. In order to express the approach to estimating the limit of particle stability simply, we take as an example particle with only one sextupole-driven resonance. In this system, the equation of particle motion is as the follows:

$$\frac{d^2u}{d\theta^2} + v^2 u = -v\beta^{3/2}\sigma\delta(\theta) . \tag{2}$$

This equation of motion can be represented in action-angle variables, and the Hamiltonian of the motion can be written as

$$H(\phi, J, \theta) = vJ + \frac{\sigma}{3}v\beta^{3/2}\delta(\theta)(\frac{2J}{v})^{3/2}\cos^3\phi , \tag{3}$$

where

$$u = (\frac{2J}{v})^{1/2}\cos\phi , \tag{4}$$

$$p = -(2Jv)^{1/2}\sin\phi . \tag{5}$$

The next step is to do the Fourier transformation, neglecting the fast varying terms, which do not drive the third-order resonance. Then the following transformation was done to make the Hamiltonian independent of θ and obtain an integral of the motion

$$\psi = 3\phi - \theta \quad \text{and} \quad K = J , \tag{6}$$

where the Hamiltonian is explicitly θ independent,

$$H(\psi, K) = (v - \frac{1}{3})K + \frac{\sigma}{24\pi}v\beta^{3/2}(\frac{2K}{v})^{3/2}\cos 3\psi . \tag{7}$$

The contours of H in (ψ, K) phase space provide the characteristics of the particle motion. Normally, it is convenient to return to (x, P_x) phase-space by inverse transformations and to plot the curves of function H, to obtain better information on particle motion. Figure 2 shows an example of a phase-space plot.

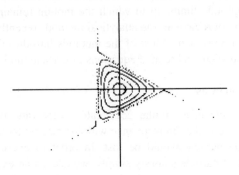

Figure 2: Example of a phase-space plot

The methods introduced above can be developed to include the case of two or three dimensions. More information related to the dynamic aperture can be gained from this analysis, such as the tune shift with amplitude, the stop-band width, the effect of crossing a resonance, etc.

3.1.2   Phase-space distortion through the Hamiltonian

We start from the Hamiltonian of single-particle motion, which includes nonlinear driving terms,

$$H(u, p, \theta) = \frac{1}{2}(p^2 + v^2 u^2) + F(\theta, u, p) . \tag{8}$$

The nonlinear driving terms $F(\theta, u, p)$ can be written as

$$F(\theta, u, p) = \sum_{k=3}^{\infty} A_k(\theta) u^k . \tag{9}$$

This Hamiltonian is not invariant as it still is time (θ) dependent. We should make the Hamiltonian θ independent, then it could provide a constant of the motion to define the invariant curves in phase-space. If the curves were closed, the motion would be stable, but if the curves were open, the motion would be unstable. We then have to try to eliminate the θ dependence in the Hamiltonian by a series of canonical transformation order by order. After certain number of transformation, the Hamiltonian has the following form:

$$\overline{H} \sim (p^2 + u^2) + B_3 u^3 + B_4 u^4 + \cdots + B_n u^n + A_{n+1}(\theta) u^{n+1} . \tag{10}$$

All of the terms lower than n are θ independent; if we assume the terms higher than $n+1$ to be negligible, then we have a θ independent invariant,

$$\overline{H} \sim (p^2 + u^2) + B_3 u^3 + B_4 u^4 + \cdots + B_n u^n .\tag{11}$$

This invariant can be used to plot curves in phase space to study the stability of the particle motion. As an example, the invariant distortion with sextupoles is shown in Figure 3.

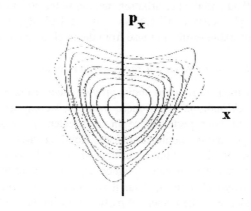

Figure 3: An example of invariant distortion

3.1.3 Calculation of phase-space distortion from the equation of motion

The calculation of phase-space distortion can also be carried out from the equation of particle motion. The phase-space ellipses corresponding to the Courant-Snyder invariant are distorted by nonlinearity. The calculation progresses by transporting the linear invariant through the lattice and kicking each point on the invariant at each sextupole. The nonlinear kicks are calculated under the assumption that the motion is purely linear. At the end of one turn, the distortion can be compared with the linear one.

3.1.4 Transfer-map methods using Lie transformations

Many studies using the Lie transformation for the dynamic aperture have been done. The work of A.J. Dragt and E. Forest at Maryland University gives detailed descriptions. This method first deals with the computation of nonlinear lattice functions and provides a connection between the conventional methods discussed above and the group theoretical tools in the Lie transformation description, and then analyzes the phase-space distortion and the quantification of the nonlinear effects.

This method is based on the perturbation in power of displacements expanding from the center orbit. The advantage of this method is that it tries to give explicit expressions for the generating function and the orbit. Their papers provide further details.

3.1.5 Other methods

Other methods of analyzing the dynamic aperture include the successive linearization method and the perturbation method in the equation of particle motion. The accelerator physicists at CERN studied the LEP dynamic aperture and got results quite comparable results to these from the tracking method.

3.2 *Numerical particle tracking using computer codes*

The analytical methods introduced above provide much useful information about the behavior of particle motion. Using these methods, we can understand the sources and mechanism of poor performance in an accelerator, and may find possible solutions to compensate the nonlinear effects. However, these methods can not take into account all of the nonlinear effects in a real accelerator, and several approximations have to be introduced, which neglect a few of the nonlinear terms, in order to get an analytical solution. Thus it is impossible to give an exact estimation of the dynamic aperture. The numerical particle tracking method then can be used to calculate the dynamic aperture for an accelerator. The high orders of all types of perturbations and errors of all kinds in the accelerator can be involved in the numerical method.

The process of a tracking simulation is that one set of particles is launched at a certain position along the ring with given initial conditions in the phase-space, and is the subject of a transfer map describing the particle motion. The launch position along the ring is chosen at an interesting point such as the interaction point in a collider, the injection point in a synchrotron, etc. The initial conditions in the phase space of the particle may be (x, x'), (y, y'), and (z, z') or (x, P_x), (y, P_y), and (z, P_z). After one turn, the positions of the particles are recorded and used as initial conditions for the next turn tracking. By iterating this process over many turns using different sets of initial conditions of the particles, we can observe the stability of the particles in different regions in phase-space including smear, tune shift, and detuning phenomena in the particle motion. The maximum amplitude of the particle corresponding to the initial condition is the dynamic aperture. Obviously the more turns of particle tracking, the better the determination of the stability of the motion compared with the real situation of particle motion, but this is limited by the computer capability. Normally, at least one damping time is necessary for tracking in an electron ring, and the necessary response time for the beam diagnostic and feedback control system for a proton synchrotron is, say, 10^5 to 10^6 turns.

Various codes have been developed to perform the numerical iterations for particle tracking simulation, such as SIXTRACK, MAD, TEAPOT, DIMAD and SAD, which can be used to execute six-dimensional tracking, and PATRACIA, RACETRACK and some others, which can be used to execute four-dimensional tracking. The main differences between these codes are in the methods used to represent the nonlinear transfer map, and the dimensions that can be tracked, such as pure transverse motion, coupled betatron motion, and coupled synchro-betatron motion.

The tracking codes have various ways of representing the transfer map for describing the particle motion around the accelerator, as follows:

• Thin-lens model

This model is used in some programs such as PATRACIA, RACETRACK, TEAPOT and MAD. The linear transformation matrices are used to represent the linear elements such as dipoles and quadrupoles. The nonlinear elements such as sextupoles, octupoles and higher multipoles (field errors) are treated in the impulse approximation as a kick in the simulation.

• High-order matrix methods

These method are used by the programs MAD and DIMAD. The six-dimensional vectors specifying the phase-space are to be used. These methods treat the transfer map in terms of a Taylor series representation, which is normally truncated at some order.

• Lie transformation technique

The transfer map is presented by Lie transformations which are generated from the Hamilton equation. The Lie algebra is used in the fomalism. This group-theoretic technique is employed in the programs MARYLIE and MAD.

• Generating function methods

These methods are based on the fact that the transfer map in an accelerator can be represented in terms of a generating function. The generating function may be obtained from the equation of motion, from matrix methods, or from Lie transformation. This technique is used in the programs DIMAD, MARYLIE and MAD.

• Canonical interaction method

The equations of motion for particles moving through the accelerator are integrated numerically in this method. An integration algorithm is employed as a canonical transformation in each step. This method is used in the program MAD.

3.3   Experimental measurement of dynamic aperture

During the operation of an accelerator, to improve the beam performance, we may need to try to increase the dynamic aperture of the machine. We can measure the dynamic aperture by a beam test experiment for an operating accelerator, and this experiment can also be used as a check to see how well the design estimation of dynamic aperture is realized. This kind of measurement was done by the CERN accelerator physicists in 1991 at the SPS machine.

4   How to increase the dynamic aperture in an accelerator

Analytical methods offer a better understanding of the mechanism of nonlinear field effects that influence particle motion, and numerical simulation can give an estimation of the dynamic aperture. It is certainly an important goal for accelerator physicists to increase the dynamic aperture during the design of a machine to obtain a dynamic aperture larger than the physics aperture. It is therefore important to compensate the nonlinear field effects and to control the tolerances in the manufacture of the magnets.

4.1   Optimum arrangement of nonlinear elements

For chromaticity correction, the strategy of the sextupole arrangement in a lattice is very important. Depending on the type and the scale of the accelerator, different compensation schemes have to be considered. For large-scale synchrotrons or colliders, most of the parts of the lattice are built up of regular cells. There is enough space to place the sextupoles in a regular way. In addition to these for chromaticity correction, more sextupoles should be introduced to compensate the beta-beta and tune shift with momentum. For small-scale colliders, the lattice is more complex and irregular than for large-scale accelerators. There is not enough space to place the sextupoles regularly, threfore more families of sextupoles are needed to compensate the nonlinear effects. For a synchrotron light source, the focusing effect is very strong in the lattice as a very small emittance is required. The harmonic sextupoles, which put on the dispersion region, have to be introduced to compensate, simultaneously, the driving terms of the most harmful harmonics and the tune shift with amplitude.

Octupoles also need to be installed in the lattice to allow tune spread for beam instability, but they have to be arranged in a suitable position in an accelerator to reduce their contribution to the nonlinear effects on the dynamic aperture. The Twiss parameters at the positions of the octupoles are important influences.

4.2 Tolerance control

The field errors in the magnet are another important source of nonlinear effect except the nonlinear elements. The errors are both systematic and random. Tolerance control is dominant for an accelerator using super-conducting magnets as the hole of the magnet is rather small and the errors are not easy to control. The different sets of magnet error fields are represented as a series of Taylor components written into the simulation code, the dynamic aperture is calculated by particle tracking, then the fabrication is repeatedly reviewed to get the optimum error distribution and to make the dynamic aperture as large as possible. Fixing the magnet tolerance by the results of particle tracking is a big issue in building an accelerator. A great deal of computer time is needed for a large-scale machine. The code for particle tracking is a powerful tool for guiding magnet fabrication.

References

1. D.R. Douglas, Dynamic Aperture Calculations for Circular Accelerators and Storage Rings, AIP No. 153, 1987.
2. J. Hagel and H. Moshammer, Analytic Approach of Dynamic Aperture by Secular Perturbation Theory, CERN 88-04, 1988.
3. A. Ropert, Dynamic Aperture, CERN 90-04, 1990.
4. Z.Y. Guo, T. Risselada and W. Scandale, Dynamic Aperture of the CERN-LHC with injection Optics, AIP No. 255, 1992.
5. W. Scandale, Dynamic Aperture, CERN 95-06, 1995.

SYNCHROTRON RADIATION AND BEAM DIMENSIONS

ZHENTANG ZHAO

Shanghai Synchrotron Radiation Center, Shanghai 201800, P.R. China
E-mail: zhaozt@ssrc.ac.cn

In this lecture, synchrotron radiation and its basic properties are introduced with emphasis on understanding the physics. The impact of synchrotron radiation on beam dynamics, such as radiation damping and quantum excitation, are discussed, as the beam dimensions are determined by the equilibrium between damping and excitation.

1   Synchrotron Radiation

Relativistic charged particles orbiting on curved trajectories radiate electromagnetic waves. This is called synchrotron radiation because its visible part was first observed at the General Electric 70 MeV synchrotron in 1947.

Synchrotron radiation was initially considered to be an undesirable by product of high energy electron circular accelerators because it limits their maximum attainable energy, but later it was realized that synchrotron radiation is an indispensable research tool for basic and applied sciences. Now tens of synchrotron radiation light sources are operated routinely for thousands of scientists around the world.

Synchrotron radiation is produced mainly by electrons traveling through three kinds of magnetic fields, generated by bending magnets, wiggler magnets and undulator magnets. Bending magnets force electrons to move on a circular path and cause electrons to emit synchrotron radiation with a continuous spectrum. Wiggler and undulator magnets, known as insertion devices, composed of $2N$ short bending magnets with alternating polarities, are used to generate synchrotron radiation for specially required photon energy, photon flux and photon brightness as well as radiation polarization. Wiggler magnets produce the same synchrotron radiation spectrum as bending magnets but with $2N$ times the enhancement of flux intensity, whereas undulator magnets generate partially coherent synchrotron radiation with a discrete spectrum and N^2 times the enhancement of flux intensity.

Synchrotron radiation has a significant impact on electron or positron motion in circular accelerators. The emission of synchrotron radiation provides radiation damping and quantum excitation to the transverse and longitudinal particle motion, and the equilibrium between the damping and excitation processes determines the beam dimensions in circular accelerators. Synchrotron radiation also has a strong influence on the hardware systems of circular accelerators, especially affecting the choice of the accelerator size and the designs of the RF and vacuum systems.

Synchrotron radiation has a broad range of applications besides its natural use for diagnosing the electron motion in accelerators. Its unique properties, such as the wide electromagnetic spectral range from infrared to hard X-ray, high intensity and high collimation, polarization and pulsed time structure as well as spatial interference, make it a very powerful light source for studies in materials, biology, physics, chemistry, medicine and micro-fabrication.

Numerous reports and lectures document synchrotron radiation and the relevant theoretical derivations. Here we focus only on the basic characteristics of synchrotron radiation from bending magnets and its effects on beam dynamics in electron storage rings; the theoretical details can be found in the references listed.

2   Basic Characteristics of Synchrotron Radiation from Bending Magnet

Electrons with relativistic energy γ moving in bending magnets emit synchrotron radiation in the tangential direction within a highly collimated angle of $\pm 1/\gamma$ at any point along the circular orbit. Therefore this radiation fan from bending magnets scans all around the circumference of the accelerator in the horizontal plane, but in the vertical plane the radiation stays within the angle $\pm 1/\gamma$.

2.1   Radiated Power

The power radiated from a relativistic electron was given, from calculation of the relevant electromagnetic fields and Poynting vector, first by Lienard in 1898, as

$$P_\gamma = \frac{2}{3} \frac{e^2}{4\pi\varepsilon_0 c} \gamma^6 \left[\left(\frac{d\vec{\beta}}{dt} \right)^2 - \left(\vec{\beta} \times \frac{d\vec{\beta}}{dt} \right)^2 \right] , \qquad (1)$$

which can be transformed into

$$P_\gamma = \frac{2}{3} \frac{e^2}{4\pi\varepsilon_0 m_0^2 c^3} \left[\left(\frac{dp_{//}}{dt} \right)^2 + \gamma^2 \left(\frac{dp_\perp}{dt} \right)^2 \right] . \qquad (2)$$

Here $\beta = (1 - \gamma^{-2})^{1/2}$ is the electron's relativistic speed, and $p_{//}$ and $p_\perp$ are the electron's longitudinal and transverse momenta. For circular motion, the radiated power becomes

$$P_\gamma = \frac{2}{3} \frac{e^2 c}{4\pi\varepsilon_0} \frac{\beta^4 \gamma^4}{\rho^2} . \qquad (3)$$

The above equations show that, if the same magnitude accelerating force is exerted on electrons, the transverse acceleration will produces γ^2 times as much radiation power as that produced by longitudinal acceleration. Furthermore the light particles will radiate more energy because the radiation power is proportional to the 4th power of the relativistic energy γ, for instance an electron emits about 10^{13} times as much power as that from a proton at the same energy. However, synchrotron radiation places a strong limitation on the maximum attainable energy of electron circular accelerators with regard to technical feasibility and economics, so that linacs will be the main accelerators for accelerating electrons to even higher energy in the future.

Integrating the radiated power over the accelerator circumference yields the energy loss in one turn:

$$U_0 = \oint P_\gamma \, dt = \frac{e^2}{3\varepsilon_0} \frac{\beta^3 \gamma^4}{\rho} . \tag{4}$$

For electrons, in engineering units, this energy loss becomes

$$U_0[\text{keV}] = 88.5 \frac{E^4[\text{GeV}]}{\rho[\text{m}]} . \tag{5}$$

For an electron beam circulating in a storage ring with an average current of I_b, the total radiated power from bending magnets per turn is

$$P_b = \frac{I_b U_0}{e} . \tag{6}$$

2.2   Spectral and Angular Power Distributions

When a charged particle travels on a circular path in a bending magnet, an observer at a fixed point outside the orbit circle can receive the synchrotron radiation only from a short arc of the electron trajectory; this short radiation pulse represents a broad frequency spectrum of synchrotron radiation.

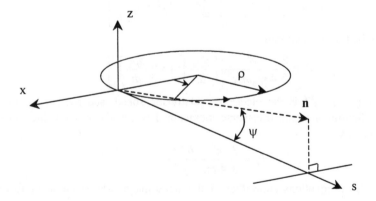

Fig. 1. Cartesian coordinate system

The complete spectral distribution of synchrotron radiation can be calculated using the Fourier transform of the radiation electric field; in the following we present it without derivation. In the coordinate system shown in Fig.1, the spectral angular power distribution is

$$\frac{d^2 P}{d\Omega d\omega} = \frac{P_\gamma \gamma}{\omega_c} F(\omega, \psi) \ . \tag{7}$$

Here Ω represents the solid angle, and ω_c is the critical frequency, which divides the integrated power into two equal parts. $F(\omega, \psi)$ is the spectral angular distribution function, $F_\sigma(\omega, \psi)$ and $F_\pi(\omega, \psi)$ indicate the horizontal and vertical polarization components:

$$\omega_c = \frac{3c\gamma^3}{2\rho} \ , \tag{8}$$

$$F(\omega, \psi) = F_\sigma(\omega, \psi) + F_\pi(\omega, \psi) \ , \tag{9}$$

$$F(\omega, \psi) = \left(\frac{3}{2\pi}\right)^3 \left(\frac{\omega}{2\omega_c}\right)^2 (1+\gamma^2\psi^2)^2 \left(K_{2/3}^2(\xi) + \frac{\gamma^2\psi^2}{1+\gamma^2\psi^2} K_{1/3}^2(\xi) \right) \ , \tag{10}$$

$$F_\sigma(\omega, \psi) = \left(\frac{3}{2\pi}\right)^3 \left(\frac{\omega}{2\omega_c}\right)^2 (1+\gamma^2\psi^2)^2 K_{2/3}^2(\xi) \ , \tag{11}$$

$$F_\pi(\omega, \psi) = \left(\frac{3}{2\pi}\right)^3 \left(\frac{\omega}{2\omega_c}\right)^2 \gamma^2\psi^2 (1+\gamma^2\psi^2) K_{1/3}^2(\xi) \ , \tag{12}$$

$$\xi = \left(\frac{\omega}{2\omega_c}\right)(1+\gamma^2\psi^2)^{3/2} \ . \tag{13}$$

Integrating the spectral angular power distribution over Ω gives the spectral distribution of the synchrotron radiation,

$$\frac{dP}{d\omega} = \frac{P_\gamma}{\omega_c} S\left(\frac{\omega}{\omega_c}\right) \ , \tag{14}$$

here

$$S\left(\frac{\omega}{\omega_c}\right) = S_\sigma\left(\frac{\omega}{\omega_c}\right) + S_\pi\left(\frac{\omega}{\omega_c}\right) \ , \tag{15}$$

$$S\left(\frac{\omega}{\omega_c}\right) = \frac{9\sqrt{3}\omega}{8\pi\omega_c}\left[\int_{\omega/\omega_c}^\infty K_{5/3}(\zeta)\,d\zeta\right] \ , \tag{16}$$

$$S_\sigma\left(\frac{\omega}{\omega_c}\right) = \frac{9\sqrt{3}\omega}{16\pi\omega_c}\left[\int_{\omega/\omega_c}^\infty K_{5/3}(\zeta)\,d\zeta + K_{2/3}\left(\frac{\omega}{\omega_c}\right)\right] \ , \tag{17}$$

$$S_\pi\left(\frac{\omega}{\omega_c}\right) = \frac{9\sqrt{3}\omega}{16\pi\omega_c}\left[\int_{\omega/\omega_c}^\infty K_{5/3}(\zeta)d\zeta - K_{2/3}\left(\frac{\omega}{\omega_c}\right)\right] . \tag{18}$$

Next we integrate the spectral angular power distribution over the frequency, which gives the angular power distribution as follows:

$$\frac{dP}{d\Omega} = \frac{21}{32}\frac{P_\gamma}{2\pi}\frac{\gamma}{(1+\gamma^2\psi^2)^{5/2}}\left(1+\frac{5}{7}\frac{\gamma^2\psi^2}{1+\gamma^2\psi^2}\right) . \tag{19}$$

3 Quantum Emission

Synchrotron radiation is emitted in discrete photons or quanta, with the photon energy of

$$u = \hbar\omega , \tag{20}$$

where $\hbar$ is Planck's constant. Then the radiated power and the number of photons emitted per unit time can be expressed as

$$P_\gamma = \int_0^\infty un(u)du , \tag{21}$$

$$N_\gamma = \int_0^\infty n(u)du = \frac{15\sqrt{3}}{8}\frac{P_\gamma}{u_c} , \tag{22}$$

where $n(u)$ is the number of photons emitted per unit time in the energy interval from u to $u+du$; from Equ. (14), (20), (21) $n(u)$ can be written as

$$n(u) = \frac{P_\gamma}{u_c^2}\frac{u_c}{u}S(u/u_c) ; \tag{23}$$

and u_c is the critical energy, which is defined as

$$u_c = \hbar\omega_c . \tag{24}$$

Two useful quantities related to radiation damping and quantum excitation are mean photon energy and mean square photon energy, which are

$$\langle u \rangle = \frac{1}{N_\gamma}\int_0^\infty un(u)du = \frac{8}{15\sqrt{3}}u_c , \tag{25}$$

$$\langle u^2 \rangle = \frac{1}{N_\gamma}\int_0^\infty u^2 n(u)du = \frac{11}{27}u_c^2 . \tag{26}$$

4 Radiation Damping

The energy loss due to synchrotron radiation is replaced by RF cavities in electron storage rings; this energy loss and compensation process provides damping of the electron synchrotron and betatron oscillations, i.e. radiation damping.

4.1 Damping of Synchrotron Oscillation

In storage rings synchrotron oscillation occurs because the off-energy electrons travel on different paths and arrive at the RF cavities at different times from the synchronous particles: the higher energy electron travels on a longer path and arrives at the RF cavities later, and vice versa. Furthermore, considering the radiation property that the higher the energy of an electron, the more energy it radiates, we find that there is an additional damping effect on the so-called synchrotron oscillation. The damped energy oscillation equation can be derived as

$$\frac{d^2\varepsilon}{dt^2} + 2\alpha_\varepsilon \frac{d\varepsilon}{dt} + \omega_s^2\varepsilon = 0 , \tag{27}$$

where ε is the deviation of the electron energy from the energy of a synchronous particles, ω_s is the synchrotron frequency, and α_ε is the damping coefficient. In terms of the storage ring revolution time T_0 and the change in radiated energy, α_ε can be expressed as

$$\alpha_\varepsilon = \frac{1}{2T_0}\frac{dU}{d\varepsilon} . \tag{28}$$

Considering the change in radiated energy with electron energy and the magnetic field experienced by the electron on the off-energy trajectory, we find that the radiated energy changes as the electron energy according to

$$\frac{dU}{d\varepsilon} = \frac{U_0}{E}(2+D) , \tag{29}$$

with

$$D = \oint \frac{\eta(s)}{\rho(s)}[\Delta - 2k(s)]ds \bigg/ \oint \frac{ds}{\rho^2(s)} , \tag{30}$$

where Δ is determined by the type of dipole magnet,

$$\Delta = \begin{cases} 1/\rho^2(s) & \text{for sector dipole} \\ 0 & \text{for rectangular dipole} \end{cases} .$$

From Eqs. (28) and (29), the damping coefficient can be written as

$$\alpha_\varepsilon = \frac{U_0}{2T_0 E}(2+D) . \tag{31}$$

Normally D is a small quantity; therefore the damping coefficient can be simply approximated as

$$\alpha_\varepsilon = \frac{U_0}{T_0 E} = \frac{\langle P_\gamma \rangle}{E} \ . \tag{32}$$

The above formula means that the synchrotron radiation damping time, the inverse damping coefficient, is the time it takes an electron to radiate all its energy away.

4.2 Damping of Betatron Oscillation

At high energy the electron can be considered to emit photons in the tangential direction, therefore this emission does not change either the position or the direction of the electron motion, i.e. x, x', z and z' do not change. During the emission the electron loses its momentum in both the longitudinal and the transverse directions, but the RF cavities compensate only the longitudinal loss. This combined process results in the damping of transverse motion.

Usually there is no dispersion in the vertical plane of the storage ring; the change in electron momentum does not influence the vertical equilibrium orbit. Therefore the electron makes pure betatron oscillations on the vertical plane; its trajectory and the corresponding slope as well as the vertical betatron oscillation invariant can be expressed as

$$z(s) = A_z \sqrt{\beta_z(s)} \cos[\psi_z(s) + \psi_{zn}] \ ,$$

$$z'(s) = -\frac{A_z}{\sqrt{\beta_z(s)}} \left(\alpha_z(s) \cos[\psi_z(s) + \psi_{zn}] + \sin[\psi_z(s) + \psi_{zn}] \right) , \tag{33}$$

$$A_z^2 = \gamma_z z^2 + 2\alpha_z z z' + \beta_z z'^2 \ . \tag{34}$$

Where $\beta(s)$ is the beta function with $\alpha_z(s) = -\beta'_z(s)/2$, $\gamma_z(s) = (1+\alpha_z^2(s))/\beta(s)$ and

$$\psi_z(s) = \int_0^s \frac{dl}{\beta_z(l)} \ .$$

After the electron is accelerated in RF cavities, its longitudinal momentum $p_{//}$ gets an increment δp while its vertical momentum p_z stays constant. Therefore the corresponding trajectory slope change can be derived as follows:

$$z' = p_z / p_{//} \ , \tag{35}$$

$$z' + \delta z' = \frac{p_z}{p_{//} + \delta p} \approx z'(1 - \frac{\delta p}{p}) \ . \tag{36}$$

Thus

$$\delta z' = -z' \frac{\delta p}{p} = -z' \frac{\delta \varepsilon}{E} \ . \tag{37}$$

From Eqs. (33) and (34) we get

$$\frac{\langle \delta A_z \rangle}{A_z} = -\frac{1}{2}\frac{\delta \varepsilon}{E} \ , \tag{38}$$

and averaging over one turn,

$$\frac{\Delta A_z}{A_z} = -\frac{U_0}{2E} \ . \tag{39}$$

Therefore the vertical damping coefficient is

$$\alpha_z = -\frac{1}{A_z}\frac{dA_z}{dt} = \frac{U_0}{2ET_0} \ . \tag{40}$$

In the horizontal plane, the electron orbit can be decomposed into a betatron part and a dispersion part, which is expressed as

$$x = x_\beta + x_\varepsilon = x_\beta + \eta \frac{\varepsilon}{E} \ , \tag{41}$$

where the horizontal betatron oscillation is similar to the vertical one, which can be expressed as

$$x(s) = A_x \sqrt{\beta_x(s)} \cos[\psi_x(s) + \psi_{xi}] \ ,$$

$$x'(s) = -\frac{A_z}{\sqrt{\beta_x(s)}}\left(\alpha_x(s)\cos[\psi_x(s) + \psi_{xi}] + \sin[\psi_x(s) + \psi_{xi}]\right) \ . \tag{42}$$

The betatron oscillation invariant is also similar to the vertical one; it can be written as

$$A_x^2 = \gamma_x x_\beta^2 + 2\alpha_x x_\beta x_\beta' + \beta_x x_\beta'^2 \ . \tag{43}$$

In the horizontal plane, in addition to the same damping process as in the vertical plane, there is an off-energy effect on the electron motion caused by the finite dispersion. Although the radiation does not change x and x', the energy loss will create a change in betatron oscillation amplitude to compensate the energy orbit displacement. Combined with the effect from the RF cavities, the increment of the horizontal betatron oscillation and its slope can be expressed as

$$\delta x_\beta = -\delta x_\varepsilon = -\eta \frac{\delta \varepsilon}{E} \ ,$$

$$\delta x_\beta' = \delta x' - \delta x_\varepsilon' = -x_\beta' \frac{\delta \varepsilon}{E} - \eta' \frac{\delta \varepsilon}{E} \ . \tag{44}$$

Considering the change in the radiated power with the electron horizontal trajectory and averaging over one turn of the storage ring, we get

$$\frac{\Delta A_z}{A_z} = -\frac{U_0}{2E}(1 - D) \ ; \tag{45}$$

then the total damping coefficient is

$$\alpha_x = -\frac{1}{A_z}\frac{dA_z}{dt} = \frac{U_0}{2ET_0}(1 - D) \ . \tag{46}$$

The damping constants in the longitudinal, horizontal and vertical planes can be summarized as

$$\alpha_i = \frac{J_i U_0}{2ET_0} \qquad i = x, z, \varepsilon \ , \tag{47}$$

where J_i is called the partition number, with

$$J_x = 1 - D \ , \qquad J_z = 1 \ , \qquad J_\varepsilon = 2 + D \tag{48}$$

and their sum is a constant:

$$J_x + J_z + J_\varepsilon = 4 \ . \tag{49}$$

5 Quantum Excitation

In dealing with the damping effect, the synchrotron radiation energy loss is assumed to be a continuous process, but in fact the radiation energy is emitted in photons of discrete energy. Furthermore the emitted photon energy and its emission time very randomly during the radiation process. This randomness causes the growth of synchrotron and betatron oscillation amplitudes, which is called quantum excitation.

5.1 Excitation of Synchrotron Oscillation

When an electron performs synchrotron oscillations with constant amplitude in storage rings, its energy oscillation can be described as

$$\varepsilon(t) = A_\varepsilon \cos[\omega_s(t - t_0)] \ . \tag{50}$$

After a photon with energy u is emitted at time t_i, the above energy deviation becomes

$$\varepsilon(t) = A_\varepsilon \cos[\omega_s(t - t_0)] - u \cos[\omega_s(t - t_i)] \ , \tag{51}$$

which results in a change in energy oscillation amplitude, i.e.

$$\langle \delta A_\varepsilon^2 \rangle = \langle u^2 \rangle \ . \tag{52}$$

Equation (52) shows that the quantum emission of a photon causes the amplitude of the energy oscillation to increase. By taking the emission rate into account, we can find the excitation rate as follows:

$$\frac{d\langle A_\varepsilon^2 \rangle}{dt} = N\langle u^2 \rangle \ . \tag{53}$$

5.2 Excitation of Horizontal Betatron Oscillation

In the horizontal plane, the electron orbit consists of the energy orbit and the horizontal betatron orbit, which is written as

$$x = x_\varepsilon + x_\beta \ , \tag{54}$$

where the horizontal betatron oscillation invariant is similar to the vertical one; it can be expressed as

$$A_x^2 = \gamma_x x_\beta^2 + 2\alpha_x x_\beta x_\beta' + \beta_x x_\beta'^2 . \tag{55}$$

When an electron emits a photon of energy u, it casuses in a change in A_x^2 as

$$A_x^2 + \delta A_x^2 = \gamma_x (x_\beta - \eta \frac{u}{E})^2 + 2\alpha_x (x_\beta - \eta \frac{u}{E})(x_\beta' - \eta' \frac{u}{E}) + \beta_x (x_\beta' - \eta' \frac{u}{E})^2 , \tag{56}$$

so that

$$\langle \delta A_x^2 \rangle = (\gamma_x \eta^2 + 2\alpha_x \eta \eta' + \beta_x \eta'^2) \frac{\langle u^2 \rangle}{E^2} . \tag{57}$$

Here we introduce a definition of a lattice function

$$H(s) = \gamma \eta^2 + 2\alpha \eta \eta' + \beta \eta'^2 ; \tag{58}$$

then the quantum excitation to the horizontal betatron oscillation can be written as

$$\frac{d\langle A_x^2 \rangle}{dt} = \frac{\langle N \langle u^2 \rangle H \rangle}{E^2} . \tag{59}$$

6   Beam Dimensions

The above results show that synchrotron radiation produces two competing effects, radiation damping and quantum excitation, on the electron motion in storage rings, and the equilibrium between these two effects determines the electron beam dimensions.

6.1   Energy Spread and Bunch length

From the energy damping and fluctuation rates, we get the following equilibrium equation:

$$\frac{d\langle A_\varepsilon^2 \rangle}{dt}\bigg|_{rad} + \frac{d\langle A_\varepsilon^2 \rangle}{dt}\bigg|_{qua} = 0 , \tag{60}$$

then

$$\sigma_\varepsilon^2 = \langle \varepsilon^2 \rangle = \frac{\langle A_\varepsilon^2 \rangle}{2} = \frac{\langle N \langle u^2 \rangle \rangle}{4\alpha_\varepsilon} . \tag{61}$$

Hence the energy spread is

$$\left(\frac{\sigma_\varepsilon}{E}\right)^2 = C_q \frac{\gamma^2}{J_\varepsilon} \frac{\langle 1/\rho^3 \rangle}{\langle 1/\rho^2 \rangle} , \tag{62}$$

where

$$C_q = \frac{55}{32\sqrt{3}} \frac{\hbar}{mc} . \tag{63}$$

For an isomagnetic lattice, ρ is a constant; therefore the energy spread of a storage ring is determined mainly by the electron energy and the bending radius of the dipole magnets. From the relation of bunch length and energy spread, i.e.

$$\sigma_t = \frac{\alpha_p}{\omega_s} \frac{\sigma_\varepsilon}{E} , \tag{64}$$

we get the bunch length

$$\sigma_t^2 = C_q \left(\frac{\alpha_p}{\omega_s} \right)^2 \frac{\gamma^2}{J_\varepsilon} \frac{\langle 1/\rho^3 \rangle}{\langle 1/\rho^2 \rangle} . \tag{65}$$

6.2   Natural Emittance

The equilibrium between damping and excitation of the horizontal betatron oscillation results in the natural emittance of the electron beam in storage rings, and the equilibrium equation can be written as

$$\left. \frac{d\langle A_x^2 \rangle}{dt} \right|_{rad} + \left. \frac{d\langle A_x^2 \rangle}{dt} \right|_{qua} = 0 , \tag{66}$$

based on the relation to natural emittance

$$\varepsilon_{x0} = \frac{\langle A_x^2 \rangle}{2} . \tag{67}$$

The natural emittance can be derived as

$$\varepsilon_{x0} = C_q \frac{\gamma^2}{J_x} \frac{\langle H/\rho^3 \rangle}{\langle 1/\rho^2 \rangle} . \tag{68}$$

In the case of an isomagnetic lattice,

$$\varepsilon_{x0} = C_q \frac{\gamma^2}{J_x} \frac{\langle H \rangle}{\rho} . \tag{69}$$

6.3   Coupling Factor and Beam Size

We can neglect the quantum excitation of the vertical betatron oscillation because its effect on vertical beam size is very weak compared with the component coupled from the horizontal plane. The coupling between vertical and horizontal motions of electrons is generated by errors of magnetic fields and the closed orbit; this coupling is measured by a coupling factor k, defined as the ratio of the vertical emittance ε_z to the horizontal one ε_x, and their relation to natural emittance is

$$\varepsilon_x = \frac{1}{1+k} \varepsilon_{x0} ,$$

$$\varepsilon_z = \frac{k}{1+k} \varepsilon_{x0} . \tag{70}$$

From Eqs. (68) the beam sizes are

$$\sigma_x = \sqrt{\beta_x \varepsilon_x} = \left[\frac{\beta_x C_q}{1+k} \frac{\gamma^2}{J_x} \frac{\langle H/\rho^3 \rangle}{\langle 1/\rho^2 \rangle} \right]^{\frac{1}{2}} ,$$

$$\sigma_z = \sqrt{\beta_z \varepsilon_z} = \left[\frac{k\beta_z C_q}{1+k} \frac{\gamma^2}{J_x} \frac{\langle H/\rho^3 \rangle}{\langle 1/\rho^2 \rangle} \right]^{\frac{1}{2}} . \tag{71}$$

At the point of dispersion, the horizontal beam size is

$$\sigma_x = \left[\beta_x \varepsilon_x + \eta^2 \left(\frac{\sigma_\varepsilon}{E} \right)^2 \right]^{\frac{1}{2}} . \tag{72}$$

The above damping and excitation processes show that electron motion has no long-term memory of its history, so that the initial conditions of an electron beam in storage ring do not influence its equilibrium dimensions.

7   Synchrotron Radiation Integrals

Five integrals are defined and widely used for electron storage rings, especially for light source machines, from which the key beam parameters can be simply expressed. The integrals are

$$I_1 = \oint \frac{\eta}{\rho} ds , \qquad I_2 = \oint \frac{1}{\rho^2} ds , \qquad I_3 = \oint \frac{1}{|\rho^3|} ds ,$$

$$I_4 = \oint \frac{(1-2n)\eta}{\rho^3} ds , \qquad I_5 = \oint \frac{H}{|\rho^3|} ds . \tag{73}$$

From these integrals we can get the momentum compaction factor

$$\alpha = \frac{I_1}{C} , \tag{74}$$

the energy loss per turn in a storage ring

$$U_0 = \frac{2 r_e E^4 I_2}{3(mc^2)^3} \quad \text{or} \quad U_0[\text{keV}] = \frac{88.5 E^4[\text{GeV}]}{\rho[\text{m}]} , \tag{75}$$

the damping partion number

$$J_x = 1 - \frac{I_4}{I_2} \quad \text{and} \quad J_\varepsilon = 2 + \frac{I_4}{I_2} , \tag{76}$$

the damping time

64

$$\tau_i = \frac{3m^3c^5}{2\pi r_e} \cdot \frac{C\rho}{J_i E^3} \quad , \qquad \tau_i[\text{ms}] = \frac{C[\text{m}]\rho[\text{m}]}{13.2 J_i E^3[\text{GeV}]} \quad , \tag{77}$$

the energy spread

$$\left(\frac{\sigma_\varepsilon}{E_0}\right)^2 = C_q \gamma^2 \frac{I_3}{2I_2 + I_4} = \frac{C_q \gamma^2}{J_\varepsilon} \cdot \frac{I_3}{I_2} \quad , \tag{78}$$

and the natural emittance

$$\varepsilon_{x_0} = C_q \gamma^2 \frac{I_5}{I_2 - I_4} = \frac{C_q \gamma^2}{J_x} \cdot \frac{I_5}{I_2} \quad . \tag{79}$$

References

1. M. Sands, The Physics of Electron Storage Rings, SLAC-121 (1970).
2. A. Hofmann, Characteristics of Synchrotron Radiation, CERN 98-04, p. 1 (1998).
3. K. Hubner, Synchrotron Radiation, CERN 90-03, p. 24 (1990).
4. K. J. Kim, Characteristics of Synchrotron Radiation, AIP Conf. Proc. 184, Vol. 1, p. 565 (1989).
5. R. P. Walker, Synchrotron Radiation, CERN 94-01, p. 437 (1994).
6. R. P. Walker, Radiation Damping, CERN 94-01, p. 461 (1994).
7. R. P. Walker, Quantum Excitation and Equilibrium Beam Properties, CERN 94-01, p. 481 (1994).
8. A. Ropert, Synchrotron Radiation and Equilibrium Beam Size, CERN 95-06, p. 783 (1996).
9. H Wiedemann, Particle Accelerator Physics I and II Springer-Verlag (1994 and 1995).

ELECTRON-POSITRON COLLIDERS

ZHANG CHUANG

Institute of High Energy Physics, Chinese Academy of Sciences
p.o.Box 918, Beijing 100039, China
e-MAIL: zhangc@mail.ihep.ac.cn

The physics of electron-positron colliders is described in four sections: 1. an overview on colliders; 2. luminosity; 3. beam-beam effects; and 4. design of an electron-positron collider. In section 1, introductory and basic questions about high energy physics and colliders as why do we need colliders? what are colliders? how do colliders work?, and electron-positron colliders, are discussed. Section 2 focuses on the issues related to a principle parameter for colluders – luminosity – by describing high energy physics and luminosity, luminosity in beam-beam collisions, average and integrated luminosity, menochromatization scheme, and methods to enhance luminosity. In section 3, an important phenomenon limiting the luminosity of colliders, – the beam-beam effect – is discussed with five subsections: on the beam-beam force, beam-beam instability, long-range beam-beam effects, beam-beam simulation, and beam-beam compensation schemes. Section 4, on the design of an electron-positron collider, is more practical, with a general description and discussion of lattice design, linear and nonlinear perturbations and their correction, intensity-dependent phenomena, beam-beam effects, and subsystem design.

1. An Overview on Colliders

1.1 High energy physics and colliders

The world is made of matter. The aims of science are discovery, insight and understanding of the natural environment and the laws that govern it. Figure 1 shows the size of natural objects in relation to subjects in the sciences sketched by Prof. A. Glashow.

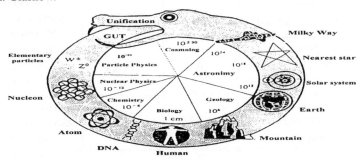

Figure 1. The Glashow sketch of the nature and the sciences

As a part of basic science, high energy physics studies "elementary particles", the building blocks of all matter. High energy physics, or particle physics, plays an essential role in basic science for it tries to answer the following fundamental questions: What are the primal constituents of all matter and energy in the universe? What are the laws governing the behavior of these constituents that let them combine to form the matter we observe?

The development of science has shown that matter is made of atoms, which are comprised of nuclei and electrons, and the nuclei are composed of protons and neutrons which consist of even smaller particles called quarks. As a microscope for observing this micro-world, the accelerator has a resolution limited by the related de Broglie wavelength λ:

$$\lambda = \frac{h}{p} = \frac{hc}{E\beta},\tag{1}$$

where $h=3.507\times10^{-15}$ eV·s is Planck constant, and p, E and β are the momentum, energy and relativistic speed of the beam used by the "microscope". Equation (1) shows that the smaller the substance the higher the required beam energy. The methods of investigating the micro-world are listed in Table 1.

Table 1. Methods of investigating the micro-world

Observed Substance	Size (cm)	Beam Energy $E = hc/\lambda\beta$	Method
Cell/Bacterium = *Aggregate of molecules*	$10^{-3}\sim10^{-5}$	0.1~10 eV	Optical microscope
Molecule = *Aggregate of atoms*	$\sim10^{-7}$	~1 keV	Electron microscope
Atom = *Nucleus + electrons*	$\sim10^{-8}$	~ 10 keV	Synchrotron radiation
Nucleus e. g. Oxygen = $8p+8n$	$\sim10^{-12}$	>100 MeV	Low-energy electron or proton accelerators
Hadron = *Aggregate of quarks* e. g. $p=u+u+d, J/\psi=c+\bar{c}$	$\sim10^{-13}$	>1 GeV	High-energy proton accelerators
Quark, Lepton ... $(u,d)(s,c)(b,t)$ $(e,v_e)(\mu,v_\mu)(\tau,v_\tau)$	$<10^{-16}$	>1000 GeV	High-energy electron or proton colliders
......	......	......	......

The history of accelerators has followed a continuous path towards higher energy. This has succeeded by repeated use of the cycle of "new principle → improved technology → saturation". There are many types of accelerators along the path: Cockcroft-Walton accelerators, electrostatic accelerators, cyclotrons, betatrons, synchro-cycrotrons, linacs, weak-focusing synchrotrons, strong-focusing synchrotrons, and colliders that produce higher "effective" energy than fixed-target machines.

1.2   Why do we need colliders ?

Table 1 shows that the smaller the substance being studied, the higher the center-of-mass energy needed. In 1954 Fermi proposed an accelerator with a center-of-mass energy of 3 TeV (1 TeV=1000 GeV=10^{12} eV). At that time, there was no concept of a collider. A fixed-target accelerator with beam energy of 5000 TeV is needed to create an E_{cm} of 3 TeV (Why? See below.) The radius of a 5000 TeV synchrotron is 8000 km for the magnetic field of 2 Tesla, i.e. its circumference is about the earth's equator. Its cost was estimated as 170 Billion \$US, and it would take 40 years to build. Clearly, this was only a dream.

Colliders would make Fermi's dream come true. The idea of a collider is to produce collisions between oppositely directed beams in order to generate high center-of-mass energy interactions between them.

The center-of-mass energy for the collision of two particles with energy E_1 and E_2, and crossing angle θ is expressed as

$$E_{cm} = \left[2E_1E_2 + (m_1^2 + m_2^2)c^4 + 2\sqrt{(E_1^2 - m_1^2c^4)(E_2^2 - m_2^2c^4)} \cos\theta \right]^{1/2}, \quad (2)$$

where m_1 and m_2 are the rest masses of the two particles. In fixed-target experiments, where $E_2 = mc^2 = E_0$, Eq. (2) is written as

$$E_{cm} \approx (2EE_0)^{1/2}. \quad (3)$$

In the case of colliders, $E_{1,2} >> E_0$; while in most existing colliders, $m_1 = m_2$, $E_1 = E_2 = E$ and $\theta = 0$, and thus Eq. (2) becomes

$$E_{cm} \approx 2(E_1E_2)^{1/2} \xrightarrow{E_1=E_2=E} 2E. \quad (4)$$

Combining Eqs. (3) and (4), we have

$$E_{cm,collider} / E_{cm,fixed target} \approx \sqrt{2E/E_0}. \quad (5)$$

It is clear that $E_{cm, collider} >> E_{cm,fixed target}$ when $E >> E_0$, which is the case in high energy physics experiments.

1.3 What are colliders ?

A collider is defined as a special type of accelerator which can produce high-energy collisions between particles of approximately oppositely directed beams. According to the types of particles, colliders are classified as electron-positron colliders (circular or linear), proton-proton (anti-proton) colliders, electron-proton colliders, muon colliders and so on. The 27 colliders in the history of the world are listed in Table 2 [1].

Table 2   Colliders in the world (DR: Double ring; SR: Single ring, LC: Linear collider)

Location	Name	$E_{cm,max}$(GeV)	Start-up
Stanford/SLAC, USA	CBX (e^-e^- DR)	1.0	1963
	SPEAR (e^+e^- SR)	5.0	1972
	PEP (e^+e^- SR)	30	1980
	SLC (e^+e^- LC)	100	1989
	PEP-II (e^+e^- SR)	10.6	1999
Frascati, Italy	AdA (e^+e^- SR)	0.5	1962
	ADONE (e^+e^- SR)	3.0	1969
	DAΦNE (e^+e^- DR)	1.0	1997
Novosibirsk, Russia	VEP-1 (e^+e^- DR)	0.26	1963
	VEPP-2/2M (e^+e^- SR)	1.4	1974
	VEPP-4 (e^+e^- SR)	14	1979
Cambridge, USA	CEA Bypass (e^+e^- SR)	6	1971
Orsay, France	ACO (e^+e^- SR)	1.0	1966
	DCI ($e^\pm e^\pm$ SR)	3.6	1976
DESY, Germany	DORIS (e^+e^- DR)	6.0	1974
	PETRA (e^+e^- SR)	38	1978
	HERA ($e^\pm p$ DR)	160	1992
CERN, Europe	ISR (pp DR)	63	1971
	Sp$\bar{p}$S ($p\bar{p}$ SR)	630	1981
	LEP (e^+e^- SR)	190	1989
	LHC (pp DR)	14000	2004
Brookhaven, USA	RHIC (heavy ions DR)	200/u	1999
	RHIC (pp DR)	500	
Cornell, USA	CESR (e^+e^- SR)	12	1979
KEK, Japan	TRISTAN (e^+e^- SR)	60	1986
	KEK-B (e^+e^- DR)	10.6	1999
Beijing, China	BEPC (e^+e^- SR)	5.6	1989
Fermilab, USA	Tevatron (pp SR)	1000	1987

1.4   How do colliders work ?

A (circular) collider usually consists of an injector, transport lines and a colliding storage ring. The injector provides the initial particles (e^+, e^-; $p, \bar{p}$; e, p, μ^+, μ^- ...) for the storage ring. The beam lines transport the injected beam to the storage ring. The beams are injected into the ring by means of septa, kickers and separators. After the beams are filled, they are accelerated to the desired energy, squeezed, and brought into collision. As the beam loses intensity over its finite lifetime, after several hours, the ring needs to be refilled. Figure. 2 shows the change in beam current with time during a regular operation of the Beijing Electron-Positron Collider (BEPC) [2].

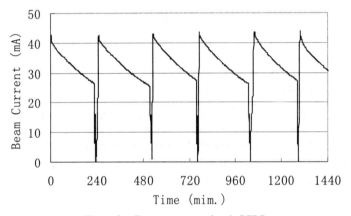

Figure 2.   Beam current vs. time in BEPC

1.5   Electron-positron colliders

In an electron-positron collider, high-energy collisions take place between e^+ and e^- beams. As seen in Table 2, most colliders ever built in the world are electron-positron colliders. The basic reason for this preference is that they provide "clean" collisions between two lepton beams. As a pair of anti-particles with the same properties but opposite charges, electrons and positrons can be injected and accelerated in a single storage ring. Another advantage is that the strong radiation damping in the e^+-e^- storage rings makes beams more stable than in other types of colliders. The synchrotron radiation also makes the e^+-e^- colliders into light sources after the high energy physics experiment is finished; DCI, SPEAR and DORIS are examples.

However, the energy loss, which scales with the fourth power of beam energy because of synchrotron radiation, limits the beam energy in e^+-e^- circular colliders. This has motivated the development of large hadron colliders [3], e^+-e^- linear colliders [4] and μ^+-μ^- colliders [5]. Figure 3 shows the layout of a linear collider in design.

The design and construction of a linear collider in the TeV center-of-mass energy range involves many physics issues and technical challenges. The design principles and technologies for linear colliders are those for linear accelerators and storage rings. In this lecture, we will concentrate on e^+-e^- circular colliders.

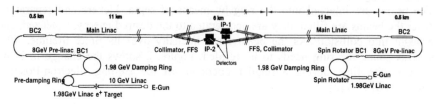

Figure 3. Layout of a electron positron linear collider for E_{cm}=1 TeV

2. Luminosity

2.1 High energy physics and luminosity

Accuracy is a major question for any experiment. The statistic error of a high-energy physics experiment depends on the number of the events to be collected, the larger the size of the data, the smaller the error. The event rate of the experiments is proportional to a parameter called luminosity (L):

$$\frac{dN}{dt} = L \cdot \sigma \tag{6}$$

where σ is the cross section of the experiment.

Luminosity is the most important parameter in evaluating a collider at a given energy. Our efforts in the design, construction and operation of a collider are focused on how to enhance its luminosity.

2.2 Luminosity in beam-beam collisions

Now, let us turn to discussing the luminosity in beam-beam collisions. When two bunches (e^+ and e^-) having $N_{\pm}$ particles and normalized distributions $\rho_{\pm}(x,t)$ collide, the luminosity is given by the overlap integral [6]:

$$L = \frac{1}{c} \int d^3 x dt \, \rho_+(x,t)\rho_-(x,t)\sqrt{c^2\left(\vec{V}_+ - \vec{V}_-\right)^2 - \left(\vec{V}_+ \times \vec{V}_-\right)^2} \, , \tag{7}$$

where $\vec{V}_+$ and $\vec{V}_-$ are the velocities of the e^+ bunch and e^- bunch. The general formula for the luminosity of short bunches is

$$L = \frac{N_+ N_- f_c}{2\pi\sqrt{(\sigma_{x,+}^2 + \sigma_{x,-}^2)(\sigma_{y,+}^2 + \sigma_{y,-}^2)}} \, , \tag{8}$$

where f_c is frequency of collisions. In the case of $\sigma_{x,+} = \sigma_{x,-} \equiv \sigma_x$ and $\sigma_{y,+} = \sigma_{y,-} \equiv \sigma_y$, Eq. (8) is simplified as

$$L = \frac{N_+ N_- f_c}{4\pi\sigma_x\sigma_y}. \tag{9}$$

It is easy to understand in the Eq. (9) why the luminosity is proportional to the number of e^+ and e^-, while the numbers of particles allowable for collision is restricted by the linear beam-beam tune shift, defined as beam-beam parameter $\xi_{x,y}$:

$$\xi_{x,y} = \frac{r_0 N \beta^*_{x,y}}{2\pi\gamma\sigma_{x,y}(\sigma_x + \sigma_y)}, \tag{10}$$

where r_0 is the classic electron radius, N is the number in the opposite bunch, and γ is the relativistic energy of the bunch. Combining Eqs. (9) and (10), we have

$$L = \frac{\pi f_c \gamma^2 \varepsilon_x \xi_x \xi_y (1+r)^2}{r_0^2 \beta^*_y}, \tag{11}$$

where $r = \sigma_y/\sigma_x$. Choosing parameters such that $\xi_x = \xi_y$ is called optimal coupling, which requires

$$r \equiv \frac{\sigma_y}{\sigma_x} = \frac{\beta^*_y}{\beta^*_x} = \frac{\varepsilon_y}{\varepsilon_x}. \tag{12}$$

Combining Eqs. (10), (11) and (12) with $r_0 = 2.818\times10^{-15}$ m, we get a formula that is often used in designing an e^+-e^- ring collider with an assumption of optimal coupling:

$$L(\text{cm}^{-2}\text{s}^{-1}) = 2.17\times10^{34}(1+R)\xi\frac{E\,(\text{GeV})k_b I_b\,(\text{A})}{\beta^*_y\,(\text{cm})}. \tag{13}$$

For bunches of finite length, luminosity is reduced because of the hourglass shape of β-function curve in the interaction region (geometric effect). The reduction factor for Gaussian beams is [7]

$$R(t_x, t_y) = L/L_0 = \int_{-\infty}^{\infty} \frac{dt}{\sqrt{\pi}} \frac{\exp(-t^2)}{\sqrt{(1+t^2/t_x^2)(1+t^2/t_y^2)}}, \tag{14}$$

with

$$t_{x,y}^2 = \frac{2\left(\sigma_{x,y,+}^2 + \sigma_{x,y,-}^2\right)}{(\sigma_{z,+}^2 + \sigma_{z,-}^2)(\sigma_{x,y,+}^2/\beta^{*2}_{x,y,+} + \sigma_{x,y,-}^2/\beta^{*2}_{x,y,-})}. \tag{15}$$

In symmetric colliders with $\sigma_y \ll \sigma_x$, the geometric luminosity reduction due to finite bunch length σ_z and crossing angle ϕ_c of collision is [8]

$$\frac{L_g}{L_0} = \sqrt{\frac{2}{\pi}} \, a \, e^b K_0(b) \xrightarrow{\sigma_z \ll \beta_y^*} \left[1 + \left(\frac{\sigma_z}{\sigma_x} \tan \phi_c\right)^2\right]^{-1/2} , \qquad (16)$$

where

$$a = \frac{\beta_y^*(x)}{\sqrt{2}\sigma_z} \cos \phi_c \quad \text{and} \quad b = a^2 \left[1 + \left(\frac{\sigma_z}{\sigma_x} \tan \phi_c\right)^2\right] , \qquad (17)$$

and K_0 is a zero-order Bessel function.

2.3   Average and integrated luminosity

After beams are brought into collision, luminosity decays with time because of particles are lost for various reasons. Let $L(t) = \exp(-t/\tau)$, where τ is luminosity lifetime. The average luminosity is

$$\langle L \rangle = \frac{\int_0^{t_c} L(t)dt}{t_c + t_f} = \frac{1 - \exp(-t_c/\tau)}{t_c + t_f} L_0 \cdot \tau , \qquad (18)$$

where t_f and t_c are beam filling time and colliding time respectively. The maximum average luminosity is reached when $d\langle L \rangle /dt_c = 0$,

$$\exp(t_c/\tau) = 1 + \frac{t_c}{\tau} + \frac{t_f}{\tau} . \qquad (19)$$

An approximate solution of Eq. (19) is

$$t_c \approx \tau \cdot \ln\left(1 + \sqrt{\frac{2t_f}{\tau}} + \frac{t_f}{\tau}\right) . \qquad (20)$$

The maximum average luminosity is obtained by replacing t_c in Eq. (18) with Eq. (20) :

$$\langle L \rangle_{\max} = L_0 e^{-t_c/\tau} \approx \frac{L_0}{1 + \sqrt{2t_f/\tau + t_f/\tau}} . \qquad (21)$$

The integrated luminosity serves as the figure of merit for a collider. Taking into account the down time and the average luminosity of the collider, an experimental year is about 10^7 s. The integrated luminosity for a collider with a peak luminosity of $L=10^{33}$ cm^{-2} s^{-1} is expected to be $\int L dt = 10^{40}$ cm$^{-2} = 10$ fb^{-1}. Thus, a essential question for a collider is how to guarantee effective operation with a low fault rate, short shutdown time, long beam lifetime, etc. As an example, for $\tau = 6$ hr, $t_f = 0.5$ hr, the optimized colliding time calculated from Eq. (20) is $t_c = 2.4$ h., the maximum average luminosity from Eq. (21) is $\langle L \rangle_{max} = 0.67 \times 10^{33}$ cm^{-2}s^{-1} for $L_0 = 10^{33}$ cm^{-2} s^{-1}, and the integrated luminosity per day is 57.9 pb^{-1}.

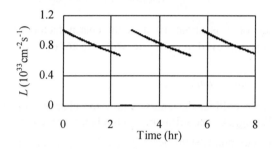

Figure 4.   Luminosity curve with $\tau = 6$ hr, $t_f = 0.5$ hr

2.4   Menochromatization scheme

In experiments such as J/ψ with a narrow resonance whose width-to-energy ratio $\Gamma/E = 2.8 \times 10^{-5}$ is much smaller than the energy spread of the beam ($\sigma_E \sim 10^{-3}$), in order to enhance the production of events, a monochromatization optics is designed in tau-charm factories with large vertical dispersion such that $D^*_{y,+} = -D^*_{y,-}$. In this way, higher-than-average-energy positrons meet lower-than-average-energy electrons at the collision point, making the center-of-mass energy deviation smaller.

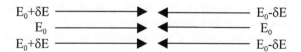

Figure 5.   Principle of the menochromator

The cross section $\sigma(w)$ for the process $e^+ + e^- \rightarrow J/\psi$ near the resonance varies significantly with center-of-mass energy w; the event rate is given by

$$dN / dt = \int_0^\infty \Lambda(w)\sigma(w)dw \quad , \tag{22}$$

where $\Lambda(w)$ is the differential luminosity, defined as [9]

$$\Lambda(w) = \frac{L_0}{\sqrt{2\pi}\sigma_w} \exp\left[-\lambda^2 \left(w - 2E_0\right)^2 / 2\sigma_w^2 \right] \quad , \tag{23}$$

where $\sigma_w = \sqrt{2}\ \sigma_E E_0$ and λ is the monochromatization factor,

$$\lambda = \sqrt{1 + \frac{(D_y^* \sigma_E)^2}{\beta_y^* \varepsilon_y}} \ .$$

(24)

The luminosity becomes

$$L = \int_0^\infty \Lambda(w)dw = L_0 / \lambda \ll L_0 \ .$$

(25)

Typical value of λ is about 10 in a tau-charm factory, and center-of-mass energy resolution is improved to σ_w / λ, i.e. with a factor of λ^{-1}.

2.5 Methods to enhance luminosity

A basic question for collider designers is how to enhance luminosity. Luminosity formula, Eq. (13), suggests the following methods for a given energy E.

- Enhance the maximum beam-beam parameter $\xi_{y,max}$. This ranged from 0.03 to 0.05 for e^+e^- colliders and it is about a constant for any specific ring. There are some ways to increase $\xi_{y,max}$, such as optimizing operating tunes, adjusting coupling, correcting closed orbit distortion, and so on.

- Increase bunch current I_b. This is limited by single-beam phenomena, maximum beam-beam tune shift, RF power, etc. For a given beam-beam parameter ξ, I_b is proportional to the natural beam emittance. Careful control of emittance will help the luminosity.

- Reduce β_y^*. Equation (13) shows that the luminosity increases linearly with the reduction of β_y^* without increasing bunch current. This is why mini-β ($\beta_y^* \sim 3$ cm) or even micro-β schemes ($\beta_y^* \leq 1$ cm) are widely used in the design of colliders. For such schemes, the detector should be designed in a way to provide space for installing insertion quadrupoles near the interaction point (IP), and the bunch length should be well controlled in order to satisfy the condition of $\beta^* \geq \sigma_z$.

- Increase the bunch number k_b. The "pretzel" scheme was developed to increase the bunch number by avoiding unwanted collisions at positions other than IP's by using electrostatic separators in the single storage ring. A straightforward way to increase k_b is to use a double ring for two species of particles. This idea was applied in the early colliders shown in the Table 2 with no success. However, recently developed impedance control and feedback effectively suppress the coupled bunch instability in multibunch operation and use of zero or small crossing angle at IP's which make the modern double-ring factories, PEP-II [10] and KEK-B [11], successful.

- Try a round beam scheme with $\sigma_x = \sigma_y$, $\beta_x^* = \beta_y^*$ and $\varepsilon_x = \varepsilon_y$, where Eq. (9) becomes

$$L = \frac{N f_c \gamma \xi}{r_0 \beta^*} = \frac{N^2 f_c}{4\pi \varepsilon \beta^*} \ . \tag{26}$$

There is some theoretical and experimental evidence that using round beams would significantly increase the beam-beam parameter and enhance luminosity, but more studies are needed before applying this scheme.

3. Beam-Beam Effects

As discussed above, the colliding beam current, and thus the luminosity, are limited by the beam-beam parameter ξ, defined as the linear beam-beam tune shift. For ξ =0.04 and β_y^*=1 cm, the equivalent quadrupole strength experienced by one bunch due to the colliding bunch is calculated as $k_{eff} = 4\pi\xi/\beta_y^* = 50 \text{ m}^{-1}$, which is a very strong quadrupole. This focusing together with its nonlinear parts will significantly influences the beam behavior in colliders.

3.1 Beam-beam force

Beam-beam interactions are nonlinear and collective effects. Let us start with the simplest case: a head-on collision between two round Gaussian bunches of length l with n particles per unit length. The transverse charge distribution is given by

$$\rho(r) = \frac{ne}{2\pi\sigma^2} e^{-r^2/2\sigma^2} \ . \tag{27}$$

Applying Gauss' theorem and Ampere's law, we obtain the radial electric field E_r and the solenoidal magnetic field B_ϕ as

$$E_r = \frac{ne}{2\pi r \varepsilon_0} \left(1 - e^{-r^2/2\sigma^2}\right), \tag{28}$$

and

$$B_\phi = \frac{ne\mu_0 \beta c}{2\pi r} \left(1 - e^{-r^2/2\sigma^2}\right) \ . \tag{29}$$

The radial force applied to the test particle with charge $-e$ at radius r is

$$F_r = -\frac{ne^2}{2\pi r \varepsilon_0} \left(1 \pm \beta^2\right)\left(1 - e^{-r^2/2\sigma^2}\right), \tag{30}$$

where the "minus" in "$\pm$" corresponds to a particle in the same bunch and the "plus"

76

to in the opposite bunch.

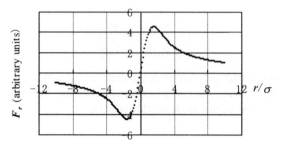

Figure. 6 Beam-beam force

The beam-beam kick in the round beam case is obtained from Eq. (30) as

$$\Delta r' = \frac{dp_r}{p} = \frac{F_r \cdot (l/2)/\beta c}{E\beta/c} = -\frac{2Nr_e}{\gamma r}(1 - e^{-r^2/2\sigma^2}) \xrightarrow{r \ll \sigma} -\frac{Nr_e}{\gamma\sigma^2}r \ . \tag{31}$$

In the general case ($\sigma_x \neq \sigma_y$), the beam-beam kick can be obtained by solving Poisson's equation for the electromagnetic potential of an elliptical Gaussian bunch [12]:

$$V(x,y) = \frac{ne}{4\pi\varepsilon_0}\int_0^\infty dt \frac{1 - e^{-\left(\frac{x^2}{2\sigma_x^2+t} + \frac{y^2}{2\sigma_y^2+t}\right)}}{\sqrt{(2\sigma_x^2+t)(2\sigma_y^2+t)}} \ , \tag{32}$$

and the beam-beam kick is

$$\Delta x' = -\frac{\partial V}{\partial x} \quad \text{and} \quad \Delta y' = -\frac{\partial V}{\partial y} \ . \tag{33}$$

3.2 Beam-beam instability

For small-amplitude particles, the beam-beam force can be approximated as a focusing lens in both x and y planes, with focal length f given by

$$(f_{x,y})^{-1} = \frac{2Nr_e}{\gamma\sigma_{x,y}(\sigma_x + \sigma_y)} \ . \tag{34}$$

The incoherent linear beam-beam tune shifts are written as

$$\Delta v_{x,y} = (f_{x,y})^{-1}\frac{\beta_{x,y}^*}{4\pi} = \frac{Nr_e\beta_{x,y}^*}{2\pi\gamma\sigma_{x,y}(\sigma_x + \sigma_y)} = \xi_{x,y} \ . \tag{35}$$

The perturbation to β^* can be derived from the matrix formalism

$$\frac{\varDelta\beta^*}{\beta^*} = -2\pi\xi \cdot \text{ctg }\mu, \qquad (36)$$

where μ is the betatron phase between two IP's. It can be seen from Eq. (36) that β^* is reduce ($\varDelta\beta^* < 0$) when

$$0 < \mu < \pi/2 \quad \text{or} \quad \pi < \mu < 3\pi/2, \quad \text{i.e.} \quad 0 < \mu/2\pi < 1/4 \text{ or } 1/2 < \mu/2\pi < 3/4. \qquad (37)$$

The motion of a particle perturbed by the opposite beam becomes unstable if the absolute value of the trace of a one-turn matrix is larger than 2:

$$\cos\mu - 2\pi\xi \cdot \sin\mu \geq 1 \quad \text{or} \quad \xi \geq \frac{\text{ctg}(\mu/2)}{2\pi}. \qquad (38)$$

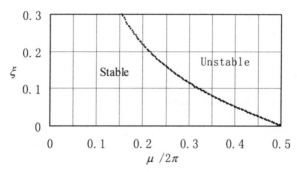

Figure 7. Stability of a weak beam executing small oscillation

In the beam-beam case, if one beam is slightly displaced with respect to the other, coherent oscillations will be induced, from which instability can result under certain condition.

With m bunches per beam, there are $2m$ modes of oscillation; with one bunch per beam there are two modes. The zero-mode is defined as two beams moving up and down together, and the π-mode as two beams moving in opposite directions.

So far we have only considered the linear part of the beam-beam force. In reality, beam-beam interaction is a nonlinear phenomenon, which gives rise to two effects.

First, it introduces dependence of the tune on the betatron amplitude, i.e. tune spread. Second, the nonlinear kick drives nonlinear resonance when the following relation is satisfied:

$$m\nu_x + n\nu_y = k. \qquad (39)$$

There is evidence, both theoretical and experimental, that nonlinear effects play an important role in beam-beam interactions.

Because of the strong damping mechanism in e^+-e^- storage rings due to synchrotron radiation, only low-order resonance needs to be taken into account.

3.3 Long range beam-beam effects

The beam-beam potential and kick are given in Eqs. (31) and (32). When two beams are widely separated, with $d = x^2 + y^2 >> \sigma_x, \sigma_y$, the beam seems like a point charge and the kick is very linear (see Fig. 6). The long-range beam-beam kick is written as

$$i\Delta x' + \Delta y' = -\frac{2Nr_0}{\gamma}\frac{1}{d^2}(ix + y).$$ (40)

The long range beam-beam force produces a perturbation with a tune-shift of

$$\Delta v_x = -\frac{\beta_x}{4\pi}\frac{\partial \Delta x'}{\partial x} = \frac{Nr_0}{2\pi\gamma}\frac{\beta_x(x^2 - y^2)}{d^4},$$

$$\Delta v_y = -\frac{\beta_y}{4\pi}\frac{\partial \Delta y'}{\partial y} = \frac{Nr_0}{2\pi\gamma}\frac{\beta_y(y^2 - x^2)}{d^4}.$$ (41)

Some empirical criteria are used in the design phase [13]:

$$d > (6{\sim}7)\max(\sigma_x, \sigma_y); \quad \xi_{xi}, \xi_{yi} < 10^{-3}{\sim}10^{-4}; \quad \Sigma\,\xi_{xi}, \Sigma\,\xi_{yi} < 10^{-3};$$ (42)

and

$$B = \frac{6.5 \times 10^{-7} n_b}{\gamma}\sqrt{\sum_i^{n_{pc}}\left(\frac{\beta_y \sigma^2}{4r^2}\right)_i^2} \leq 10.$$ (43)

3.4 Beam-beam simulation

Computer simulation techniques have been developed to study the complex process of beam-beam interaction. The simulations are of two types, weak-strong and strong-strong, illustrated in Figure 8.

Figure 8. Weak-strong (a) and strong-strong (b) interactions

Weak-strong mode: only the distribution of the weak beam is changed;
Strong-strong mode: distributions of both beams are changed.

Several computer codes have been developed for beam-beam simulation, such as BEAMBEAM, MAD [14], SIXTRACK, LIFETRAC, BBC [15], and some others. The simulation codes consist of modules that read the lattice parameters and track the particle motion in a machine with the beam-beam force, betatron and synchrotron oscillations, radiation damping and its fluctuation, and even machine imperfection. However, simplification has often taken place in order to focus on the major problem and to save computing time.

For example, in the code for beam-beam interaction with crossing angle (BBC), the simulation algorithm of the code is traced as [16]:

$$x(0) \xrightarrow{\;L\;} x^*(0^*) \xrightarrow{\;SBM\;} x^*(0^*) \xrightarrow{\;L^{-1}\;} x(0) \xrightarrow{\;Ring\;} x(0) \xrightarrow{\;L\;} \cdots$$

Tracking in "Ring" is based on linear unperturbed betatron oscillation with synchrotron radiation damping and its fluctuation:

$$\begin{pmatrix} u \\ u \end{pmatrix}_{n+1} = \lambda \begin{pmatrix} \cos\mu & \beta_0 \sin\mu \\ -\dfrac{\sin\mu}{\beta_0} & \cos\mu \end{pmatrix} \begin{pmatrix} u \\ u \end{pmatrix}_n + \sqrt{(1-\lambda^2)} \begin{pmatrix} r_1 \\ r_2 \end{pmatrix}, \qquad (44)$$

where $\mu = 2\pi\nu$, $\lambda = \exp(-\tau/T_0)$, and r_1 and r_2 are Gaussian random numbers with unit r.m.s. value. In Synchrotron Beam Mapping (SBM) [17], L and L^{-1} are actions for Lorenz and inverse Lorenz transformations applied to treat the beam-beam interaction with a crossing angle. Figure 9 shows the result of a beam-beam simulation in the ν_x-ν_y plane using BBC for the Beijing Tau-Charm Factory (BTCF) [18].

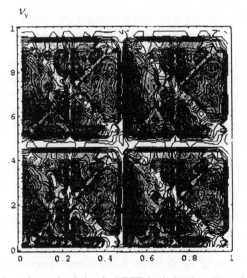

Figure 9. Beam-beam simulation for BTCF, the darker the pattern, the higher the luminosity

3.5 Beam-beam compensation schemes

Beam-beam effects limit the maximum intensity of beams to be brought into collision — see Eq. (10) —, and thus the achievable luminosity, — see Eq. (11). Therefore the question arises: can we do something about beam-beam compensation in order to improve the performance the colliders?

As discussed above, the beam-beam force is an electromagnetic interaction between two beams in collision. Can it be compensated by an external electromagnetic field? Unlike a regular quadrupole, the beam-beam "lens" applies focusing in both the x and the y planes. This defeats the simple compensation schemes with magnets. There are some schemes to compensate the beam-beam force by using a field generated by the space charge of the beams themselves or by foreign particle beams.

A space-charge compensation scheme was tested in DCI at Orsay [19] as illustrated in Figure 10. In this scheme, two pairs of bunches are stored in a double ring-collider. If each bunch has exactly the same intensity and distribution, then the particles in the bunches will encounter no beam-beam force when they meet.

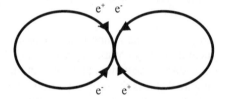

Figure 10. Four beam compensation scheme in DCI

However, this scheme showed no advantage in reality. The experiments indicated that the beam-beam limit with four beams was not significantly different in comparison with that in the two-beam case. Why did the scheme not succeed? The answer was "Nonlinear resonance still strongly affects the four-beam interaction, "with coherent signals due to the collective effects observed [20].

Another space-charge compensation scheme was later proposed for CESR at Cornell. In this proposal, a transverse secondary low-energy electron beam with matched profile intercepts the circulating beam once a turn and then is discarded. Some technical issues still need to be studied, and this scheme has not yet been tested.

The beam-beam force may cause resonance, as shown in Eq. (39). Carefully choice of the parameters of a collider may cancel some of the resonance lines. A set of beam-beam resonances can be eliminated by adjusting the phase advance between IP's such that $\Delta\phi_x=\Delta\phi_y=2\pi(p/N+2k+1)$, then all resonances of order N are canceled [21]. This is a partial compensation scheme and has not been tried in a storage ring.

Much work remains to be done on beam-beam compensation. Readers are encouraged to make efforts on this aspect.

4. Design of an Electron-Positron Collider

4.1 General description

Accelerator physics is applied science rather than pure physics, and is applied to design, construction, operation and improvement of accelerators, etc. The design of an electron-positron collider, like that of other types of accelerators, involves almost all issues of accelerator physics. Figure 11 illustrates the issues and their interrelations in the design.

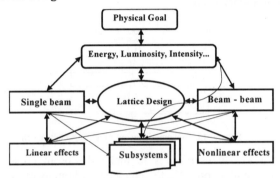

Figure 11. The issues and their interrelations in design a collider.

As shown Figure 11, beam energy, beam intensity, and the luminosity of the collider are determined according to the required physical goal. The design efforts for a collider are aimed at maximizing the luminosity, and many subjects should be studied. Most of the issues in Figure 11 have been discussed in this school and it need be only briefly mentioned here.

4.2 Lattice design

Lattice design is at the center of Figure 11. The design of the lattice for proton and ion machines has to concentrate on a particle-beam optics problem, where the transverse and longitudinal emittances are determined by their injectors and injection schemes, i.e. independent of the lattice. In electron (positron) rings, the synchrotron radiation determines the particle distribution in six-dimensional phase space. The characteristics of the synchrotron radiation, and then the beam emittances, are manipulated and influenced by such lattice parameters as bending radius, focusing strength, and Twiss parameters, described with synchrotron integrals [22].

For the single purpose of an electron-positron storage ring for high energy physics, the lattice should be optimized such that the beam transverse emittance is reasonably large and β_y^* possibly small in order to maximize luminosity, as shown in Eq. (11). The design and optimization criteria for synchrotron radiation storage rings, where small beam emittance is preferred, are different. For a collider such as BEPC with the dual purposes of high energy physics and synchrotron-radiation applications, the lattice should be designed to provide corresponding flexibility.

The lattice in a collider usually consists of three parts, i.e. arc cells, dispersion suppressors, and low-β (mini-β or even micro-β) insertions.

Most high energy accelerators and storage rings apply FODO lattices in their arcs with a periodic sequence of quadrupole magnets of alternating polarity (focusing and defocusing) and bending magnets between the quadrupoles. Figure 12 shows the layout of a FODO cell.

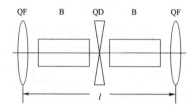

Figure 12. Schematic layout of a FODO lattice cell

The goal of a low-β insertion is to make the beam cross section at the interaction point as small as possible to achieve high luminosity in a collider. The limitations to smaller β^* are strengths and apertures of the low-β quadrupoles, and space needs for the experimental detector, chromaticity, tolerances and closed orbit. The idea of mini-β or even micro-β was to making β^* smaller and smaller by placing the insertion quadrupoles inside the detector without increasing the β-values at the quadrupoles.

The function of the dispersion suppressors is to match the dispersion from its finite value in the arcs to zero at the interaction points. Various types of dispersion schemes are used in colliders. Some keep the FODO structure the same as in the arcs while more independently powered quadrupoles are applied, others change the distribution of bending magnets in the matching section.

Designing the lattice of a collider is a heavy task that needs large amounts of man power and computer time. Many interplays and iterations with other design tasks and subsystems are necessary in determining lattice parameters, defining the space for ring components, chromaticity correction, dynamic aperture study, etc. Fortunately, a number of computer codes are available for lattice design, such as MAD, SAD, MAGIC [23], SYNCH [24], DIMAD [25], MARYLIE [26], TEAPOT [27], TRANSPORT [28], TURTLE [29], PARTRICIA [30], RACETRACK [31], HARMON [32] and some others. The design of a modern collider and its optimization are possible only with computer codes.

As an example, the lattice of the BEPC storage ring has a basic four-fold mirror symmetry with respect to the interaction points. The e^+ and e^- transport lines are launched into the ring near the west and east symmetrical points respectively. The lattice consists of two interaction region insertions, two injection region insertions, and four "trunks" that connect two half insertions in each quadrant. The so-called trunk is a group of FODO quasi-cells in which the geometrical arrangement of the magnets is regular while the strength of each F or D may be different. This kind of lattice pattern differs from the lattice structure mentioned earlier in this subsection. The operation of BEPC shows that the lattice is reliable and flexible. The beam emittance can be adjusted in a wide range with a constant tune, and the phase advance between two injection kickers stays ~ 180° for both the colliding mode (v_x=5.83) and the dedicated SR mode (v_x =9.38).

4.3   Linear end nonlinear perturbations and their correction

Designing an unperturbed lattice under ideal conditions is only the first step toward the complete design of a collider. More sophisticated design needs to take into account magnetic errors, misalignment of accelerator components, and other perturbations. The linear and nonlinear perturbations and their correction involve linear and nonlinear perturbed dynamics as discussed in other lectures at this school [33][34].

The issue of chromaticity correction is mentioned in the last subsection as it is closely related to the lattice design. The chromaticity in an electron-positron collider must be corrected to a slightly positive value in order to avoid head-tail instability. At least two families of sextupoles are required to correct the chromaticity in both the horizontal and the vertical planes. Severe chromatic aberrations exist in colliders due to the low-β insertions. To deal with this problem, more families of sextupoles are often employed.

Another systematic perturbation in a collider is due to solenoids in the detectors needed for secondary-particle identification. The solenoid field gives rise to linear coupling between horizontal and vertical motions and degrades performance of the collider. There are basically two solenoid compensation schemes: anti-solenoid or skew quadupoles. A pair of anti-solenoid magnets placed in the same straight section of the detector can effectively compensate the effects of the detector solenoid. The question is how to find space for the compensation magnets as it is always very crowded inside the detector. Four pairs of skew quadupoles carefully arranged symmetrically around the interaction regions can compensate the solenoid effects. In order to avoid excitation of vertical dispersion, the skew quadupoles should be placed in dispersion-free positions.

For such a small machine as BEPC, the dispersion-free sections in the interaction regions are too short to find right betatron phases for skew quadupoles. One pair of skew quadupoles at the so called magic phase of

$$\phi_x = k_1\pi \quad \text{and} \quad \phi_y = k_2\pi + \pi/2, \tag{45}$$

k_1 and k_2 being integers, has been applied, and the results are in good agreement with the theoretical prediction [35].

The dipole errors due to field deviation in bending magnets and horizontal displacement of quadrupoles cause horizontal closed orbit distortion (COD); and the non-zero roll angle of bending magnets and vertical displacement of quadrupoles cause vertical COD. These random errors need to be simulated and corrected with computer programs so that the corresponding tolerances can be determined.

The quadrupole errors due to field deviation in quadrupole magnets and displacement of sextupole magnets perturb the betatron tune and Twiss parameters. These are also random errors, which need to be studied with computer simulation in order to determine their tolerances.

Multipole errors in all magnets cause nonlinear resonance and degrade the performance of a collider. These are basically the systematic errors due to the finite dimension of the magnetic pole. The tolerances for these magnetic errors should be such that they do not significantly reduce the dynamic aperture dominated by nonlinear effects of the chromaticity sextupoles. Based on the simulation, the field quality requirement and tolerances of the multipole components in magnets can be determined.

Several computer codes have been developed for the study of linear end nonlinear errors, such as MAD, SAD, PETROS [36], DIMAD, MARYLIE, etc. There are always conflicts between desired performance and the reasonable tolerances, and the designers of a collider should discuss and negotiate with hardware experts in order to reach the best compromise.

4.4 Intensity-dependent phenomena

The higher the beam current in a collider, the higher the luminosity that can be obtained. Beam-beam effects limit the current allowable for collisions in a collider, as detailed in the section 3. Another current limitation is due to single-beam collective effects. As electron-positron colliders are operated in the relativistic region, direct interactions among the particles in the same bunch, or space-charge effects, are negligible. There are two types of single-beam effects in colliders. Coherent effects are defined as the interactions between beams and metal environment in the accelerators; incoherent effects include intrabeam scattering, beam-ion effects, and beam-photoelectron interactions. The physics of the single-beam collective phenomena has been well described in this school [37]. Understanding the nature of the phenomena is important for appropriate design to minimize the collective effects. I mention them here only in relation to collider design.

As a measure of interactions between particle beams and their environment, coupling impedance is an essential parameter in the design of a collider. The impedance of the collider is due to all the vacuum components along its circumference. In the design stage, one should make a detailed budget of the impedances of all components. In the design of BTCF for example, the impedance and loss factors of most vacuum components, such as RF cavities, resistive wall, beam position monitors, shielding bellows, masks, slots of antechambers, electrostatic separators, Y-shaped junctions, chambers in the interaction range, collimators, feedback kickers, etc., were calculated. Table 3 summarizes the results [18].

Table 3. Calculated impedance and loss factors for BTCF

Item	Quantity	L (nH)	K_L(V/pC)
RF cavities	3		3.63
Resistive wall			0.51
Beam position monitors	136	5.03	0.45
Shielding bellows	76	3.95	0.17
Masks	52	3.29	0.044
Slots of antechambers	28	0.7	
Chamber in IR	1	3.78	0.047
Y-shaped chambers	2	1.12	1.44
Separators	2	1.34	1.0
Collimators	2	3.14	0.04
Feedback kickers	2	27.0	0.66
Total	304	49.4	7.97

Computer codes such as SUPERFISH [38], URMEL [39], TCBI [40], ABCI [41] and MAFIA can be applied for impedance calculation. The design of the components should be optimized to satisfy not only their functional specification but also the requirements of the impedance budget. The impedance of some components needs to be measured also in the design and R & D stages.

Knowing the impedance of the collider, we can study the thresholds of a variety of instabilities and their growth rate at the designed beam current. Knowing the broadband impedance, we can study single-bunch effects such as head-tail effect, mode coupling, and bunch lengthening. Bunch length is a crucial issue for colliders on the scale of BEPC and BTCF. The impedance should be well controlled so that the threshold of bunch lengthening is above the design bunch current. Higher-order mode (HOM) electromagnetic field in RF cavities will excite coupled bunch instabilities. The dangerous modes and the growth rate of the instabilities they cause should be checked. The resistive wall impedance causes unstable coupled-bunch motion. The growth time of the most dangerous mode of resistive wall instability is estimated to be 5 msec at the design tunes of BTCF [18].

Ions from residual gas ionization or dust may cause ion trapping and fast-ion instability. Photoelectrons generated by synchrotron radiation may interact with positron beams and cause photoelectron or electron cloud instability. These foreign particle instabilities become significant in factory-type machines, where electron and positron beams move in two individual rings. Vacuum components should be designed to avoid these instabilities.

The intensity of beams in a collider decreases exponentially with time. The coherent and incoherent collective effects influence beam lifetime. The beam-gas scattering lifetime has contributions from elastic scattering and bremsstrahlung on nuclei as well as elastic and inelastic scattering on the electrons of gas atoms. The beam-gas lifetime needs to be estimated for the design vacuum pressure and gas composition. When particles execute betatron oscillation with different speeds, Coulomb scattering occurs, and some of them will be lost when their energy deviation becomes larger than the RF bucket height. This intra-beam scattering in a bunch, or Touschek effect, reduces beam lifetime. Depending on beam energy, bunch density and RF voltage, the Touschek lifetime should be carefully checked in the design stage, especially for lower energy machines.

The computer codes BBI [42], ZAP [43] were developed to estimate the growth rates of various types of instabilities. For more sophisticated studies one needs to carry out dedicated simulations.

4.5 Subsystem design

The components of collider include subsystems: RF system, magnet system, power supply system, vacuum system, beam instrumentation system and control system, and etc. Detailed discussion of individual subsystems is beyond the scope of this lecture. Here, I will discuss only the subsystems closely related to the design of an electron-positron collider, taking BTCF as an example.

Electron and positron beams circulating in the vacuum chamber of a collider undergo energy loss, which scales with the fourth power of beam energy:

$$U_0 (\text{MeV}) = 0.0885 \frac{E^4 (\text{GeV}^4)}{\rho (\text{m})}, \tag{46}$$

where U_0 is energy loss per turn and ρ is the bending radius. The lost energy is compensated by RF power. Along with an increase in beam energy, the required RF voltage ($\propto E^4$) and cavity power lost ($\propto E^2$) are steeply increased. As the particles move almost with the speed of light, the RF frequency stays constant in electron-positron colliders.

In designing the RF system of a collider, one needs to choose its parameters and components carefully. The frequency of the RF system is chosen taking into consideration the beam property requirements and industrial availability, which is in the range 50 MHz to 500 MHz. The higher the RF frequency, the smaller the cavity size. On the other hand, the beam-cavity interactions get stronger at higher frequency. However, most recently constructed colliders have 500 MHz RF systems. The HOM issue is of overwhelming importance for storage rings, especially in high luminosity colliders. Single-cell cavities are often adopted to avoid possible modes due to coupling between cells. The computer codes SUPERFISH, ULTRAFISH,

URMEL and MAFIA have been developed for cavity design and optimization. RF windows, a high power source and low level control are also important for stable and reliable operation of the RF system.

Accelerators are often categorized by the weight of iron and copper in their magnets. The magnet system is an essential part of an accelerator. In a collider it usually consists of two types of magnets: general magnets and special magnets. In BTCF, the former contain bending dipoles, normal quadrupoles, large-bore quadrupoles, small-size quadrupoles, sextupoles and dipole correctors; and the latter contain injection kickers, Lambertson septum magnets, wigglers, separators, and vertical deflecting magnets in the interaction region. All the magnets should be carefully designed to meet their specification and required field quality. Mathematics is used in the design of a magnet is to find optimized boundary conditions by solving the magnetostatic part of Maxwell's equations. Computer programs, MARE [44], MAGNET [45], POISSON [46] and TOSCA [47], can be applied for this purpose. Their accuracy is such that in most cases no prototype is required, which saves considerable time and expense.

The power supply system should be designed to meet the requirement of operating the collider with various modes. The specifications of the power supply system, which include DC stability and current ripple, are based on the allowable beam energy stability, tune shift and Twiss parameter perturbation. Furthermore, the design of the power supply system should provide the flexibility to adjust the operation modes, but with consideration of manufacturing cost. Table 4 lists the major requirements for the power supply system of BTCF [18].

Table 4. BTCF power supplies and their current tolerance

Powered Magnets	Number of Power Supplies	Stability & Reproducibility	Current Ripple
Dipoles	4	1×10^{-4}	1×10^{-5}
Quadrupoles	55	1×10^{-4}	1×10^{-5}
Sextupoles	60	1×10^{-4}	3×10^{-5}
Wigglers	1	3×10^{-4}	3×10^{-5}
Correctors	200	3×10^{-4}	3×10^{-5}

The vacuum system in an electron-positron collider presents a technical challenge. To satisfy the requirements of sufficient beam lifetime, low background, and high luminosity, each vacuum device must satisfy demanding design criteria. A very large flux of synchrotron radiation has to be absorbed by photon absorbers to avoid excessive heating or gas generation. For good beam lifetime, the design vacuum pressure for a collider is usually around 1×10^{-9} Torr in the arcs. While in the interaction region it should be better than 5×10^{-10} Torr to assure a low background in the detector. The design should be reliable so that beam lifetime is achieved soon after the initial stored beam is injected on start-up and so that the vacuum system recovers quickly after sections are vented or opened for alteration. It is also important, as mentioned in the previous subsection, that the chamber wall be designed as smooth as possible to minimize coupling impedance.

The beam instrumentation system, which consists of various beam monitors and signal processing electronics, is an important part of a collider. This system must provide precise and sufficient information about the beams so that machine operators can improve the performance of the collider by monitoring the beam behavior, optimizing the lattice parameters, correcting the closed orbit, increasing the injection efficiency, and enhancing the beam current and luminosity. The approach to designing the beam instrumentation system may differ from one collider to another, but the basic considerations are the same. The design philosophy for the beam instrumentation system of BTCF [18] is as follows:

- The beam parameters should be measured with appropriate precision and speed;
- The dynamic range of the measurement must be large enough to cover a broad range of beam intensities and energies;
- The coupling impedance of all beam monitors needs to be as small as possible;
- Interference and disturbance to beams must be minimized during measurement;
- Processes of all measurements should be controlled by a computer as well as operated manually;
- Commercial products should be used wherever possible to reduce the R & D cost and efforts.

The beam instrumentation system in a collider usually consists of a DC current transformer (DCCT) for average beam current measurement, beam position monitors (BPM) for measuring closed orbit and single-pass trajectory, fluorescent screens for monitoring the injected beam position and profile, stripline electrodes for tune measurement, synchrotron radiation monitors for beam profile measurement, and transverse and longitudinal feedback to suppress beam instability.

The control system is the "brain and nerve" of a collider, from which the operators know the status of every part of the collider and the behavior of the beam so that they can operate the machine properly. The control system must be capable of operating the collider with separate missions and controlling devices in central and local control rooms; controlling and monitoring all the equipment in the collider including RF system, magnet power supplies, vacuum units, and beam instruments; and providing a user-friendly man-machine interface for operators. The control system usually consists of computers, network, interfaces, timing, video and communication, and interlock. With the rapid development of computer and network technology, the control systems for colliders have become more powerful and more efficient. A common used control architecture is called "standard model, " i.e. workstation with UNIX system – VME with real-time operating system – low-level device interface. However, any individual collider involves many special requests to the control system. And a control system with high reliability, high performance and high accuracy is a precondition of the success of a collider.

Up to now, cryogenic and superconducting technology, the major subject of this school, has not yet been involved. This is a relatively new technology, which has been widely applied in accelerators in the 1990s. Superconductivity eliminates Joule effects so that the running cost can be reduced, and makes higher electromagnetic fields available. In electron-positron colliders, superconducting magnets are used in high gradient quadrupoles for squeezing β^*, to get a strong solenoid field to compensate coupling due to the detector, and in wigglers which need a higher magnetic field. Superconducting RF cavities can provide a higher accelerating field so that the number of cavities is reduced, resulting in a lower machine impedance. Because of the high Q-value the superconductng cavities can be shaped into a very smooth structure with a large beam port, thereby reducing the parasitic impedance due to the HOM's and improving the stability of the high intensity bunches. Furthermore, the large-bore beam tube of superconducting cavities is specially designed to enable the transmission of all HOM's, and the HOM power leaking out into the pipe will be absorbed by ferrite located outside the cryostat at the room temperature beam tube. Consequently, one obtains a nearly mono-mode cavity and gets ride of the HOM coupler at the cavities. Along with the application of supercondcting technology in the field of accelerators, the performance of electron-positron colliders will be further improved. The successful school of AAS'99 will certainly promote the application of supercondcting technology in accelerators.

References

1. Rees J., Colliders, in Handbook of Accelerator Physics and Engineering, World Scientific (1998), ISBN 9810235003, pp.11–13.
2. Fang S.X., Chen S.Y., Beijing Electron-Positron Collider, Proc. 14th Int. Conf. on High Energy Accelerators (1989), p. 51.
3. The Large Hadron Collider – Conceptual Design, CERN/AC/95-05 (LHC) (1993).
4. Tigner M., Nuovo Cimento 37 (1965) 1228.
5. μ^+-μ^- Collider, A feasibility study, BNL52503, Fermilab-Conf-96/092, LBNL-38946 (1996).
6. Moller C., Danske Vidensk K., Selsk. Mat.-Fys. Medd. 23 (1945) 1.
7. Furman M.A., Proc. Particle Accelerator Conf. 1991, p. 422.
8. Napoly O., Particle Accelerators 40 (1993) 180.
9. Jowett J., Springer-Verlag Lecture Notes in Physics 425 (1992) 79.
10. PEP-II, An Asymmetric B-Factory, LBLPUB5379, SLAC418 (1993).
11. KEK B-Factory Design Report, KEK Report 95-7 (1995).
12. Montague B., CERN 68-38, (1968).

13. Tenneson J.L., Long-Range Forces in APIARY-6, ABC Note 28 (1991).
14. Iselin F.C., The MAD program, Reference Manual, CERN LEPTH/85-15 (1985).
15. Hirata K., BBC User's Guide, CERN SL-note 97-57 AP (1997).
16. Hirata K., Phys. Rev. Lett. 74 (1995) 2228.
17. Hirata K., Moshammer H., Ruggiero F., Particle Accelerators 40 (1993) 205.
18. Feasibility Study Report on Beijing Tau-Charm Factory, IHEP-BTCF Report03 (1996).
19. Augustin J.E. et al, Proc. 7th Int. Conf. on High Energy Accelerators (1969), p. 113.
20. Zyngier G., AIP Proc. 57 (1979) p. 136.
21. Peggs S., Proc. Workshop on AP Issues for SSC, UM HE 84-1 (1984) p. 58.
22. Zhao Z.T., These proceedings.
23. King A.S., Lee M.J., Lee W.W., SLAC183 (1995).
24. Garren A.A., Eusebio J.W., UCID10153, (1975).
25. Servranckx R., User's guide for program DIMAD, SLAC285, UC28 (1985).
26. Dragt A., et al., MARYLIE, A program for charged particle beam transport systems based on Lie algebraic methods, University of Maryland (1985).
27. Schachinger L., Talman R., SSC52 (1985).
28. Brown K.L., Carey D.C., Iselin F.C., Rothacker F., TRANSPORT, A computer program for designing particle beam transport systems, CERN 80-04 (1980).
29. Brown K.L., Carey D.C., Iselin F.C., TURTLE, A computer program for simulating particle beam transport systems, CERN 74-2 (1974).
30. Wiedemann, PEP Note 220 (1976).
31. Wrülich, Particle tracking in accelerators with higher multipole fields, Springer-Verlag Lecture Notes in Physics 215 (1984).
32. Donald M.H.R., PEP Note 311 (1979).
33. Qin Q., These proceedings.
34. Guo Z.Y., These proceedings.
35. Fang S.X., Huang N., Zhang C., Compensation of solenoid effects in BEPC, Proc. 4th China-Japan Joint Symp. on Accelerators for Nuclear Science and Their Application, (1990) p. 24–26.
36. Steffen K., Kewisch J., DESY PET 76/09 (1976).
37. Chao A.W., These proceedings.
38. Halbach K., Holsinger R.F., Particle Accelerators 7, 213 (1976).
39. Weiland T., Nucl. Instrum. Methods, 216, 329 (1983) p. 329–348.
40. Weiland T., Particle Accelerators 15, 245, 292 (1984).
41. Chin Y.H., User's guide for ABCI (Azimuthal Beam Cavity Interaction) Version 8.8, CERN SL/94-02(AP), LBL-35258, UC414 (1994).
42. Zotter B., BBI – a program to compute bunched beam instabilities in high energy accelerators and storage rings, CERN LEP/TH 89-74, (1989).
43. Zisman M.S., Chattopadhyay S., Bisognano J.J., ZAP user's manual, LBL21270, UC28 (1986).

44. Perin E., van der Meer S., The program MARE for computation of two-dimensional static magnetic fields, CERN 67-7 (1967).
45. Iselin F.C., Two-dimensional magnetic fields including saturation, CERN Program Library, T600 (1971).
46. Iselin F.C., Solution of Poisson's or Laplace's equation in two-dimensional regions, CERN Program Library, T604 (1976).
47. Armstrong A.G.A.M., Riley C.P., Simkin C.P., TOSCA User Guide, Version 3.1, RAL Report R1-81-070 (1980).

INTRODUCTION TO THE INJECTOR LINAC FOR ELECTRON RING ACCELERATORS

SHU-HONG WANG

IHEP/KEK

1 Introduction

An RF linac is usually chosen to be the injector for ring accelerators (Collider and Synchrotron Radiation Facility), since in a dc machine it is very difficult to have a stable voltage higher than a few MeV.

1.1 What Injector Linac Performance is Required by a Ring Machine?

It is most important for an electron ring machine to have integrated luminosity (for e^+- e^- colliders) or integrated brightness (for SRFs) as high as possible. Thus its injector linac is usually required to have the following:

♦ A high electron/positron current from the linac, and a high injection rate into the ring, to minimize injection time. Thus a high bunch current, multi-bunch in a bunch train, and/or a high repetition rate are preferred. Particularly for e^+-e^- colliders, a high current of primary e^- beam is needed to yield high current of e^+ beam at the conversion target.

♦ Full energy injection is preferred, either with an injector linac only, or with the addition of a booster ring to pre-ramp the e^+/e^- energy before injection into the main ring.

♦ The energy spread and phase spread of the beam from the linac should be smaller than the longitudinal acceptance of the ring machine.

♦ The transverse emittance of the ring beam is dominated by the equilibrium of the SR damping effect and the quantum excitation effect of electrons (or positrons) in the ring, and can be well adjusted by tuning the ring parameters. Thus the ring beam emittance is not directly dominated by the linac beam emittance. However, a smaller emittance of the linac beam is preferred to facilitate beam transport and injection into the ring.

♦ Stable and reliable operation of the injector linac is important for highly effective operation of the ring machine. Thus, in addtion to having high quality hardware of linac elements, some feedback systems (energy and phase) are required.

1.2 Topics of this Lecture

Because of the limited time (50 min.) for this lecture, we have to concentrate on the elementary concepts of the commonly used room-temperature, disk-loaded, traveling -wave (TW) rf linac, including the concept of particle acceleration with an rf linac, and the essential accelerating structure. Other important elements of linac, such as the electron gun and the rf pulse compressor are also briefly described.

As an injector linac is operated at high beam current, the interaction between bunch charge and accelerating structure may become an important issue. The high current bunch excites a wakefield, which affects the electron's motion longitudinally and transversely, and thus may dilute the beam emittance and energy spread. Some advanced techniques used to cure such dilution have been developed in the past years, readers are referred to other lectures and/or related papers.

2. Elementary Principles of the Electron Injector Linac

2.1 Acceleration with the RF Linac

Assuming an electric-magnetic field travels in a uniform cylindrical waveguide, its fundamental mode TM_{01} has the EM components, longitudinal electric field E and azimuthal magnetic field B, as shown in Fig.1. Their distributions are analytically described in the following expressions :

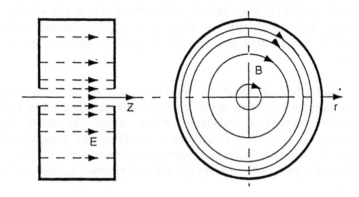

Figure 1. EM field pattern of TM_{01} mode

$$E_z(r,z,t) = E_0 J_0(k_c r) e^{j\omega t - k'z},$$

$$E_r(r,z,t) = jE_0[1 - (\frac{\omega_{cr}}{\omega})^2]^{1/2} J_1(k_c r) e^{j\omega t - k'z},$$

$$E_\theta = 0,$$

$$B_\theta(r,z,t) = j\mu_0 E_0 J_1(k_c r) e^{j\omega t - k'z},$$

$$B_r = B_z = 0.$$

J_0 and J_1 are zero- and first-order Bessel functions respectively. $k_c = \omega_{cr}/\omega$ is the wave number that corresponds to the cutoff frequency of the waveguide, ω_{cr}. The value of k_c is obtained from the first root of $J_0(k_c r) = 0$,

$$k_c r = \frac{\omega_{cr}}{c} \cdot r = 2.405 \; ;$$

$k' = \alpha + jk_0$, where α is the field attenuation factor due to the rf loss on a resistive wall and $k_0 = \dfrac{\omega}{v_p}$ is a wave number with frequency ω and phase velocity v_p, which is equal to the velocity of light, c.

Let us first consider the case of no power loss (ideal conductor, $\alpha = 0$), then its propagation property (dispersive relation) is as follows:

$$k_0^2 = (\frac{\omega}{c})^2 - k_c^2 = (\frac{\omega}{c})^2 - (\frac{\omega_{cr}}{c})^2 .$$

It describes the relations among k_c, ω and k_0 in the form $J_0(k_c r) = e^{j(\omega t - k_0 z)}$, as shown in Fig. 2.

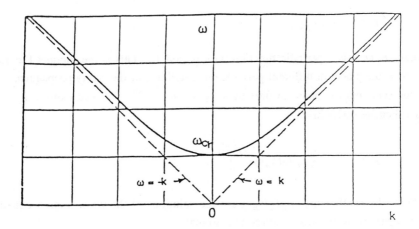

Figure 2. Dispersion curve for a uniform waveguide

For TM$_{01}$ to exist in the waveguide, k_0 should be a real number, so that $\omega \geq \omega_{cr}$. This means that only the waves with $\omega \geq \omega_{cr}$ can be propagated in the waveguide. But their phase velocity is

$$v_p = \frac{\omega}{k_0} = \frac{c}{\sqrt{1 - (\omega_{cr}/\omega)^2}} \geq c \quad .$$

Obviously, these waves can not resonantly accelerate electrons.

To have an accelerating structure in which the propagated waves have $v_p \leq c$, we must modify the structure to slow down the v_p, for instance, by introducing a periodic disk-loaded structure, as shown in Fig. 3. Then the wave amplitude is periodically modulated:

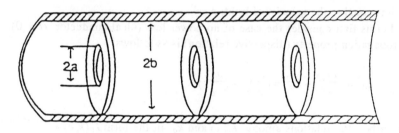

Figure 3.   Disk-loaded TW structure

$$E_z(r,z,t) = E_L(r,z)e^{j(\omega t - k_0 z)},$$

where $E_L(r,z)$ is a periodic function with period L_c. This is the Floquet theorem: at the same place in different periods, the amplitude of the stably propagated field is the same but the phase differs by a factor of $e^{jk_0 L}$. We can express $E_L(r,z)$ as a Fourier series in z:

$$E_L(r,z) = \sum_{n=-\infty}^{\infty} E_n J_0(k_n r)e^{-j\frac{2\pi n}{L}z},$$

where the coefficients $E_n J_0(k_n r)$ are the solutions of the wave motion equation with cylindrical boundary condition, so that

$$E_z(r,z,t) = \sum_{n=-\infty}^{\infty} E_n J_0(k_n r)e^{j(\omega t - k_n z)},$$

where k_n is the wave number of the nth space harmonic wave, $k_n = k_0 + 2\pi n/L$ which has the phase velocity of $v_{np} = \dfrac{\omega}{k_n} = \dfrac{\omega}{k_0(1 + 2\pi n/k_0 L)} \leq c.$

With the above expressions we find that :

♦ A traveling wave consists of infinite space harmonic waves, as shown in Fig. 4.

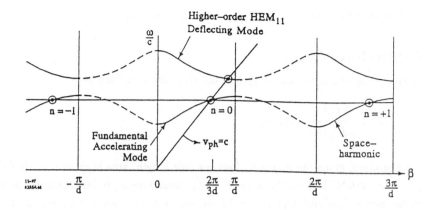

Figure 4. Brillouin diagram for a periodically loaded structure

♦ Harmonic waves with $n > 0$, propagating in the $+z$ direction, are forward waves; those with $n < 0$, propagating in the $-z$ direction, are back waves.

♦ If each forward wave has the same amplitude and phase velocity as a back wave, then they form a standing wave. Therefore a method of analysis using space harmonic waves can be used to describe both a standing wave and a traveling wave.

♦ Because the various space harmonic waves have different phase velocities, only one of the harmonic waves can be used to resonantly accelerate particles. The fundamental mode $(n = 0)$ generally has the largest amplitude and hence is used for acceleration.

♦ When a TW is used to accelerate particles, a particle that "rides" on the wave at phase φ_0 and moves along the axis has an energy gain per period (L_c, cell length) of

$$\Delta W = e\, E_0 L_c \cos\varphi_0,$$

where E_0 is the field on axis averaged over a period: $E_0 = \dfrac{1}{L_c}\int_0^{L_c} E_z(0,z)dz$

♦ Figure 4 also shows a second upper branch, which is one of an infinity of such high-order modes (HOMs), and intercepts the $v_p = c$ line. These modes are so-called wake-fields, which can be excited by the transversely off-set beam.

2.2 Essential Parameters of a TW Accelerating Structure

2.2.1 Shunt-Impedance Z_s

The shunt-impedance per unit length of the structure is defined as

$$Z_s = \frac{E_a^{\ 2}}{-dP_w/dz} \quad (\text{M}\Omega/\text{m}).$$

It expresses, given the rf power loss per unit length, how high an electric field E_a can be established on the axis. $P_w \propto E_a^2$, therefore Z_s is independent of E_a and the power loss and depends only on the structure itself: its configuration, dimension, material and operating mode.

2.2.2 Quality Factor Q

The unloaded quality factor of an accelerating structure is defined as

$$Q = \frac{\omega U}{-dP_w/dz},$$

where U is the stored energy per unit length of structure. The Q also describes the efficiency of the structure. With this definition one can see that, given the stored energy, the higher the Q, the less the rf loss; or given rf loss, the higher the Q, the higher the E_a (since $U \propto E_a^2$).

2.2.3 Z_s/Q

With the definitions of Z_s and Q, we have $Z_s/Q = E_a^2/\omega U$. This defines, for establishing a required electric field E_a in a structure, the minimum stored energy required. Obviously Z_s/Q is independent of power loss in the structure.

2.2.4 Group Velocity v_g

Group velocity is the velocity at which the field energy travels along the waveguide, therefore it is given by

$$v_g = P_w / U \ ,$$

where P_w is the power flow, defined by integrating the Poynting vector over a transverse plane. For TM_{01} mode,

$$P_w = \int_0^a E_r H_\theta \, 2\pi r dr \ ,$$

where a is the iris radius. For TM_{01} mode, $E_r \propto r$ and $H_\theta \propto r$, so that $v_g \propto a^4$.

2.2.5 Attenuation Constant τ_0

We define the attenuation constant to as

$$\tau_0 = \int_0^{L_s} \alpha(z) dz \ ,$$

where $\alpha(z)$ is the attenuation per unit length of the structure, as mentioned in section 2.1. This is one of the most important parameters for a TW structure, since it defines the ratio of output power to input power for an accelerating section (of length L_s), and determines the power loss per unit length

$$P_{out} = P_{in} e^{-2\tau_0} \ , \qquad \frac{dP_w}{dz} = \frac{P_{in}}{L_s}(1 - e^{-2\tau_0}).$$

It is clear that, the larger the τ_0, the smaller the output power, and hence the higher the rate of power use. On the other hand, a smaller τ_0 gives a larger group velocity of the structure and thus a larger iris radius ($v_g \propto a^4$) and a larger transverse acceptance. This is important in the positron injector design, since the positron beam usually has a large transverse emittance at the beginning of acceleration. Finally, τ_0 should be chosen by a compromise between these two effects. The output power is usually absorbed by a load installed at the end of the section, as shown in Fig. 5.

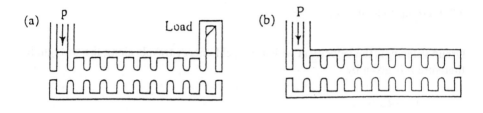

Figure 5. Power absorber at the end of TW section
(a) Disk-loaded TW Structure   (b) SW Structure

2.2.6 Working Frequency f_0

The working frequence is one of the basic parameters of the structure, since it affects most of the other parameters according to the following scaling laws:

♦ Shunt-impedance $Z_s \propto f_0^{1/2}$ ♦ Quality Factor $Q \propto f_0^{-1/2}$

♦ Total rf peak power $P_{tot} \propto f_0^{-1/2}$ ♦ Minimum energy stored $Z_s/Q \propto f_0^{-1}$

♦ RF energy stored $U \propto f_0^{-2}$ ♦ Power filling time $t_F \propto f_0^{-3/2}$

♦ Transverse dimension of structure, a and $b \propto f_0^{-1}$

The final choice of f_0 is usually made by adjusting all of the above factors and by considering the available rf source as well. Most electron injector linac work at a frequency of about 3000 MHz (S-band), e.g. 2856 MHz ($\lambda \approx 10.5$ cm) for the SLAC linac and many others.

2.2.7 Operation Mode

Here we define the operation mode, which is specified by the rf phase difference between two adjacent   accelerating cells. For instance 0-mode, $\pi/2$-mode, $2\pi/3$ - mode and π-mode are the operation modes that have the phase differences 0, $\pi/2$, $2\pi/3$ and π respectively between two adjacent cells, as shown in Fig. 6.

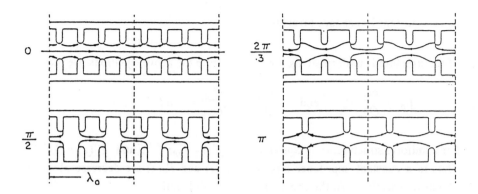

Figure 6. Operation modes

For a SW structure, 0-mode or π-mode has the highest shunt-impedance but the lowest group velocity, thus it has high accelerating efficiency but may not be stable in operation. On the other hand, the π/2-mode has the lowest shunt-impedance but the largest group velocity, and thus has lower accelerating efficiency but high operation stability. For a structure with both high efficiency and high stability, the solution is to use a so-called bi-periodic structure, e.g. a side-coupled cavity structure, which combines the advantages of π-mode and π/2-mode. For a disk-loaded TW structure the optimum operation mode is the 2π/3-mode, because it has the highest shunt-impedance.

2.3 Disk-Loaded Waveguide Structure

With the definitions of structure parameters mentioned above, the rf power distribution along the linac section is

$$\frac{dP_w}{dz} = -\frac{\omega P_w}{Qv_g} = -2\alpha_0 P_w,$$

where α_0 is the attenuation per unit length of structure, $\alpha_0 = \dfrac{\omega}{2Qv_g}$.

2.3.1 Constant-Impedance Structure

If the structure is uniform along the z axis, from the above equations we have

$$E_a^2 = \frac{\omega Z_s}{Q v_g} P_w \quad \text{and} \quad \frac{dE_a}{dz} = -\frac{\omega E_a}{2Q v_g} = -\alpha_0 E_a.$$

For a uniform structure, α_0 = constant, $E_a(z) = E_0 e^{-\alpha_0 z}$, and $P_w(z) = P_0 e^{-2\alpha_0 z}$. Thus in a constant-impedance structure, $E_a(z)$ and $P_w(z)$ are decrease along the z axis in a section. At the end of a section with length L_s,

$$E_a(L_s) = E_0 e^{-\tau_0} \quad \text{and} \quad P_w(L_s) = P_0 e^{-2\tau_0},$$

where $\tau_0 = \alpha_0 L_s = \dfrac{\omega L_s}{2Q v_g}$ is the section's attenuation. The energy gain of an electron that " rides " on the crest of the accelerating wave and moves to the end of section is

$$\Delta W = e \int_0^{L_s} E_a(z) dz = e E_0 L_s \frac{1 - e^{-\tau_0}}{\tau_0},$$

Using $E_0^2 = 2 Z_s \alpha_0 P_{in}$ (P_{in} = input power) $\quad \Delta W = e \sqrt{2 Z_s P_{in} L_s} \cdot (\dfrac{1 - e^{-\tau_0}}{\sqrt{\tau_0}}).$

For an optimized design of a constant impedance structure, we should maximize ΔW. Given P_{in} and L_s we make

$$Z_s \quad \text{maximum} \quad \text{and} \quad (\frac{1 - e^{-\tau_0}}{\sqrt{\tau_0}})_{max} \Rightarrow \tau_0 = 1.26.$$

Given L_s and Q, we can obtain the optimized group velocity v_g. Obviously, the smaller v_g, the bigger τ_0; and the bigger v_g, the lower E_0. An effective way to control v_g is to adjust the iris radius a along the section. On the other hand, the power filling time of a waveguide is

$$t_F = L_s / v_g = 2\pi \tau_0 / \omega.$$

To decrease t_F, then, τ_0 should be adjusted to be < 1.26.

2.3.2 Constant Gradient Structure

To keep $E_a = E_0 =$ constant along the structure, the structure is not made uniform, and α_0 has a z dependence, $\alpha_0 = \alpha_0(z)$. The question is how to determine $\alpha_0(z)$. Let us change the radii of the structure, a and b, to vary v_g along the section, and keep the variations of Q and Z_s along z so small that they can be neglected, then we have

$$dP_W / dz = -2\alpha_0(z)P_w \quad \text{and} \quad P_{L_s} = P_0 e^{-2\tau_0},$$

where $\tau_0 = \int_0^{L_s} \alpha_0(z)dz$ is a section's attenuation. Since $E_a^2 = -Z_s \dfrac{dP_w}{dz}$, to keep $E_a =$ const. We need $dP_w / dz =$ const. so that

$$P_w(z) = P_0 + \frac{P_{L_s} - P_0}{L_s} z = P_0 \left[1 - \frac{1 - e^{-2\tau_0}}{L_s} z \right].$$

Thus in a constant gradient structure, P_w should be linearly decreased along the structure.

With $dP_W / dz = -2\alpha_0(z)P_W$ and $v_g(z) = \omega / 2Q\alpha_0(z)$, we have

$$\alpha_0(z) = \frac{1}{2L_s} \cdot \frac{1 - e^{-2\tau_0}}{1 - \dfrac{z}{L_s}(1 - e^{-2\tau_0})} \quad \text{and} \quad v_g(z) = \frac{\omega L_s}{Q} \cdot \frac{1 - \dfrac{z}{L_s}(1 - e^{-2\tau_0})}{1 - e^{-2\tau_0}}.$$

Thus in a constant gradient structure, the $v_g(z)$ also decreases along the structure in the same way as $P_W(z)$. The energy gain for an on-crest particle is

$$\Delta W = e \int_0^{L_s} E_a(z)dz = eE_0 L_s.$$

Since $E_0^2 = -Z_s \dfrac{dP_{L_s}}{dz} = \dfrac{Z_s P_0}{L_s}(1 - e^{-2\tau_0})$, then $\Delta W = e\sqrt{Z_s P_0 L_s (1 - e^{-2\tau_0})}$.

To have ΔW_{max}, we should have

Z_s maximum and τ_0 maximum $\Rightarrow$ all power should be lost in the structure.

On the other hand we should also consider the filling time, $t_F = 2Q\pi / \omega$, and τ_0 should be chosen by a compromise among some effects.

An example of a SLAC constant gradient structure is shown in Fig. 7. Each section is designed to be a tapered structure: $2b$ (8.4 to 8.2 cm), $2a$: (2.6 to 1.9 cm), $v_g / c \approx 0.021$ to 0.007, $L_s = 3.05$ m, and $Z_s \approx 57$ MΩ/m. The advantages of the constant gradient structure are its uniform power loss and lower average peak surface field, thus most injector linacs are designed as constant gradient structures.

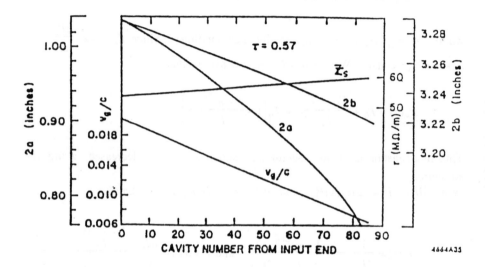

Figure 7. Parameters of a SLAC constant gradient structure

2.4 Electron Gun and Preinjector

Two types of electron preinjector are commonly used: a dc high voltage gun with a bunching system and an rf gun followed by a short accelerating structure.

The dc electron gun has a cathode (thermionic or photo-cathode) and an anode. It produces electrons with pulse lengths of 1 μs to several μs and a beam energy of 50 keV to 500 keV. If the gun uses a thermionic cathode, then a wire-mesh control grid is needed to form a beam pulse, which normally works at a voltage of about minus 50V with respect to the cathode; if the gun uses a photo-cathode, then the electrons are produced by the photo-electric effect, using a laser pulse incident on the cathode, and no grid in the gun.

Since the electrons from the gun have $v < c$ (e.g. ~ 0.5 c), the electron bunch can be shortened by using a bunching system that modulates the electron's velocity with an rf field in the cavity or in the waveguide (bunchers) , followed by a drift space.

The first stage of bunching the beam from the dc high voltage gun is commonly accomplished by using standing wave (SW) single-cavity bunchers, followed by some traveling wave (TW) bunchers for further bunching and acceleration. Usually the first few cells of the TW buncher have $v_p < c$ in order to synchronize with the beam. Figure 8 shows a SLAC preinjector system. It consists of a dc high voltage gun (80 kV), two sub-harmonic SW bunchers (178.5MHz), a four-cell TW fundamental frequency buncher (2856 MHz, $v_p = 0.75$ c), and a fundamental frequency accelerator (2856 MHz, 3 m long, $v_p = c$). This preinjector can bunch a beam of 7×10^{10} electrons with bunch length 2.5 ns (FWHM) and energy 120 keV to a bunch of 5×10^{10} electrons in 20 ps at 40 MeV.

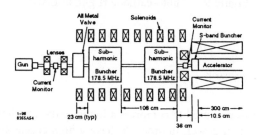

Figure 8. SLAC Preinjector System

The rf gun is followed by an accelerating structure, since the electrons from the cathode are soon bunched by the rf field. The rf gun consists of one or more SW cavities with the cathode installed in the upstream wall of the first cavity. Compared with the dc gun, the rf gun has the advantage of quickly accelerating electrons to relativistic velocity (about 5 to 10 MeV), which avoids collective effects such as the space-charge effect and provides a shorter bunch length and lower beam emittance at the cathode. However, the rf gun has some time-dependent effects due to its time-dependent rf field, which may dilute the performance of electron bunches.

Like the dc gun, the rf gun can have a thermionic cathode or a photo-cathode. As an example, Fig. 9 shows an L-band photo-cathode rf gun at LANL, which consists of a single half-cell cavity at L-band, a Cs_3Sb cathode, and a laser pulse train with wavelength 522 nm. The electron pulse pattern produced from the cathode is a bunch train of 6 µs, with micro-bunches of 53 ps (FWHM), separated by 9.2 ns. A high bunch charge of 27 nC was produced in each micropulse. At the bunch charge of 10 nC, the measured emittance is about 3×10^{-5} m.

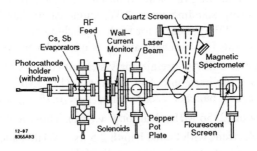

Figure 9.   Photo-cathode rf gun at LANL

2.5 RF Pulse Compressor

The rf pulse compressor is used to compress a longer pulse at a lower power into a shorter pulse with higher power, in order to have a high accelerating gradient. So far, three methods of rf pulse compression are bejing used: the SLAC Energy Development (SLED), which stores rf energy in rf cavities; the SLED-II, which stores rf energy in long resonant delay lines; and the Binary Pulse Compressor (BPC), which stores rf energy in TW delay lines. Most injector linacs use the SLED compressor, while the SLED-II and BPC are used for linear colliders. Here we decribe only SLED, as an example.

The principle of SLED can be seen in Fig. 10, where part (a) shows non-SLED operation, with the rf power from klystron directly transmitted to the linac. Part (b) shows the SLED system, which has two major components: a π-phase shifter on the drive side of the klystron, and two high-Q ($Q_0 = 100,000$) cavities on the output side of the klystron, operated at TE_{015}-mode and connected to a 3-db coupler. During the first part of the pulse, the phase of the rf drive signal is

reversed and the rf cavities fill up with energy at that phase. The fields emitted by
the cavities (E_e) add to the field reflected by the cavity coupling irises (RE_{in}),

$$E_{out} = E_e + RE_{in} ,$$

and the power flows toward the accelerator sections.

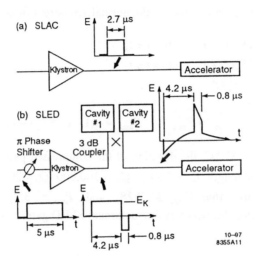

Figure 10.   Principle of SLED: (a) operation without SLED, (b) with SLED

Based on the principle of energy conservation, we have the following power
equilibrium equation:

$$P_{in} = P_{out} + P_c + \frac{dU_c}{dt},$$

where P_{in}, P_{out} and P_c are the input power, output power and power lost in
cavities respectively, and U_c the stored energy in the cavities. Starting with this
equation we can deduce the basic SLED differential equation, which relates the
emitted field E_e to the input field E_{in}, as follows:

$$T_l \frac{dE_e}{dt} + E_e = \alpha E_{in} ,$$

where $\quad T_l = Q_l / \pi f_0$, $\quad \alpha = 2\beta /(1 + \beta)$, $\quad Q_l = Q_0 /(1 + \beta)$ and $\quad \beta$ is the cavity coupling coefficient $\beta = Q_0 / Q$. If $E_{in}(t)$ is normalized as follows

$$E_{in}(t) = \begin{cases} 1 & 0 \le t < t_1 \\ -1 & t_1 \le t \le t_2 , \\ 0 & t > t_2 \end{cases}$$

then by solving the equation, we get the normalized output field,

$$E_{in}(t) = \begin{cases} E_e(t) - 1 & 0 \le t < t_1 \\ E_e(t) + 1 & t_1 \le t \le t_2 , \\ E_e(t) & t > t_2 \end{cases}$$

At $t = t_1$, $E_{out} = E_{max}$: $E_{max} = \alpha(1 - e^{-t_1 / T_l}) + 1$.

For the SLED at SLAC, the duration of the compressed pulse $T_{cp} = t_2 - t_1 = 0.8$ µs, the input pulse length $T_k = 3.5 T_{cp}$, $Q_0 = 100{,}000$, $Q_e = 20{,}000$, $\beta = 5$, $\alpha = 1.667$, $T_l = 1.86$ µs, thus $E_{max} \le 2.28$, and $P_{max} \le 5.2$. The beam energy multiplication factor M, which is the maximum SLED gain, is about 1.614.

3. Bibliography

1. P.M.Lapostolle and A.Septier, Eds., Linear Accelerators, North Holland and Wiley, 1970.
2. G. A. Loew and R. Talman, Elementary principles of linear accelerators, AIP Conf. Proc. 105, 1983.
3. T. P. Wangler, Introduction to Linear Accelerators, Los Alamos Report, LA-UR-94-125, 1994.
4. D.H. Whittum, Introduction to Electrodynamics for Microwave Linear Accelerators, SLAC-PBU-7802, April 1998.
5. A. W. Chao and M. Tigner, Eds., Handbook of Accelerator Physics and Engineering, World Scientific, 1998.

INJECTION INTO THE STORAGE RING

XIAOAN LUO

Institute of High Energy Physics, Chinese Academy of Sciences, Beijing, China
E-mail: luoxa@mail.ihep.ac.cn

More and more large accelerators will be needed for high energy physics. Large accelerators consist of many smaller ones. The beams need to be injected from one accelerator to another one, therefore it is important to increase the efficiency of injection. This lecture introduces the principles of injection into electron storage rings. It describes the beam transverse oscillation during the injection process, injection elements, and the choice of the optimum injection time.

1 Introduction

Normally in a circular machine, such as a collider or a synchrotron light source, the beams cannot be produced, but are extracted from another smaller accelerator, passed through a transfer line, and injected into the circular accelerator. Injection is the final stage of the transfer of beams from one accelerator to another, either from a linear to a circular machine or from one circular accelerator to another.

As there are dynamic aperture limits, orbit oscillation in the transverse plane, and longitudinal acceptance in the whole storage ring, the beams cannot be injected into the ring smoothly without any measurements. Some special injection elements must be used to fulfill these requirements. The parameters of the beams at the end of the transfer line should be matched with those at the injection point of the storage ring in transverse and longitudinal phase space. Then the beams can be injected with high efficiency, minimum beam loss, and low dilution of the emittance.

2 Injection of electron and positron beams

Electron (or positron) beams have some special properties, such as radiation damping and quantum excitation. We can make good use of these properties in injecting beams into an electron machine. Usually single-turn injection is chosen to accumulate electron beams in a storage ring.

2.1 Injection process

Here we discuss only transverse injection. At the beginning, the beams from the transfer line enter the acceptance of the transverse phase space with large amplitude compared with that of the central closed orbit because there is some distance between the transfer line and the storage ring. Under radiation damping and quantum excitation, the injection beams will reach the equilibrium state as accumulated beams, which circulate along the central closed orbit. The beam size will be damped too. Then the phase space that was occupied by the initial injected beams is "empty", leaving room for a second pulse from the transfer line to be injected into the phase space. This process is repeated during the whole injection.

Therefore electron beams can be injected multi-times. After repeated injections, the beam current can reach a high intensity. To obtain the high current in a shorter time, the repetition frequency of injection can be increased, but it cannot be too high. Usually, the interval between injections should be longer than one damping time of the storage ring because the beam size and the amplitude of the beam oscillation are larger than damped ones during the injection. In this period, the beam is easy to be lost if it is kicked by the magnet.

The radiation damping effect in the ring leads to damping of the betatron and synchrotron oscillations. The horizontal damping time can be calculated as

$$\tau_x[ms] = \frac{C[m]\rho[m]}{13.2 J_x E^3[GeV]}$$

where C is the circumference of the storage ring, ρ is the bending radius, J_x is the damping partition function, and E is the beam energy.

Sometimes special wiggler magnets are used in the ring to reduce the damping time and allow more frequent injections.

2.2   Aperture

In order to inject beams into the acceptance of the machine and avoid hitting any place within the vacuum chamber of the storage ring, we must consider the dynamic aperture carefully. The dynamic aperture must be large enough to let the beams be injected into the ring with minimum beam loss. Figure 1 shows the transverse acceptance at the end of the septum magnet.

When the first pulse of injected beams reaches the entry of the ring, the first-turn beam is injected into the ring via the septum magnet. Then it is acted on by a fast local orbit bump produced by the kicker magnets in the ring so that the aperture can accept the beam from the transfer line. Otherwise a large dynamic aperture would be needed in the storage ring, which is difficult to have and unnecessary. After the first-turn injection, the local orbit bump is rapidly eliminated so that the kicker magnets won't interfere with the second-turn beam when it passes the injection region. This prevents the beam from hitting the vacuum chamber. After some time, the first injected beams become the circulating beams whose trajectory is

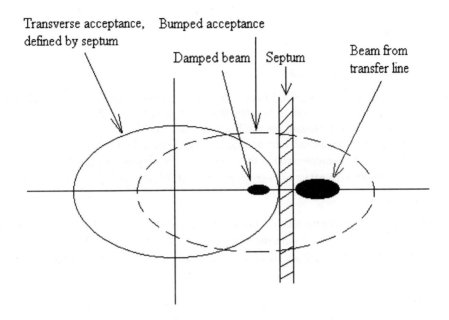

Transverse acceptance, Bumped acceptance
defined by septum

Damped beam   Septum   Beam from
transfer line

Figure 1. The dynamic aperture of the injected and the circulating beams in the transverse plane at the end of the septum magnet.

the central orbit. When the second pulse of injected beams enters the ring, it will be kicked by the kicker magnets, and the circulating beams will also be kicked from the kicked by the kicker magnets, and the circulating beams will also be kicked from the central orbit at the same time. Thus there is enough dynamic aperture for both injected beams and circulating beams.

The beams can be injected into the ring in both transverse planes, horizontal or vertical.

For injection into the horizontal phase space, the central closed orbit is bumped in the horizontal plane so that the transverse acceptance can accommodate the injected beams from the exit of the septum magnet. The horizontal plane is favored because the horizontal acceptance is normally larger than the vertical one in a conventional machine.

However in some e^+e^- colliders, horizontally injected beams in one ring which make large oscillations before damping will interact parasitically with the circulating beams in the other ring in the region where the two beams share the same pipe.

Therefore vertical injection will be chosen. Then the orbit will be bumped in the vertical plane.

2.3   Transverse oscillation

In the whole ring, the injected beams oscillate coherently around the closed orbit during the injection process. It is difficult to find an analytic solution to express the oscillation. Simply we regard the oscillation as two parts:
1.   The injected beams oscillate around the central closed orbit X_c.
2.   The particles oscillate around the trajectory of the injected beams X_β.
    If we neglect the effect of energy spread $\eta(\Delta P/P)$, the orbit can be written as

$$X = X_c + X_\beta$$

After damping, the beam reaches the equilibrium state, $X_c = 0$. Thus we can regard the attenuation of X_c as a pure damping process. Assuming τ_x is the damping time and X_0 is the initial amplitude of the injected beam, then X_c can be expressed as

$$X_c = X_0 e^{-t/\tau_x}$$

For X_β, we assume the initial emittance of the injected beam is ε_i. It will be the same as the emittance of the circulating beam, ε_c, in the final equilibrium state.

If $\varepsilon_i \gg \varepsilon_c$, then damping is the main cause of the change in emittance, approximately

$$\varepsilon_i(t) = \varepsilon_i e^{-2t/\tau_x}$$

If $\varepsilon_i \ll \varepsilon_c$,

$$\varepsilon_c(t) = \varepsilon_c(1 - e^{-2t/\tau_x})$$

Then the emittance can be regarded as

$$\varepsilon(t) = \varepsilon_i(t) + \varepsilon_c(t) = \varepsilon_c + (\varepsilon_i - \varepsilon_c)e^{-2t/\tau_x}$$

Similarly, the energy spread is

$$\delta(t) = \delta_c + (\delta_i - \delta_c)e^{-t/\tau_x}$$

Thus the beam size at time t is changed as

$$\sigma(t) = \sqrt{\varepsilon(t)\beta_0 + \eta^2\delta^2(t)}$$

During the injection, the beam oscillates according to the different phase advances and beta-functions. The oscillating amplitude of the injected beam is

$$X(s) = \sqrt{\frac{\beta(s)}{\beta_0}} X_0$$

where β_0 is the beta-function at the injection point, $\beta(s)$ is the beta-function at any place s, and η is the dispersion at the injection point.

So we must consider the physical aperture carefully at some special places where the beta-functions are larger during the injection process. The initial amplitude of the injected beam, X_0, must be satisfied as

$$X_0 > n\sigma_i + m\sigma_c + D_x\left[\left(\frac{\Delta P}{P}\right)_i + \left(\frac{\Delta P}{P}\right)_x\right] + X_s$$

where σ_i is the horizontal beam size of the injected beam, σ_c is the horizontal beam size of the ring, $(\Delta P/P)_i$ is the momentum spread in the transfer line, $(\Delta P/P)_x$ is the momentum spread in the ring, D_x is the dispersion in the ring, and X_s is the effective thickness of the septum.

Zero dispersion at the injection point is normally chosen in the storage ring. Then the beam size is smaller and the amplitude of the offset at the injection point can be reduced. Thus the orbit oscillation and the aperture needed are smaller.

2.4   Parameter matching

The parameters of the injected beams at the end of the transfer line should be matched with those at the injection point in the ring. This means that the beta, alpha and dispersion functions, β_x, β_y, α_x, α_y, D_x, D_y, D'_x and D'_y at the end of the transfer line must be matched with the machine lattice parameters at the injection point. Also the phase space contours containing defined percentages of the beam must be approximately the elliptical contours in transverse phase space given by the Courant-Snyder invariant:

$$\beta\gamma - \alpha = 1$$
$$\alpha = -\frac{1}{2}\beta'$$
$$\gamma\beta^2 + 2\alpha yy' + \beta y'^2 = \varepsilon$$

2.5   Longitudinal acceptance

For the longitudinal plane, the energy spread should be matched with the longitudinal acceptance:

$$\varepsilon_{RF} = \pm\left[\frac{2U_0}{\pi\alpha hE_0}[\sqrt{q^2 - 1} - \cos^{-1}(1/q)]\right]^{1/2}$$

$$q = \frac{eV_{RF}}{U_0}, \qquad\qquad U_0 = eV_0 \sin\phi_s$$

where U_0 is the energy loss per turn, α is the momentum compaction, and V_{RF} is the total voltage of the radio-frequency cavities.

The longitudinal accelerating phase ϕ_s must be in the stable region when the beams reach the radio-frequency cavity.

3   Injection elements

Injection elements consist of kicker magnets, bump magnets, septum magnets, and separators (for the collider machine). There are many special requirements for these elements.

3.1   Kicker magnet

The fast bump orbit must be realized by the kicker magnets. The bump orbit should be satisfied as follows:
- A certain displacement X_k and injection angle X'_k at the injection point;
- No closed orbit distortion outside the bump orbit.

In order to meet these demands, four kicker magnets are used:

$$\sum_{i=1}^{4} \sqrt{\beta_0 \beta_i}\, \sin\Delta\phi_{0i} \cdot \theta_i = X_k$$

$$\sum_{i=1}^{4} \sqrt{\frac{\beta_0}{\beta_i}}(\cos\Delta\varphi_{0i} - \alpha\sin\Delta\varphi_{0i}) \cdot \theta_i = X'_k$$

$$\sum_{i=1}^{4} \sqrt{\beta_0 \beta_i}\, \sin\Delta\phi_{1i} \cdot \theta_i = 0$$

$$\sum_{i=1}^{4} \sqrt{\frac{\beta_1}{\beta_i}}(\cos\Delta\varphi_1 - \alpha_1\sin\Delta\varphi_{1i}) \cdot \theta_i = 0$$

The angular deflection θ that the kicker magnet imparts to the beam to make it move subsequently on the central equilibrium orbit is

$$\theta = X / \sqrt{\beta_0 \overline{\beta}_x}\, \sin\mu_x$$

where β_0 and $\overline{\beta_x}$ are the beta-functions of the injection point and the kicker magnets respectively, μ_x is phase advance between the kicker and the injection point.

A high value of $\overline{\beta_x}$ is advantageous to reduce the kicker strength, and a high β_0 helps reduce the relative contribution to θ due to the septum thickness. A reasonably large value of the beta-function in the transverse plane is needed at the point of the septum magnet, but it should not be so large as to give rise to significant extra chromaticity.

If the phase advance between the two kickers is just π and there are no constraints on the injection angle or other special conditions, two kickers are enough for the injection.

Some types of kicker magnets are
1. current loop inside the ring vacuum,
2. terminated transmission line inside the vacuum, and
3. ferrite magnet outside the vacuum.

In the injection region, the kickers should be well matched as the beams already stored in the ring will see the rising and falling edges of the kicker waveform during the injection process. Fast kicker magnets require rapid rise and fall times of the pulse, which are functions of the thyratron characteristics and the kicker magnet design.

3.2   Bump magnet

To decrease the strength of the kicker magnets, slow bump magnets are needed. During one injection, the slow bump magnets can be used to produce a local bump orbit. When the injected pulse enters the injection point, the kicker magnets will add another fast local bump orbit. Thus the kicker magnets needn't be too strong. The slow time means that the bump orbit will be eliminated before the next injection pulse arrives.

3.3   Septum magnet

Septum magnets are used to connect the transfer line and the storage ring. The injected beams pass through a double window of stainless-steel foil that is used in the septum magnet to isolate the ultrahigh vacuum needed in the storage ring from the high vacuum in the transfer line. Then the beams enter the storage ring. To avoid the radiation, the injection pipe is normally at the inside of the storage ring.

If the transfer line and the storage ring are at different heights, a Lambertson-type septum magnet is usually used. If they are in the same plane, horizontal septum magnet can be used.

If the injection energy is relatively low, it is important to have adequate shielding of the beam from the septum leakage field. The sources of this stray field for a magnetic septum are the magnetomotive loss field from the yoke, the gaps between the yoke and the septum, the gaps between the septum turns, and the cooling channels in the septum conductors.

3.4   Separator

For a circular collider, before two beams collide, more beam current should be accumulated. During the injection process, the two beams must be separated at the crossing points and the parasitic interaction points so that the two beams can reach high intensity without interaction. After finishing injection, the separators are turned off to allow collision at the special interaction .point. Normally electrostatic separators are chosen for a circular electron-positron collider.

4   Optimum time between two injections

For a collider or a factory machine, the injection efficiency needs to be high. We must choose the optimum time between injection cycles so that the average luminosity can reach the maximum value.

We assume here that the lifetime of luminosity is τ and each injection begins with the same initial luminosity L_0. Then the luminosity will decay as

$$L(t) = L_0 e^{-\frac{t}{\tau}}$$

Assuming that the data-taking period for a high physics experiment is T, the total integrated luminosity accumulated during time T is given by

$$\int_0^T L dt = \frac{T}{t_i + t_e} \int_0^{t_e} L_0 e^{-t/\tau} dt$$

Thus the average integrated luminosity $\overline{L}$ can be inferred as

$$\overline{L} = \frac{\tau(1 - e^{-t_e/\tau})}{t_i + t_e} L_0$$

where t_i and t_e are the injection time and physics experiment time respectively. If the injection time and the lifetime of the luminosity are known, the optimum time for experimental physics can be inferred as

$$\frac{t_i + t_e}{\tau} = e^{\frac{t_e}{\tau}} - 1$$

Then the maximum average luminosity is

$$\overline{L}_{\max} = L_0 e^{-\frac{t_e}{\tau}}$$

We can choose the optimal period t_e for the physics experiment, and calculate the average luminosity.

Of course, it is more complicated to obtain the maximum average luminosity for a real accelerator. For different machines, we can use different injection models, such as resonant injection, longitudinal phase space injection, multi-bunch injection, and so on.

In general, we should choose the optimum injection models to obtain high average luminosity or reach high beam current in a shorter time.

References

1. G.H. Rees, Injection, CAS, Fifth general accelerator physics course, 1994, pp. 1–12.
2. PEP-II Conceptual Design Report, SLAC-418, 1993.
3. Feasibility Study Report on BTCF, 1996.
4. X. Liang and C. Zhang, Theoretical Design Study on BEPC Injection System, 1983.
5. 1-2 GeV Synchrotron Radiation Source, PUB-5172, LBL, 1986.

RF SYSTEM FOR ELECTRON STORAGE RINGS

KAZUNORI AKAI

KEK, High Energy Accelerator Research Organization, 1-1 Oho, Tsukuba, Ibaraki-ken, 305 Japan
E-mail: kazunori.akai@kek.jp

This paper presents an overview of RF systems for electron storage rings with emphasis on beam-loading and beam dynamics issues. Basic analysis of the RF system with beam-loading is discussed as well as RF system-related beam instabilities and some transient effects. Feedback control loops to cure the instabilities and stabilize the accelerating field are also discussed.

1 Introduction

The main roles of RF accelerating systems in storage rings are (1) to provide the beam with energy to compensate for one-turn energy loss or to ramp the beam energy, and (2) to bunch the beam longitudinally around the synchronous phase. Recently, increasingly high accelerating voltage has been required for large storage rings and linear colliders. Efforts have been continuously made to generate higher accelerating gradient and higher RF power. Many important technical issues are involved in the design and fabrication of the RF cavities as well as the peripheral components such as input couplers, windows and HOM dampers, the high power RF source and other high power components, and the RF control systems.

Since the RF cavities can be major sources of the impedance of the ring, beam-induced field in the cavities is an important issue to be treated, especially in high intensity accelerators. As the stored beam current is increased, it becomes more and more important to avoid beam instabilities. The RF voltage can be affected by the fundamental (accelerating) mode beam-loading, as well as by other causes such as noise, errors, and changes of operating parameters. The effect is to shift the beam position or to excite coherent longitudinal oscillations of the beam, which may lead to emittance growth or even to beam loss and unstable operation of the RF system. This problem must be solved in the RF system design. In addition, instabilities driven by higher-order modes (HOM) of the cavities must be avoided. Several types of HOM-damping schemes have been developed and used around the world.

This paper mainly presents a basic analysis of the RF system with beam-loading as well as RF system-related beam instabilities and cures. In Section 2 a brief overview of the RF system is given by comparing energy frontier e^+e^-

colliders and so-called e^+e^- factories. Section 3 presents a basic analysis of an RF cavity with an RF high power system. In Section 4 the effect of beam-loading on the RF system and beam is discussed. In Section 5 beam instabilities caused by a long-range wake of RF cavities are discussed as well as some methods to cure them. Finally, Section 6 presents an example of an RF control system, including feedback loops and other methods of stabilizing the overall RF system and beam. In this paper, the discussion relates to the RF system in e^+e^- rings unless otherwise noted.

2 Overview of the RF system

2.1 RF signal flow

Figure 1 shows major RF components and the signal flow between each component and the beam. In some cases, e. g. a large storage ring with many RF cavities, several RF stations are distributed along the ring. An output from an RF signal generator is distributed to the RF stations via reference lines such

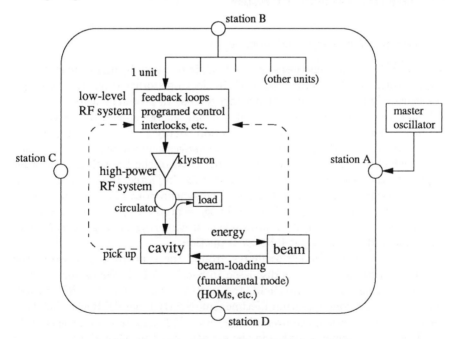

Figure 1. Schematic view of the RF system with beam.

as coaxial cables or fiber-optic lines. For accurate control of the relative phase between stations, phase-lock feedback loops are often used to compensate for changes in the electrical length of the cables due to temperature changes or other causes.

In each RF station, the RF signal is transmitted to a high power amplifier, e. g. a klystron. The RF output power from the high power source is transmitted to an RF cavity via a high power transmission line, such as a wave guide. The output power is often sent to more than one cavity via a magic-tee, hybrid, etc. Between the klystron and the cavities, a circulator is usually inserted to terminate reflection power from the cavity, to prevent the reflection power from returning to the klystron.

The cavities and the beam act in two ways: the cavities supply the beam with energy and the beam induces the wakefield in the cavity. The effect of beam loading is stronger in high current machines such as high luminosity factories. The low-level RF system includes several feedback loops for stable and accurate control of the RF voltage and other parameters. It also contains an interlock system to protect the cavities and other high power components when problems arise in the RF system.

2.2   Required RF voltage and beam power

In storage rings, the beam loses energy U_{loss} in one turn by quantum radiation and parasitic loss due to the wakefield. In electron storage rings, the radiation loss is usually the main cause of the energy loss. In proton synchrotron rings, although the radiation loss is negligibly small, energy should be delivered to the beam during ramping from the injection energy to the maximum energy.

Figure 2 shows the RF voltage, which changes sinusoidally, and the bunch phase, which is defined with respect to the crest of the RF voltage. In order to provide the beam with the longitudinal focusing force, the crest of the RF voltage should be larger than U_{loss}/e. The synchronous phase ϕ_s, where the energy gain is equal to U_{loss}, is given by $\phi_s = \arccos(U_{loss}/eV_c)$. In electron storage rings, $0 < \phi_s < \pi/2$. In proton synchrotron rings, $0 < \phi_s < \pi/2$ above transition and $-\pi/2 < \phi_s < 0$ below transition.

A particle in the bunch oscillates around the synchronous phase at the synchrotron frequency ω_s, as discussed in Section 5.1. If the oscillation amplitude is larger than a threshold, the particle goes out of the RF bucket. For a sufficiently long quantum lifetime, sufficiently large RF bucket height is needed. In many cases, the overvoltage ratio eV_c/U_{loss} must be at least 1.3.

The desired bunch length and the synchrotron tune should also be taken into account. When U_{loss} and the rms bunch length σ_z are given, the necessary

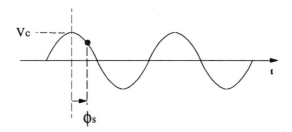

Figure 2. RF voltage and the synchronous phase.

RF voltage is given by

$$V_c = \sqrt{\left(\frac{U_{loss}}{e}\right)^2 + \left(\frac{cC\eta E\sigma_\varepsilon^2}{e\omega_{rf}\sigma_z^2}\right)^2}, \tag{1}$$

where c is the velocity of light, C the ring circumference, η the slippage factor, σ_ε the relative energy spread, ω_c the RF angular frequency, and E the beam energy. A shorter bunch length requires a larger V_c. The synchrotron tune $\nu_s = \omega_s/\omega_0$, where ω_0 is the revolution frequency, is related to the bunch length as

$$\nu_s = \frac{C\eta\sigma_\varepsilon}{2\pi\sigma_z}. \tag{2}$$

The power transferred from the RF system to the beam, P_b, to compensate for the energy loss is given by

$$P_b = I_b \times \frac{U_{loss}}{e} = I_b V_c \cos\phi_s, \tag{3}$$

where I_b is the DC beam current.

2.3 Energy frontier colliders and high luminosity factories

Figure 3 shows the design RF voltage and stored beam current in several e^+e^- storage rings, comparing energy frontier colliders and factories. RF-related parameters for some storage rings are listed in Table 1. The factories are very high luminosity colliders that are intended to produce particular particles for precise experiments. In energy frontier e^+e^- rings such as LEP2 and TRISTAN, the synchrotron radiation loss is very large and the necessary RF voltage is more than several hundred MV, which requires a high gradient and a large number of cavities. In these colliders, a large number of superconducting (SC) cavities have been operated at a gradient of 4 to 6 MV/m. In the factories, on the other hand, although the required RF voltage is relatively low, the stored beam current is of the order of an ampere, which is much higher than that in the energy frontier colliders. As a result, each cavity should deliver higher power, and heavier demands are placed on curing beam instabilities.

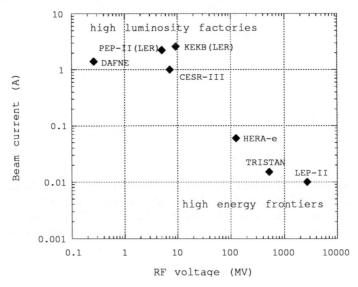

Figure 3. RF voltage and stored beam current in energy frontier colliders and high luminosity factories.

3   Fundamentals of cavities

3.1   Free and damped oscillation in a cavity

Here we discuss standing-wave accelerating structures used in e^+e^- storage rings. The solutions of Maxwell's equations in the cavity can be analyzed by expanding the electromagnetic field with orthogonal functions, each of which corresponds to a resonant mode of the cavity. [1] Here we will not show a detailed analysis using orthogonal functions, but only some important results for application.

First, we consider an isolated cavity without any input or output port. We also consider a single mode which is separated from other modes. It is characterized by resonant frequency ω_c, intrinsic Q-value Q_0, and shunt

Table 1. RF-related parameters for some e^+e^- energy frontier colliders and high luminosity factories.

	Energy frontier		High luminosity factories			
Ring	TRISTAN	LEP2	DAΦNE	CESR-III	KEKB	
Physics			ϕ	B	B/(asymmetric)	
No. of rings	1	1	2	1	2	
Beam energy (GeV)	32	100	0.51	5.3	3.5 (e^+)	8.0 (e^-)
Beam current (A)	0.015	0.006	1.4	0.5+0.5	2.6	1.1
Luminosity (10^{33}/cm^{2}s)	0.04	0.11	0.53	2	10	
Circumference (m)	3016	27000	97.69	768	3016	
Bunch length (cm)			3.0	1.3	0.4	
RF frequency (MHz)	508.6	352.2	368.3	500	508.9	
RF voltage (MV)	520	3400	0.26	7~12	9.4	16.2
Beam power (MW)	~5		0.03	1.3	4.5	4.0
Cavity type	NC+SC	SC	NC	SC	NC	NC+SC
No. of cavities	104+32	288	1	4	20	12+8
Input power (kW)	200/60			325	355	~400
No. of cav/klystron	4/4	8	1	1~2	2	2/1
No. of klystrons	26/8		1	3	10	6/8

impedance R. The solution of Maxell's equations gives the free and damped oscillation of the electromagnetic field as

$$\vec{E}, \vec{H} \propto \exp(-\omega_c t/2Q_0)\exp(j\omega_c t). \qquad (4)$$

Without any loss, the Q_0 is infinity and the oscillation does not decay. The electromagnetic energy in the cavity is dominated by electric energy or magnetic energy, each of which appears alternately at an interval of 1/4 oscillation period. It is analogus to a simple L-C circuit of resonant frequency $\omega_0 = \sqrt{1/LC}$. In an actual cavity, there is a loss due to finite conductivity σ at the wall or some lossy material inside the cavity. The Q-value represents the decay time constant of the oscillation and, equivalently, the bandwidth of the resonance. The resonant frequency ω_c is slightly decreased compared with the no loss case of ω_0. In the case of wall loss, ω_c is shifted as $\omega_c = \omega_0(1 - 1/2Q_0)$.

The power loss averaged in one oscillation period P_c is related to the stored energy in the cavity, U, as $P_c = -dU/dt$, and both of them are proportional to $\vec{E}\vec{E}^*$ or $\vec{H}\vec{H}^*$. Consequently, the Q_0 is related to U and P_c as

$$Q_0 = \frac{\omega_c U}{P_c}. \tag{5}$$

The power loss at the wall, P_c, is

$$P_c = \frac{R_s}{2} \int_{surface} \left| \vec{H} \right|^2 dS, \tag{6}$$

where R_s is the surface resistance. For a normal-conducting cavity (NC) $R_s = \sqrt{\omega_c \mu / 2\sigma}$. The R_s for a superconducting cavity (SC) comes from the BCS part and residual loss: $R_s = R_{BCS} + R_{residual}$. For example, at 500 MHz, $\sigma = 5.8 \times 10^7 / \Omega m$ and $R_s = 5.8 \times 10^{-3}\Omega$ for NC. On the other hand, $R_s \sim 10n\Omega$ for SC. Thus the SC cavities have very large Q_0 compared with NC cavities.

3.2 Accelerating voltage and shunt impedance

Let us assume that the cavity is driven by an external RF source at frequency ω_{rf}. The electric field on the axis of the cavity $E_z(z,t)$ varies sinusoidally as

$$E_z(z,t) = E_{z0}(z)\exp(j\omega_{rf}t). \tag{7}$$

When a particle of charge q passes the cavity on axis with velocity v (Fig-

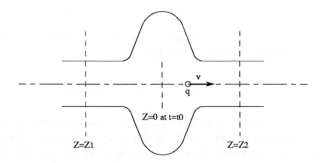

Figure 4. A particle passes in the cavity.

ure 4), it gains energy as

$$W = q\text{Re}\left\{ \int_{z_1}^{z_2} E_z(z, t - t_0 = z/v)dz \right\}$$

$$= q\text{Re}\left\{ \int_{z_1}^{z_2} E_{z0}(z)\exp\left\{ j\omega_{rf}(t_0 + z/v) \right\} dz \right\}. \tag{8}$$

The cavity voltage V_c is defined as

$$V_c \equiv \left| \int_{z_1}^{z_2} E_{z0}(z) \exp(j\omega_{rf}z/v)dz \right|. \tag{9}$$

Note that V_c is *not* equal to $\int E_{z0}(z)dz$. The ratio of V_c to this value is called the transit time factor. The energy gain changes according to the arrival time of the particle at the cavity t_0 as

$$W = q\mathrm{Re}(V_c e^{j\omega_{rf}t_0 + \phi_0}) = qV_c \cos(\omega_{rf}t_0 + \phi_0). \tag{10}$$

It changes sinusoidally and the maximum energy gain is qV_c on crest, as shown in Figure 2.

We now introduce the shunt impedance R as

$$R \equiv \frac{V_c^2}{P_c}. \tag{11}$$

From Eqs. (5) and (11) we obtain $R/Q_0 = V_c^2/(\omega_c U)$. Since $V_c^2/U \propto \omega_c$, the R/Q_0 (often written as R/Q) depends only on geometry: it does not depend on surface resistance or frequency. For NC cavities, the cell shape is usually optimized to have a high R/Q value, for example, with a nose-cone to concentrate the electric field on the beam axis. On the other hand, since SC cavities have a very high Q_0 (low P_c), it is not important to maximize the R/Q: a smoother cell shape is preferred to avoid multipacting on the surface. High Q_0 is still important to reduce the heat load for cryogenic systems.

3.3 Connection to a transmission line

Next we consider a cavity connected to an external RF source via a transmission line (Figure 5). We assume that the source is well-padded, i. e. that a circulator is inserted between the RF source and the cavity so that the power reflected from the cavity does not go back to the high power source, but goes to the load terminating the line.

The free oscillation with no input power from the source decays as $\exp(-\omega_c t/2Q_L) \times \exp(j\omega_c t)$, where Q_L is the loaded Q-value:

$$Q_L = \frac{Q_0}{1 + \beta}. \tag{12}$$

The β is the input coupling factor between the cavity and the transmission line. In the case that the field is excited by a source inside the cavity, such as beam, and no power comes from the RF source, β equals the ratio of power dissipated at the load to power at the cavity wall. The time constant, $T_f \equiv 2Q_L/\omega_c$, is called the filling time.

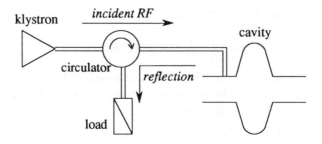

Figure 5. Cavity connected to an RF source via a transmission line.

Consider an external RF source operating at RF frequency ω_{rf}, which is not necessarily equal to the resonant frequency of the cavity ω_c. As discussed above, the source is assumed to be well-padded. Then the output power of the source is not affected by the reflection from the cavity and thus it can be treated as a current generator. This is the case in most accelerator RF systems.

Here we consider that the cavity is excited only by the RF source and no beam is stored. According to the analysis using the orthogonal functions, the admittance at ω_{rf} observed from the transmission line toward the cavity at the so-called detuned short plane is

$$Y_{in} = G_c \left\{ 1 + jQ_0 \left(\frac{\omega_{rf}}{\omega_c} - \frac{\omega_c}{\omega_{rf}} \right) \right\} \frac{G_0}{\beta G_c}, \tag{13}$$

where G_0 is the characteristic impedance of the transmission line. This expression is equivalent to a parallel resonant circuit with admittance $G = G_c$, capacitance $C = Q_0 G_c/\omega_c$, and inductance $L = 1/(\omega_c Q_0 G_c)$, transformed by an impedance transformer with a step-up ratio $n_c = \sqrt{G_0/\beta G_c}$, as shown in Figure 6a. The LCG circuit and the transformer represent the cavity and the input coupler, respectively. The G_c is chosen such that the power dissipation P_c is equal to $G_c V_c^2/2$. Then the relation between G_c and R is $G_c = 2/R$, see Eq. (11). Furthermore, looking at the line admittance and the generator current from the cavity, they are transformed into $\beta G_c = G_0/n_c^2$ and $i_g = i_{g0}/n_c$, respectively, and the transformer can be eliminated. This simplifies the circuit, as shown in Figure 6b.

In the following, we use a phasor representation that treats the phase of voltage or current with respect to a reference frame rotating with angular frequency ω_{rf}. We indicate a phasor by a tilde. Let us write the voltage $\tilde{V}_c$

(a)

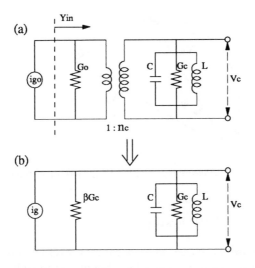

(b)

Figure 6. Equivalent circuit for the cavity and RF source: (a) with an impedance transformer, and (b) simplified circuit.

induced by the generator as $\tilde{V}_g$. From the simplified circuit,

$$\tilde{V}_g = \frac{\tilde{i}_g}{G_c\left[1 + jQ_0\left\{(\omega_{rf}/\omega_c) - (\omega_c/\omega_{rf})\right\}\right] + \beta G_c}. \tag{14}$$

Note that the quantity which can be measured is not the generator current i_g but the output power P_g traveling in the transmission line. They are related as $\left|i_g^2\right| = 8P_g\beta G_c$. Introduction of the tuning angle ψ,

$$\tan\psi \equiv -Q_L\left(\frac{\omega_{rf}}{\omega_c} - \frac{\omega_c}{\omega_{rf}}\right)$$

$$\approx 2Q_L\frac{\omega_c - \omega_{rf}}{\omega_c} \quad \text{for } |\omega_c - \omega_{rf}| \ll \omega_c, \tag{15}$$

casts Eq. (14) into useful forms:

$$\tilde{V}_g = \tilde{V}_{gr}\cos\psi e^{j\psi}, \tag{16}$$

$$V_{gr} = \frac{2\sqrt{\beta}}{1+\beta}\sqrt{RP_g}, \tag{17}$$

where the phase of $\tilde{V}_{gr}$ is the same as that of $\tilde{i}_g$. Note that ψ is the relative phase between $\tilde{i}_g$ and $\tilde{V}_g$. Equations (15) and (17) are represented in a phasor relation, as in Figure 7.

128

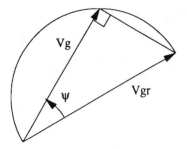

Figure 7. Phasors representing the generator voltage.

4 Beam-loading and RF system

So far we have discussed the case where a cavity is excited only by an external
RF source. The cavity is also excited by a beam due to wake. The wakefield
can be characterized according to its time constant, or equivalently, bandwidth of the impedance. Short-range wake (broad-band impedance) is important for single-bunch effects: for example, potential-well distortion, bunch
lengthening, mode-coupling instability, etc. It is usually analyzed with the
Vlasov equation and distribution function. On the other hand, long-range
wake (narrow-band impedance) dominates coupled-bunch effects: resonant
build-up of beam-loading, Robinson instability, coupled-bunch instability, etc.
A rigid bunch, macro particle model can be applied in this case.

The wakefield in the cavity is composed of many frequency components:
the fundamental (accelerating) mode, higher-order modes (HOM's) below the
cut-off frequency of the beam ducts, and wide-band spectrum above the cut-
off. The accelerating mode of a cavity usually has a very high-Q impedance.
Also the Q-values of some HOM's can be large. Consequently, the long-range
wake plays a more important rolle in the operation of the accelerating mode
and coupled-bunch phenomena related to the accelerating mode or HOM's of
cavities. We will focus on the long-range wake, assuming that each high-Q
mode is separate and therefore they do not interfere with each other.

4.1 Single-bunch beam-induced voltage

If we are interested in how the electromagnetic field builds up while a single
particle passes the cavity, we have to solve Maxwell's equations closely. In
most cases, however, what concerns us is the total effect, i. e. the beam-

induced voltage after the particle exits the cavity and the total energy change of the particle due to passing the cavity. Then the problem is greatly simplified as follows:

1. Beam-induced voltage $\vec{V_{b0}}$ just after the beam passes the cavity: For simplicity, we assume that the field is zero before the particle enters the cavity. The beam-induced voltage by a single Gaussian bunch of charge q is

$$\vec{V_{b0}} = -\frac{\omega_c}{2}\left(\frac{R}{Q}\right)qe^{-\omega_c^2\sigma_z^2/2c^2}, \tag{18}$$

which is obtained from energy conservation. The negative sign indicates that the particle loses the energy due to the wake.

2. Free and damped oscillation of the voltage: As discussed in Section 3, the voltage oscillates at ω_c and decays with time constant T_f. In the phasor representation,

$$\vec{V_b} = \vec{V_{b0}}\exp(-\frac{t}{T_f})\exp\{j(\omega_c - \omega_{rf})t\}. \tag{19}$$

3. Energy gain or loss of the particle while passing the cavity: The voltage that the particle itself feels is just half of V_{b0}. This is called the fundamental theorem of beam-loading. The proofs of 1 and 3 are given by Wilson [2].

4.2   Resonant build-up

Next we consider a bunch train with bunch spacing T_b. From Eq. (19) the voltage induced by the preceding bunch becomes, after T_b, $\tilde{V}_b = \tilde{V}_{b0} \times e^{-\tau}e^{j\delta}$, where $\tau = T_b/T_f = \omega_c T_b/2Q_L$ and $\delta = (\omega_c - \omega_{rf})T_b$. Note that $\delta = \tau \tan\psi$ from Eq. (15). By summing up all the voltage induced by the preceding bunches, we obtain the beam-induced voltage by a bunch train. As shown in Figure 8, the voltage just before the bunch passes is V_b^-. By adding to it the single-bunch beam-induced voltage, we obtain the voltage just after the bunch passed V_b^+. The voltage that the bunch itself feels is just the average of these two voltages. The result is

$$\tilde{V}_b = \tilde{V}_{b0} \times (K + e^{-\tau}e^{j\delta} + e^{-2\tau}e^{2j\delta} + e^{-3\tau}e^{3j\delta} + \cdots)$$
$$= \tilde{V}_{b0} \times \left(\frac{1}{1 - e^{-\tau}e^{j\delta}} + K - 1\right), \tag{20}$$

$$K = \begin{cases} 1, & \text{for } \tilde{V}_b^+ \; : \text{just after the bunch passes} \\ \frac{1}{2}, & \text{for } \tilde{V}_b \; : \text{beam feels} \\ 0, & \text{for } \tilde{V}_b^- \; : \text{just before the bunch passes.} \end{cases}$$

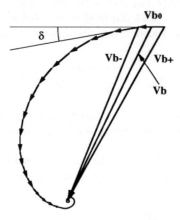

Figure 8. Resonant build-up of beam-induced voltage.

When $T_b \ll T_f$, the voltages are approximately equal:

$$\tilde{V}_b \approx \tilde{V}_b^+ \approx \tilde{V}_b^- \approx \tilde{V}_{b0} \times \frac{1}{\tau - j\delta} = \tilde{V}_{b0} \times \frac{1}{\tau} \cos \psi e^{j\psi}. \tag{21}$$

This case corresponds to continuous beam-loading. Equations (18) and (21) lead to

$$\tilde{V}_b = -\frac{q}{T_b} \left(\frac{R}{Q} \right) Q_L e^{-\omega_c^2 \sigma_z^2 / 2c^2} \cos \psi e^{j\psi}. \tag{22}$$

Since q/T_b is the DC beam current I_b, we finally obtain

$$\tilde{V}_b = \tilde{V}_{br} \cos \psi e^{j\psi}, \tag{23}$$

$$V_{br} = -\frac{RI_b}{1 + \beta} e^{-\omega_c^2 \sigma_z^2 / 2c^2}, \tag{24}$$

where the phase of $\tilde{V}_{br}$ is opposite the beam direction. This means that the beam-induced voltage decelerates the beam. These relations can be expressed in a phasor diagram, as shown in Figure 9.

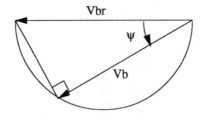

Figure 9. Phasors representing the beam-induced voltage.

4.3 Equivalent circuit and phasors

So far we have treated the RF source and the beam independently. Now we combine them by superposition of the beam current and the generator current and of the beam-induced voltage and the generator voltage. The simplified equivalent circuit including both the generator and the beam is shown in Figure 10. We should point out that the superposition can be applied to quantities that are solutions of linear equations, such as current or voltage, but not to power or energy.

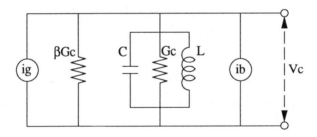

Figure 10. Simplified equivalent circuit for the cavity, RF source and beam.

Figure 11 shows the phasors for the steady state. Here we assume continuous beam-loading with bunch spacing $T_b \ll T_f$. The RF voltage $\tilde{V}_c$ is the vector sum of $\tilde{V}_g$ and $\tilde{V}_b$ as $\tilde{V}_c = \tilde{V}_g + \tilde{V}_b$. Here we have introduced the loading angle ϕ_L, which is the relative phase between $\tilde{V}_{gr}$ and $\tilde{V}_c$. When no beam current is stored, ϕ_L is equal to ψ. From the phasor diagram we obtain the

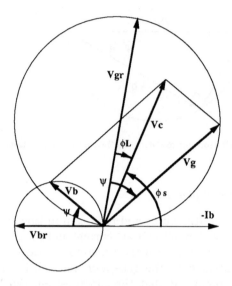

Figure 11. Phasor representation for the steady state.

following useful relations:

$$V_{gr} \cos \phi_L = V_c + V_{br} \cos \phi_s, \tag{25}$$

$$\tan \psi = (1 + \frac{V_{br}}{V_c} \cos \phi_s) \tan \phi_L - \frac{V_{br}}{V_c} \sin \phi_s. \tag{26}$$

4.4 Power, coupling and tuning angle

The generator power P_g needed to provide RF voltage V_c for the steady state is given by Eqs. (17) and (25) as

$$P_g = \frac{1}{\cos^2 \phi_L} \frac{(1+\beta)^2}{4\beta} \frac{(V_c + V_{br} \cos \phi_s)^2}{R}. \tag{27}$$

Given V_c and V_{br}, the loading angle ϕ_L and the coupling β determine P_g. First, consider P_g as a function of β, then $\partial P_g/\partial \beta = 0$ leads to $\beta = 1 + P_b/P_c$. The P_g is minimum when the coupling β is chosen to satisfy this relation. This case is called optimum coupling. The β is usually chosen to be optimum coupling at the design stored current or at a higher current.

Next we discuss the loading angle ϕ_L. In storage rings, the resonant frequency of cavities, ω_c, is usually controlled according to ϕ_L. Figure 12

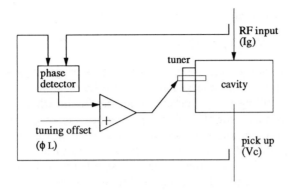

Figure 12. Block diagram of a tuning control system.

shows a typical example of a tuning system. Both the input RF signal and the cavity voltage are monitored, and their relative phase is detected by a phase detector. It is compared with the desired loading angle and their difference is amplified to control a tuner of the cavity: thus ω_c is controlled by a feedback loop. The P_g is minimum when $\phi_L = 0$, see Eq. (27). This case is called optimum tuning. With optimum coupling and optimum tuning at the same time, no reflection power $(P_r = P_g - P_b - P_c)$ comes back from the cavity. The ϕ_L is often set at a small negative value in order to increase the Robinson stabiltiy margin, which will be discussed later.

At optimum tuning $(\phi_L = 0)$, Eq. (26) becomes $\tan \psi = -(V_{br}/V_c) \sin \phi_s$, which corresponds to a detuning freqency:

$$\Delta f = \frac{1}{2\pi}(\omega_c - \omega_{rf}) = -\frac{I_b \sin \phi_s}{2V_c}\left(\frac{R}{Q}\right)f_{rf} = -\frac{P_b \tan \phi_s}{4\pi U}. \tag{28}$$

The cavities are usually thus detuned as the beam current increases.

4.5   Transient issues

When T_b is comparable to, or larger than, T_f, transient effects should be taken into account. It is important also for the continuous beam-loading case, if the driving current i_g, i_b or some other parameters, such as ω_c, change.

First, let us consider a simple case of a transient due to a change from a stationary state $\tilde{V}_i$ to another stationary state $\tilde{V}_f$. For example, the voltage $\tilde{V}_c$ changes when the beam is injected into the ring, stored beam is lost, or RF is turned on or off. Assume that the change is much faster than T_f. Then it can be considered as a superposition of the final state $\tilde{V}_f$ and the difference which decays as a free and damped oscillation as

$$\tilde{V}_c = \tilde{V}_f + (\tilde{V}_i - \tilde{V}_f)e^{-t/T_f}e^{j(\omega_c - \omega_{rf})t}. \tag{29}$$

Next we discuss the bunch-gap transient. In storage rings where a very high current beam is stored, a large number of buckets should be filled with the beam. Instead of filling many buckets uniformly in the whole ring, a bunch gap is often introduced for the following reasons: (1) to avoid instabilities caused by ion trapping, and (2) to allow for rise time of a beam-abort kicker which protects physics detectors and accelerator hardware.

The bunch gap, however, modulates the amplitude and phase of the RF voltage, since the beam-loading effect is different between the gap and the beam. The modulation then shifts the longitudinal synchronous position bunch-by-bunch: the head bunch is shifted backward and the tail bunch forward. As an example, a simulation for the KEKB HER [4] (Figure 13) shows that the amplitude and phase of the RF voltage and the beam phase are modulated by the gap.

If the gap modulation is large, the following problems can arise: (1) In a double-ring collider, such as most factory rings, the colliding point is shifted bunch-by-bunch, and the luminosity can be reduced by the hourglass effect. (2) The modulation of the cavity field can affect the RF control system by way of the feedback loops. For example, the klystron output power can be modulated by the gap, which may lead to saturation of the klystron power.

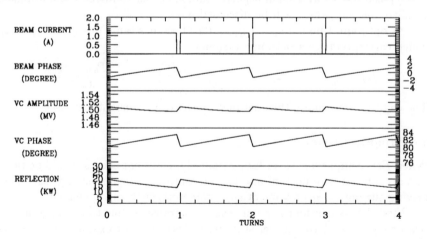

Figure 13. Simulation of a bunch-gap transient for KEKB-HER.

For optimum tuning, the beam phase modulation $\Delta\phi$ due to a gap of length $c\Delta t$ is approximately [5]

$$\Delta\phi = \frac{1}{2}\left(\frac{R}{Q}\right)\frac{\omega_{rf}}{V_c}I_b\Delta t = \frac{P_b\Delta t}{2\cos\phi_s U}. \tag{30}$$

It is seen that $\Delta\phi$ is inversely proportional to the stored energy U: a high stored energy is beneficial for reducing the bunch-gap transient.

5 RF system-related instabilities and cures

In this section, first the equation of longitudinal motion and synchrotron oscillation of a particle is presented. Next, several types of RF system-related beam instabilities are discussed, where a bunch is treated as a macro particle and the long-range wake due to cavities is involved.

5.1 Synchrotron oscillation of a particle

While a particle is travelling along a ring, its energy changes: it loses energy by synchrotron radiation and parasitic loss due to wake, and it gains energy at RF cavities. Storage rings are usually designed so that the energy loss or gain per turn is much smaller than the beam energy. Then we can use the averaged values of energy E, momentum p, and velocity v for each turn. Relations between these quantities are given from relativistic mechanics: $E = m_0c^2\gamma = \sqrt{p^2c^2 + m_0^2c^4}$, $\vec{p} = m_0\gamma\vec{v}$, and $\gamma = 1/\sqrt{1-\beta^2} = 1/\sqrt{1-(v/c)^2}$. The pass length C and travelling time T for one turn depend on E.

If the energy loss U_{loss} and gain of a particle are balanced at every turn, all the values mentioned above are constant from turn to turn. Such a particle is called a synchronous particle. The synchronous particle passes the RF cavity at a constant RF phase ϕ_s, which is called the synchronous phase, and $U_{loss} = eV_c\cos\phi_s$. It is convenient to take the synchronous particle as a reference and treat the difference between a particle under consideration and the synchronous particle: $\Delta T = T - T_0$, $\Delta E = E - E_0$, $\Delta p = p - p_0$, and so on, where T_0, E_0, p_0 are the values for the synchronous particle. Note that $\Delta E/E_0 = \beta^2\Delta p/p_0 = (\beta\gamma)^2\Delta v/v_0$.

The momentum compaction factor α_p and the slippage factor η are defined as

$$\frac{\Delta C}{C_0} = \alpha_p\frac{\Delta p}{p_0}, \tag{31}$$

$$\frac{\Delta T}{T_0} = \eta\frac{\Delta p}{p_0}. \tag{32}$$

Since $T = C/v$, $\Delta T/T_0 = \Delta C/C_0 - \Delta v/v_0 = (\alpha_p - 1/\gamma^2)p/p_0$. Then the relaton between α_p and η is

$$\eta = \alpha_p - 1/\gamma^2. \tag{33}$$

Since $\gamma \gg 1$ for electron storage rings, $E \cong pc$ and $\eta \cong \alpha_p$.

Let us make a comment on a relation between the RF frequency f_{rf} and the conditions for the synchronous particle. When the RF frequency f_{rf} is changed to $f'_{rf} = f_{rf} + \Delta f_{rf}$, p_0 also changes. Since $T_0 = h/f_{rf}$, $\Delta p/p_0 = -(1/\eta)\Delta f_{rf}/f_{rf}$. If $\eta > 0$, increasing f_{rf} decreases p_0, and vice versa. The momentum p_0 is often changed by changing the RF frequency for measuring dispersion or chromaticity of a ring.

Now we discuss synchrotron oscillation in electron storage rings. First we derive the equation of motion for nonsynchronous particles. Let $\delta_n = \Delta E/E_0$ be the energy deviation averaged from the n-th to the $(n+1)$-th turn, and τ_n be the arrival time delay at the n-th turn with respect to the synchronous particle. Then the equation of motion is given by

$$\delta_{n+1} - \delta_n = \frac{1}{E_0}\left\{eV_c\cos(\phi_s + \omega_{rf}\tau_n) - U_{loss}(\delta_n)\right\}, \tag{34}$$

$$\tau_{n+1} - \tau_n = \eta T_0 \delta_n. \tag{35}$$

The energy loss U_{loss} includes the radiation loss U_{rad} and the parasitic loss. Since the radiation loss depends on the energy,

$$U_{loss}(\delta_n) = U_{loss}(0) + \delta_n U'_{rad}(0)$$
$$= eV_c\cos\phi_s + \delta_n U'_{rad}(0). \tag{36}$$

By substituting $(\delta_{n+1} - \delta_n)/T_0 \rightarrow d\delta/dt$ and $(\tau_{n+1} - \tau_n)/T_0 \rightarrow d\tau/dt$ and eliminating δ, we obtain

$$\frac{d^2\tau}{dt^2} = \frac{eV_c\eta}{E_0 T_0}\left\{\cos(\phi_s + \omega_{rf}\tau) - \cos\phi_s\right\} - \frac{U'_{rad}(0)}{E_0 T_0}\frac{d\tau}{dt}. \tag{37}$$

For small amplitude oscillation, we can use the approximation,

$$\cos(\phi_s + \omega_{rf}\tau) \approx \cos\phi_s - \omega_{rf}\tau\sin\phi_s, \tag{38}$$

then Eq. (37) becomes

$$\frac{d^2\tau}{dt^2} = -\frac{eV_c\eta\omega_{rf}\sin\phi_s}{E_0 T_0}\tau - \frac{U'_{rad}(0)}{E_0 T_0}\frac{d\tau}{dt}. \tag{39}$$

The solution is $\tau = \tau_0\exp(j\omega_s t - t/t_d)$, where

$$\omega_s = \sqrt{\frac{eV_c\eta\omega_{rf}\sin\phi_s}{E_0 T_0}}, \tag{40}$$

$$\frac{1}{t_d} = \frac{1}{2T_0} \frac{dU_{rad}(0)}{dE}. \tag{41}$$

The solution indicates that the particle oscillates around the synchronous particle at an angular frequency ω_s and the oscillation decays with a damping time t_d. Since U_{rad} is proportional to E^2, $dU_{rad}(0)/dE = 2U_{rad}(0)/E_0$. Then $t_d \approx E_0 T_0 / U_{rad}(0)$.

When the oscillation amplitude is large, the approximation Eq. (38) can not be applied. The particle motion deviates from the linear oscillation. We will not discuss this case in detail here.

5.2   Growth rate of longitudinal coupled-bunch instability

Single-bunch case
In the following, we discuss beam instabilities caused by long-range wakes, mainly due to RF cavities. For stable operation of the accelerating mode, it is important to consider longitudinal instabilities related to it. It is also necessary to avoid instabilities caused by higher-order modes (HOM's) of cavities, both longitudinally and transversely. Although the longitudinal case is discussed here, transverse instabilities can be treated similarly. The radiation damping effect is neglected here: U_{rad} is considered independent of the energy deviation δ_n.

Let $W_\parallel$ be the longitudinal wake function of the 0-th moment accumulated over one turn. The equation of motion at the n-th turn is given by

$$\delta_{n+1} - \delta_n = \frac{1}{E_0} \left\{ eV_c \cos(\phi_c + \omega_{rf}\tau_n) - U_{rad} \right\}$$

$$- \frac{e^2 N}{E_0} \sum_{k=-\infty}^{\infty} W_\parallel \left\{ c(nT_0 + \tau_n) - c(kT_0 + \tau_k) \right\}, \tag{42}$$

$$\tau_{n+1} - \tau_n = \eta T_0 \delta_n, \tag{43}$$

where N is the number of particles in a bunch. The wake function can be devided into static (independent of τ_n) and dynamic terms as

$$W_\parallel \left\{ c(nT_0 + \tau_n) - c(kT_0 + \tau_k) \right\} \approx W_\parallel \left\{ cT_0(n-k) \right\}$$

$$+ c(\tau_n - \tau_k) W_\parallel' \left\{ cT_0(n-k) \right\}. \tag{44}$$

The static term of the wake function shifts the synchronous phase, that is, the reference point of τ. This term is omitted in the following analysis.

Assuming small amplitude oscillation, we obtain

$$\frac{d^2\tau}{dt^2} = -\omega_s^2 \tau - \frac{e^2 N \eta}{E_0 T_0} \sum_{k=-\infty}^{\infty} c(\tau_n - \tau_k) W_\parallel' \left\{ cT_0(n-k) \right\}. \tag{45}$$

138

If we assume that the solution is expressed as $\tau = \exp(j\Omega t)$, then Eq. (45) becomes

$$\Omega^2 - \omega_s^2 = \frac{e^2 N\eta}{E_0 T_0} \sum_{k=-\infty}^{\infty} c\left[1 - \exp\{-j\Omega(n-k)\}T_0\right] W_\parallel'\left\{cT_0(n-k)\right\}. \quad (46)$$

By Fourier transforming it (for a detailed derivation, see, for example, Chao [6]), we obtain

$$\Omega^2 - \omega_s^2 = -j\frac{e^2 N\eta}{E_0 T_0^2}\left[\sum_{p=-\infty}^{\infty}\left\{(p\omega_0 + \Omega)Z_\parallel(p\omega_0 + \Omega) - p\omega_0 Z_\parallel(p\omega_0)\right\}\right]. \quad (47)$$

By assuming that Ω is not far from ω_s and using the fact that $\mathrm{Re}\, Z(\omega)$ is an even function of ω, we obtain the growth rate:

$$\tau_g^{-1} = \mathrm{Im}(\Omega) = \frac{e^2 N\eta}{2E_0 T_0^2 \omega_s}\mathrm{Re}\sum_{p=-\infty}^{\infty}\left\{(p\omega_0 + \omega_s)Z_\parallel(p\omega_0 + \omega_s)\right\}. \quad (48)$$

It is represented in a more useful form as

$$\tau_g^{-1} = \frac{\eta I_b}{4\pi(E_0/e)\nu_s}\sum_{p=1}^{\infty}\left\{\omega(p)^+\mathrm{Re}Z_\parallel(\omega(p)^+) - \omega(p)^-\mathrm{Re}Z_\parallel(\omega(p)^-)\right\}, \quad (49)$$

where $\omega(p)^+ = (p - 1 + \nu_s)\omega_0$ and $\omega(p)^- = (p - \nu_s)\omega_0$. As illustrated in Figure 14, the impedance at the upper or lower synchrotron sidebands contributes to this instability.

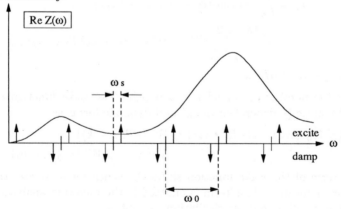

Figure 14. The longitudinal instability is excited by the impedance at the upper synchrotron side bands and suppressed by lower side bands (for $\eta > 0$).

Coupled-bunch instabilities

The above discussion can be easily extended to a many-bunch case. Assume that M equally-spaced buckets are filled with equal-charge bunches (Figure 15).

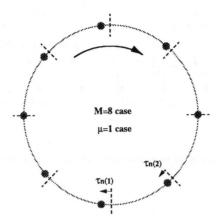

Figure 15. Longitudinal coupled-bunch motion (the case of $M=8$ and $\mu=1$)

The equation of motion for the m-th bunch is given by

$$\frac{d^2\tau^{(m)}}{dt^2} = -\omega_s^2\tau^{(m)} - \frac{e^2N\eta}{E_0T_0}$$

$$\times \sum_{k=-\infty}^{\infty}\sum_{m'=1}^{M} c(\tau_n^{(m)} - \tau_k^{(m')})W_{\parallel}'\left\{cT_0(n - k + \frac{m - m'}{M})\right\}. \qquad (50)$$

We assume that the solution is expressed as $\tau^{(m)} \propto \exp(2\pi j\mu m/M) \times \exp(j\Omega_\mu t)$, where μ is the coupled-bunch mode number ($\mu = 0, 1, 2, \ldots, M - 1$). Similarly to Eq. (49) we obtain the growth rate:

$$\tau^{-1} = \frac{\eta I_b}{4\pi(E_0/e)\nu_s}\sum_{p=1}^{\infty}\left\{\omega_\mu(p)^+\,\text{Re}Z(\omega_\mu(p)^+) - \omega_\mu(p)^-\,\text{Re}Z(\omega_\mu(p)^-)\right\}, (51)$$

where $\omega_\mu(p)^+ = \{(p-1)M + \mu + \nu_s\}\omega_0$ and $\omega_\mu(p)^- = \{pM - \mu - \nu_s\}\omega_0$. Figure 16 shows driving and damping side bands for the case of $M=5$. It is similar to the single-bunch case (Figure 14) except that there are M independent modes.

140

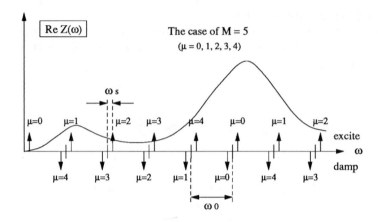

Figure 16. Driving and damping impedance for longitudinal coupled-bunch instability (the case of $M=5$).

5.3 Instabilities due to the accelerating mode

Robinson instability

It is seen from Eq. (49) or Eq. (51) that a narrow-band, i.e. high-Q, impedance can cause an instability. It is called the Robinson instability. [7] In particular, high impedance of the accelerating mode of RF cavities can give rise to a very strong instability. Let us consider the case $\eta > 0$. If $\omega_c > \omega_{rf}$, i.e. $\psi > 0$, see Eq. (15), the impedance at the upper synchrotron sideband of the RF frequency Re $Z_\parallel(\omega_{rf} + \omega_s)$ becomes larger than that at the lower sideband Re $Z_\parallel(\omega_{rf} - \omega_s)$, as shown in Figure 17. Consequently, ω_c should be detuned toward the lower side. The condition $\psi < 0$ is always fulfilled if the cavity is operated at optimum tuning ($\phi_L = 0$) or with a negative value of ϕ_L, see Eq. (26), which is a usual method of tuning, as discussed in Section 4.4.

Instability driven by detuned accelerating mode

When a high beam current is stored in a large storage ring such as a B factory, the detuning frequency Δf given by Eq. (28) is large, while the revolution frequency f_{rev} is relatively small. As a result, Δf can be comparable to, or even larger than, f_{rev}. As shown in Figure 18, the driving force of the coupled-bunch modes -1, -2, etc., can be very large because of the high impedance of the accelerating mode. A serious longitudinal coupled-bunch instability

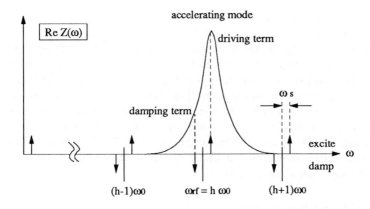

Figure 17. The exciting and damping frequencies related to the acclerating mode.

of these modes can be excited in B factories, whereas this is no problem for relatively small rings (f_{rev} is large), such as ϕ and τ-charm factories, even with a high beam current.

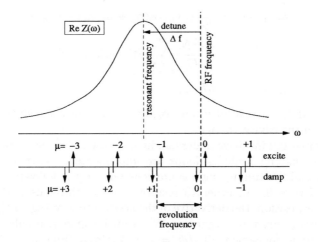

Figure 18. Instability driven by the detuned accelerating mode.

For the two B factories, KEKB at KEK and PEP-II at SLAC, completely different measures have been taken to overcome this instability. As seen in Eq. (28), Δf is inversely proportional to the stored energy U in the cavity.

A straightforward solution is to reduce Δf by increasing U. At KEKB, two types of cavities are used: the SC damped cavity [8] and the ARES NC cavity system. [9] The SC cavity can be operated at a high gradient, resulting in a large stored energy. The ARES is a three-cavity system where an accelerating cavity is resonantly coupled with an energy-storage cavity operating in a high-Q mode via a coupling cavity in between. This system increases the total stored energy by an order of magnitude compared with conventional NC cavities, while the cavity dissipation power is kept at a reasonable level.

The other way to avoid this instability, used at PEP II, is to reduce the coupling impedance selectively at upper synchrotron sidebands by means of a feedback loop with a comb filter. [10]

Static Robinson instability

Here we discuss another type of longitudinal instability related to the accelerating mode. We assume that the oscillation period is much slower than the range of wake. In this case, the arrival time τ_k does not change much within the wake range and the dynamic term of Eq. (44) can be neglected. The static term can give rise to an instability when the stored beam current is high. The equation of motion is

$$\delta_{n+1} - \delta_n = \frac{1}{E_0} \left\{ eV_g \cos(\phi_g + \omega_{rf}\tau_n) - U_{rad} \right\}$$

$$- \frac{e^2 N}{E_0} \sum_{k=-\infty}^{\infty} W_\| \left\{ cT_0(n-k) \right\}, \tag{52}$$

$$\tau_{n+1} - \tau_n = \eta T_0 \delta_n, \tag{53}$$

where V_g and ϕ_g, instead of V_c and ϕ_c, are used to distinguish the generator voltage from the cavity voltage.

For simple and intuitive understanding, we use the phasor representation. We assume that no feedback loops are implemented. Then the generator voltage $\tilde{V}_g$ and the tuning angle ψ are constant. In the following, we discuss the case of $\eta > 0$. In order to restore a nonsynchronous particle to the synchronous position, the derivative of the accelerating voltage, dV_a/dt, where $V_a = V_c \cos\phi_s$, must be negative. When an incoherent oscillation or zero current limit is concerned, $dV_a/dt = -\omega_{rf}V_c \sin\phi_s$ (Figure 19, left) and it is stable when $\sin\phi_s > 0$. On the other hand, when the bunch phase is shifted coherently, the beam-induced voltage $\tilde{V}_b$ does not contribute to the restoring force, since it shifts in the same way as the bunch phase. Then only $\tilde{V}_g$ contributes to the restoring force. In this case $dV_a/dt = dV_g \cos\theta_g/dt = -\omega_{rf}V_g \sin\theta_g$, where $\theta_g \equiv \phi_s - \phi_L + \psi$ (Figure 19, right) and stability requires

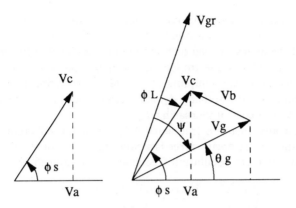

Figure 19. Restoring force for incoherent (left) and coherent oscillation (right).

$\sin \theta_g > 0$. By eliminating either ϕ_L or ψ, using Eq. (26), we obtain two equivalent representations:

$$2V_c \sin \phi_s + V_{br} \sin 2\psi > 0, \tag{54}$$

$$2V_c \sin \phi_s - V_{br} \sin 2(\phi_s - \phi_L) > 0. \tag{55}$$

Figure 20 shows the boundaries of the stable region in terms of ψ (left) and

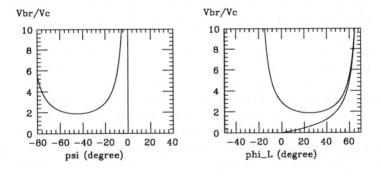

Figure 20. Robinson stability limit as a function of ψ (left) and ϕ_L (right). The case of $\phi_s = 70$ degrees is shown. The two lines indicate the static and dynamic limits. The area between them is a stable region.

ϕ_L (right). It also shows the dynamic stability criterion. As the ratio V_{br}/V_c

increases, the stable region becomes small. To operate the system stably, avoiding possible errors or transient effects, a sufficiently large margin is desired. The RF feedback or feed-forward method is effective for increasing the margin, since it reduces the impedance seen by the beam, as discussed in Section 6. Increasing the coupling β — that is, reducing Q_L — to match a higher current, and/or shifting ϕ_L for further detuning, is also effective to some extent, at the cost of increasing P_g.

We should also mention frequency shifting of the coherent synchrotron oscillation. Since the restoring force is smaller for coherent than for incoherent oscillation, the coherent oscillation frequency ω'_s is reduced as

$$\left(\frac{\omega'_s}{\omega_s}\right)^2 = \frac{V_{gr}\cos\psi\sin(\phi_s - \phi_L + \psi)}{V_c\sin\phi_s}. \tag{56}$$

For optimum tuning, ($\phi_L = 0$), it becomes [3]

$$\left(\frac{\omega'_s}{\omega_s}\right)^2 = \frac{1 - [(V_{br}/V_c)\cos\phi_s]^2}{1 + [(V_{br}/V_c)\sin\phi_s]^2}. \tag{57}$$

Note that $\omega'_s = 0$ corresponds to the static Robinson stability limit.

5.4 Instability caused by HOM's and HOM-damped cavities

The maximum beam current could be limited also by coupled-bunch instabilities caused by higher-order modes of cavities. In a few cases where relatively low beam current is stored in a small ring, the instability could be avoided by just tuning a few dangerous modes at safe frequencies, i.e. somewhere between the driving frequencies of the instabilities. With increased beam current, or in a large ring, however, it would be unrealistic to tune all these modes at the same time. Then the impedance of longitudinal and transverse HOMs must be sufficiently reduced. In particular, it is important to reduce the Q_L values. Several types of HOM couplers have been used for this purpose.

For factories or very high intensity synchrotron light sources, very strong damping is required, and the Q_L values for dangerous modes should typically be much less than 100. Several types of so-called "damped cavities" have been proposed and developed. The damping methods adopted for them can be classified as follows:

1. Use of widely-opened beam pipes.

2. Use of wave guides with cut-off frequencies above the fundamental mode but below the lowest frequency HOM.

3. Use of a coaxial or radial line with a band-rejection filter for the operating mode.

The first scheme allows for a simple structure and is favorable for SC cavities. It is used for the KEKB [8] and CESR-B [11] single-cell SC cavities. The HOMs extracted via the beam pipes are absorbed by an absorber attached inside the beam pipe. This scheme is based on the "single-mode cavity" devised by Weiland. [12] In order to damp the lowest frequency dipole modes, one beam pipe is further enlarged for the KEKB cavity, or a fluted beam pipe [13] is adopted for the CESR-B cavity. The second method is used in many NC damped cavities for factories, including PEP II [14] and DAΦNE. [15] The third scheme was originally independently proposed for a choke-mode cavity for linear colliders [16] and for a SC crab cavity. [17] A more detailed review of HOM-damping has been provided by Yamazaki. [18]

6 RF control system

The amplitude and phase of the RF voltage in the cavities can be disturbed for many reasons such as gain and phase change of the klystron, phase shift change in cables or wave guides, RF noise or AC ripple, etc. Furthermore, as we have already shown, as the stored beam current is increased, heavy beam-loading can give rise to several kinds of instabilities or transient effects, which can limit the stored beam current and/or beam quality.

Many RF systems have servo control loops to maintain the amplitude and phase of the RF voltage in the cavities. The RF voltage is monitored by a pickup antenna at the cavity. Its phase deviation with respect to the reference phase is detected and amplified to control a phase shifter, so that it establishes a phase lock loop (PLL). The amplitude of the pickup signal is detected and compared with the desired operating voltage. An RF modulator is controlled according to the error signal, which provides the auto level control loop (ALC). The resonant frequency of cavities should also be appropriately controlled. The tuning control loop was described in Section 4.4.

For a ring with high stored beam current, heavy beam-loading should be carefully taken into account. First, the ALC and PLL loops can be coupled to each other because of the imaginary component of the beam-loading: an error in amplitude signal affects not only the ALC but also the PLL loop. The same thing occurs for a phase error. The cross talk of the loops can lead to unstable operation for a high beam current, when V_{br} is much larger than V_c. [19] Second, transient issues caused by the bunch-gap or by injection or extraction of the beam can modulate the amplitude or phase. To cope with the problem, RF feedback [20] or RF feed-forward are useful and have been applied.

Figure 21 shows a block diagram of the RF feedback loop. The RF signal picked up at the cavity is directly fed back to the low-level RF circuit and combined with a driving signal. It leads to $V_c = (HI_d + I_b)Z/(1 + HZG)$ and

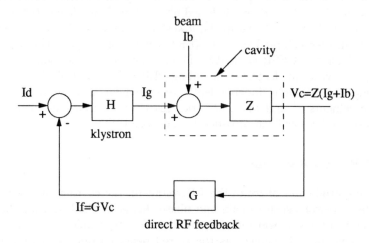

direct RF feedback

Figure 21. Block diagram of direct RF feedback.

thus the impedance seen by the beam is reduced by a factor $Z/(1 + HZG)$. With a high loop gain the beam-induced voltage is eliminated by the loop. Although the loop gain in an actual case is limited by the loop delay, including the filling time of the cavity, still some part of the beam-induced voltage is cancelled, so that the beam-loading effect is reduced. It is also effective to reduce the bunch gap transient, if the bandwidth of the loop is much larger than the revolution frequency.

Figure 22 shows an example of feed-forward compensation. The beam signal is picked up at a monitoring electrode and filtered to extract the RF frequency component. It is fed to the low-level circuit and combined with a driving signal, so that the beam-induced voltage in the cavity can be eliminated. The feed-forward method can also be used for reducing the bunch-gap transient. A signal generated in a programmed way according to the bunch filling pattern is fed to modulate the amplitude and phase of the generator RF signal so that the bunch-gap transient is eliminated.

As an example, Figure 23 shows a schematic view of the KEKB RF control system. [4] In addition to the ALC, PLL and direct RF feedback loops, PLL and

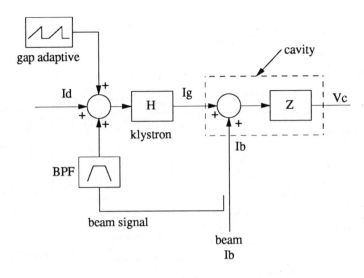

Figure 22. An example of feed-forward compensation.

auto gain control (AGC) loops for the klystron are implemented to stabilize the amplitude and phase of the klystron output. They reduce phase variations due to cathode voltage variations and eliminate power supply ripple and noise around the synchrotron frequency of the beam. The AGC loop also keeps the open loop gain of the direct feedback loop, HZG, constant, even if the klystron gain changes.

Although the phase and amplitude of each cavity voltage is stabilized with these feedback loops, random 0-mode synchrotron oscillations of about ± 0.5 degrees due to noise in the drive system were observed at the beginning of operations. The effect has been cured by adding a 0-mode damping feedback system. The beam signal picked up by a button electrode is transmitted to a high-Q cavity filter to obtain the RF frequency component. The relative phase between the beam signal and the reference RF is detected and transmitted to a band-pass filter centered at the synchrotron frequency so that the synchrotron oscillation is detected. After the output signal is rotated by 90 degrees using a low-pass filter, it is fed to a phase shifter to change the RF phase of the whole ring.

148

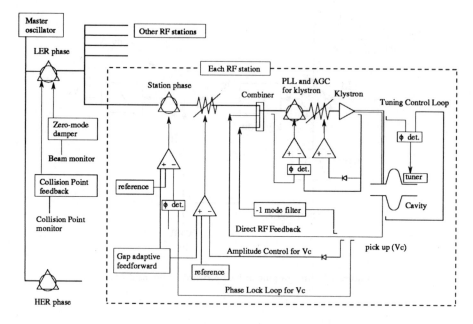

Figure 23. Feedback control loops for KEKB RF system.

7 Summary

Increasingly heavy demands are being placed on the RF system for energy frontier colliders and high luminosity factories: higher accelerating gradient, more RF power to the beam, much stronger damping of HOM's, etc. It is also important to design the RF system to cope with the problems arising from the heavy beam-loading.

References

1. J. C. Slater, Microwave Electronics, Van Nostrand (1950).
2. P. B. Wilson, AIP Conf. Proc. 87, p. 450 (1982).
3. P. B. Wilson, Proc. 9th Int. Conf. on High Energy Accel., p. 57 (1974).
4. K. Akai, et al., Proc. 6th EPAC (1998); KEK Preprint 98-82 (1998).
5. D. Boussard, Proc. 1991 Part. Accel. Conf., p. 2447 (1991).
6. A. W. Chao, Physics of Collective Beam Instabilities in High Energy Accelerators, John Wiley and Sons (1993).
7. K. W. Robinson, CEAL-1010 (1964).

8. S. Mitsunobu, et al., KEK Proc. 93-7, p. 140 (1993).

9. Y. Yamazaki and T. Kageyama, Part. Accel. 44, 107 (1994).

10. P. Corredoura et al., Proc. 4th EPAC, p. 1954 (1994).

11. H. Padamsee, et al., Part. Accel. 40, 17 (1992).

12. T. Weiland, DESY 82-24 (1982).

13. T. Kageyama, Proc. 8th Symp. on Accel. Sci. and Tech., Saitama, Japan, 1991, p. 116.

14. R. A. Rimmer, KEK Proc. 93-7 p. 123 (1993).

15. S. Bartalucci, et al., Proc. 1993 Part. Accel. Conf., Washington, DC, p. 778.

16. T. Shintake, Jpn. J. Appl. Phys. 31 (1992) L1567.

17. K. Akai, et al., Proc. B Factories, SLAC-400, p. 181 (1992).

18. Y. Yamazaki, in Frontiers of Accelerator Technology, p. 506 (1994).

19. F. Pedersen, IEEE Trans. Nucl. Sci. NS-22, 1906 (1975).

20. D. Boussard, IEEE Trans. Nucl. Sci. NS-32, 1852 (1985).

VACUUM SYSTEM OF THE ELECTRON STORAGE RING

MASANORI KOBAYASHI

High Energy Accelerator Research Organization, Institute of Materials Structure Science, Photon Factory

The lecturer expects the students to know the fundamentals of vacuum physics and vacuum engineering, that is, kinetic theory of gases, pressure units, gas flow regimes, conductance for aperture and tubes, vacuum pumps, vacuum components, and vacuum system materials[1, 2]. This knowledge is helpful for the students at this school to understand the vacuum system for the electron storage ring. The lecture includes the following subjects.
1. Beam Lifetime and Vacuum Pressure in the Electron Storage Ring
2. Photon Generation and Gas Load by Photodesorption
3. Pumping Speed and Conductance of Vacuum Duct
4. Pressure Distribution along the Vacuum Duct
5. Practical System Design of the Electron Storage Ring
6. Example of an Electron Storage Ring; Antechamber Structure

1 Beam Lifetime and Vacuum Pressure in the Electron Storage Ring

Synchrotron radiation (SR) users have the following requirements: that beam lifetime is long, SR intensity is almost constant, and beam position is stable throughout their measurements. It is not easy to satisfy their requirements rigorously. Since these requirements are closely related to vacuum system performance, the electron storage ring must be carefully designed.

First we show what determines the beam lifetime of the stored electron beam. Electrons with high energy, i.e. in GeV region, collide with the residual gas molecules in the vacuum ducts, and they are scattered from the correct electron orbit. In other words, such electrons are lost from the stored beam and thus stored beam current decreases. Beam lifetime τ is defined by the following equation:

$$- \mathrm{d}I_B/\mathrm{d}t = I_B/\tau. \tag{1}$$

The beam lifetime τ is expressed as follows:

$$1/\tau = \sum_i \{\sigma_A + \sigma_B + \sigma_C\}_i \, p_i \tag{2}$$

where the σ's are cross sections in scattering process, the subscript i indicates the residual gas component, and the summation Σ should be over all i's. The p is gas pressure [Torr]. The term σ_A corresponds to Moeller scattering between electrons and those of residual gas molecules $\left(\propto 2\pi r_0^2 / \gamma_c\right)$, the term σ_B to Rutherford scattering between electron and nuclei of residual gas atoms $\left(\propto 4\pi(r_0 Z_i / \gamma)^2 / \theta_c^2\right)$, and the term σ_C to Bremsstrahlung $(\propto 4\alpha_0^2 \, Z_i(Z_i+1) \times \{(4/3)\ln(\gamma/\gamma_c)-(5/6)\} \times \ln(183 Z_i^{-1/3}))^3$.

For example, we can obtain each cross section σ for the PF ring (E_B=2.5GeV):

$$\sigma_A = 529 Z_i / \gamma_c, \tag{3}$$

$$\sigma_B = 1058(Z_i/\gamma)^2/\theta_c^2, \tag{4}$$

$$\sigma_C = 2.46 \, Z_i(Z_i+1) \times \{(4/3)\ln(\gamma/\gamma_c)-(5/6)\} \ln(183 Z_i^{-1/3}). \tag{5}$$

The third term in Eq. (2), σ_C, is larger than σ_A and σ_B, and therefore determines the lifetime of electron beams; Bremsstrahlung is the dominant process in beam loss. Pressure is sometimes said to be a *measure* of the vacuum, therefore the "order" of vacuum pressure is of interest (see Eq. (2)). But beam lifetime is inversely proportional to vacuum pressures; $\tau \propto 1/p$, thus the "factor" of the pressure can influence beam lifetime directly.

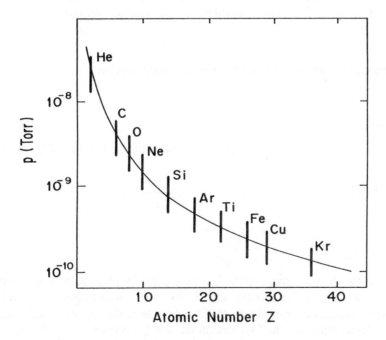

Fig.1 Required pressure for a beam lifetime of 20 hours

The beam lifetime is also proportional to $Z_i(Z_i+1)$ of the residual gas molecules, i.e., the beam lifetime is approximately proportional to the square of the atomic number Z_i. Figure 1 shows the required pressure when beam lifetime of 20 hours, and the pressure is a function of Z_i according to Eqs. (2) to (5). Total cross sections $(\sigma_A+\sigma_B+\sigma_C)$ are listed in Table 1.

Table 1　$(\sigma_A+\sigma_B+\sigma_C)$ for residual gas molecules

H_2	352	CH_4	3840
H_2O	5600	CO	8390
N_2	8260	CO_2	13600
He	495	Ar	23300

Using the values in Table 1, we can calculate τ [sec] from $\{(\sigma_A+\sigma_B+\sigma_C)p[\text{Torr}]\}^{-1}$.

It is clearly understood that hydrogen is not a serious problem as residual molecules. On the contrary, carbon monoxide gives a shorter beam lifetime. In a real electron storage ring, the dominant residual gas species are hydrogen and carbon monoxide molecules. When these are present in similar amounts in the vacuum duct, the lifetime is determined by carbon monoxide pressure.

2　Photon Generation and Gas Load by Photodesorption

2.1　Photon Generation and Photon Spectrum

An accelerated electron can generate synchrotron radiation (SR). Radiated power P_W is given by Eq. (6)[4,5]:

$$P_W[W]=\frac{27}{32\pi^3}\frac{e^2c}{\rho^3}\left(\frac{\lambda_c}{\lambda}\right)^4\gamma^8\left\{1+(\gamma\psi)^2\right\}^2\left\{K_{\frac{2}{3}}^2(\xi)+\frac{(\gamma\psi)^2}{1+(\gamma\psi)^2}K_{\frac{1}{3}}^2(\xi)\right\}$$

$$\xi=\frac{\lambda_c}{2\lambda}\left\{1+(\gamma\psi)^2\right\}^{\frac{3}{2}},\quad \lambda_c=\frac{4\pi}{3}\frac{\rho}{\gamma^3} \tag{6}$$

where e is the charge, c speed of the light, and ρ the bending radius of the electron orbit. λ is radiated wavelength and λ_c is critical wavelength. ψ is vertical divergence against the orbit plane. In the right side of Eq. (6), $K(\xi)$ is the second kind of Bessel function. The first term in the last { } corresponds to a component whose electric vector is parallel to the orbit plane and the second to be the normal component.

We can obtain the number of photons by dividing Eq. (6) by photon power hc/λ, where h is Planck's constant:

$$N_p(\lambda, \psi) = P_W/(hc/\lambda)$$

$$= \frac{27}{32\pi^3} \frac{e^2 c}{\rho^3} \left(\frac{\lambda_c}{\lambda}\right)^4 \gamma^8 \left\{1 + (\gamma\psi)^2\right\}^2 \left\{K_{\frac{2}{3}}^2(\xi) + \frac{(\gamma\psi)^2}{1+(\gamma\psi)^2} K_{\frac{1}{3}}^2(\xi)\right\} \left(\frac{\lambda}{hc}\right) \qquad (7)$$

Photon intensity is shown in Fig. 2(a), where N_p is the number of emitted photons per second per mrad in the orbit plane per 1 eV energy width, and vertical divergence Ψ is integrated. Photon spectrum [photons/s/mrad/% energy band width] is shown in Fig. 2(b).

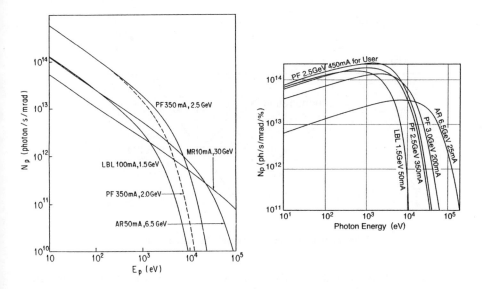

Fig. 2(a) Photon intensity vs. photon energy; (b) photon spectrum.

2.2   Gas Load by Photodesorption

The main outgassing in an operating electron storage ring is photodesorption, while thermal desorption is less.   In discussing the vacuum system of electron storage rings, we concentrate on photodesorption.

Photons irradiate the inner surfaces of the vacuum ducts. Some of the incident photons are reflected according to the incidence angle and the optical constants n and k of the duct materials. The rest of them penetrate the materials and lose their energy along their penetration path. The penetrating photons produce secondary electrons in their loss process. The secondary electrons escape and their kinetic energy changes to thermal energy. Some secondary electrons can reach the surface boundary and escape as photoelectrons into the vacuum[6]. This process is shown schematically in Fig. 3.

Photon stimulated desorption (PSD) and (photo-)electron stimulated desorption (ESD) occur at the surfaces of the vacuum ducts. Photodesorption includes both PSD and ESD. As the cross section of PSD is smaller than that of ESD, photodesorption is caused mainly by ESD of photoelectrons.

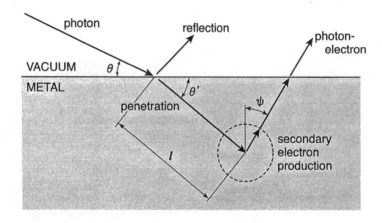

Fig. 3 Photon reflection, penetration and photoelectron production.

The amount of photodesorption (Q) is defined and measured by the pressure rise (Δp) due to photon irradiation and the pumping speed (S) per unit length of the vacuum system[7], i.e.

$$Q[\text{Pam}^3/\text{s/m}] = \Delta pS. \tag{8}$$

The photodesorption yield η [molecules /photon] is defined in practical expression as,

$$\eta[\text{molecules/photon}] = k \, Q/N_p = k \, \Delta p \, S/N_p \tag{9}$$

where k is a unit conversion factor, that is [molecules/Pam^3], and N_p is the number of incident photons [photons/s/m]. Usually outgassing by photon irradiation is evaluated by using the photodesorption yield η.

The photodesorption yield η is about 10^{-1} to 1 when photons irradiate the vacuum duct initially, i.e. they incident on the *virgin* duct for photon irradiation. After long-term irradiation by photons, the photodesorption yield η decreases to the order of 10^{-6}. The decreasing process is called "beam cleaning" or "self cleaning" of photon irradiation. Tthe decrement can be written as

$$\eta = (\text{integrated photon dose})^{-n}, \quad 0.5 < \eta \le 1.0 \ . \tag{10}$$

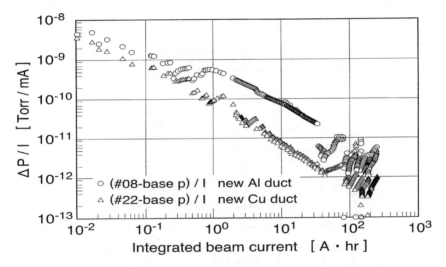

Fig. 4 Beam cleaning of newly installed vacuum ducts in the KEK Photon Factroy.

An example of pressure change in an electron storage ring is shown in Fig. 4, where open circles indicate pressure on a newly installed aluminum alloy duct and triangles pressure on a newly installed OFHC copper duct[8]. Pressures (Δp) are normalized by stored beam current. The normalized pressure decreases with integrated beam current, and the decrement gives us n in Eq. (10) as about 0.8 to 0.9.

The photodesorption q_p [molecules/cm^2/s] is expressed[9, 10] as

$$q_p = N_s Y_{pe} \sigma F_p \tag{11}$$

where N_s [molecules/cm^2] is the number of molecules in the surface layer, Y_{pe} [photo-electrons/photon] is the photoelectron yield, σ[cm^2/photoelectron] is the cross section of electron stimulated desorption, and F_p [photons/cm^2/s] is the incident photon flux. Therefore, we obtain η as

$$\eta = q_p / F_p = N_s Y_{pe} \sigma F_p / F_p = N_s Y_{pe} \sigma. \tag{12}$$

Usually Y_{pe} and σ are approximately constant, therefore decreasig η means the decreasing of N_s. In order to reduce η, that is, to reduce the nuber of molecules in the surface layer, different duct materials and surface treatments shown in Table 2 were tested experimentally. The observed experimental results[11] for η are shown in Fig. 5

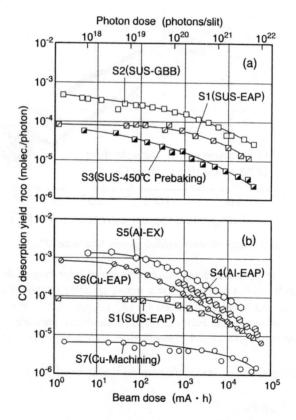

Fig. 5 Photodesorption of CO vs. beam dose. (a) Effect of stainless steel. (b) Effect of materials and surface treatment.

Table 2
Surface treatment conditions[11]

Sample	Material	Treatment
S1	SUS316	Electrolytic-abrasive polishing
S2	SUS316	Glass beads blast
S3	SUS316	Electrolytic polishing and 450°C ×48 hr pre-baking
S4	A6063	Electrolytic-abrasive polishing
S5	A6063	Extrusion
S6	OFC C-1	Electrolytic-abrasive polishing
S7	OFC C-1	Machining

In the vacuum duct of an electron storage ring, the photon flux intensity and the number of molecules in the surface layer differ from place to place. Therefore the gas load caused by photodesorption varies along the vacuum ducts of the storage ring (see Fig. 6).

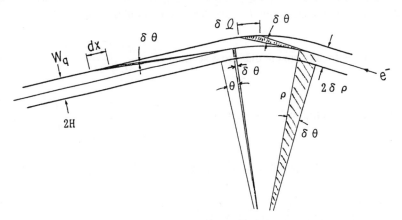

Fig. 6 Photon irradiation along the vacuum duct.

The photon irradiation intensity $F_p(L)$ [photons/s/unit length] along the vacuum duct depends on duct geometry: bending radius ρ, duct width $2H$, and the distance L from the end of the bending magnet in the straight section.

In the bending magnet section, F_p is constant:

$$F_p = 8.08 \times 10^{17} I_B E_B \{\rho/(\rho + \delta\rho)\}/2\pi. \tag{13}$$

In the straight section downstream from a bending magnet, F_p is not constant because the angular divergence $\delta\Theta$, which corresponds to unit length dx, changes with L, which is given by $L=H+\rho(1-\cos\Theta)/\tan\Theta-\rho\Theta$.

$$F_p = 8.08\times10^{17}I_B E_B(\delta\Theta/2\pi).\tag{14}$$

By calculating $\delta\Theta$ vs. L for constant dx, we can obtain the photon intensity along the vacuum duct. The photon intensity will be discussed in Section 4.

3 Pumping Speed and Conductance of Vacuum Duct

Here we review vacuum conductance in the molecular flow region (discussed in the textbook on vacuum science and engineering). In the following equations, R is the universal gas constant, M molar mass, T temperature of gases, and C the conductance of the duct. L is vacuum duct length, d is the diameter of the duct with circular cross section, and a and b is axis lengths in elliptical cross section[1,2].

Long tube with uniform circular cross section: $\quad C = \dfrac{d^3}{3L}\sqrt{\dfrac{\pi RT}{2M}}$. $\qquad(15)$

Long tube with elliptical cross section: $\quad C = \dfrac{2\pi}{3L}\dfrac{a^2 b^2}{\sqrt{(a^2+b^2)/2}}\sqrt{\dfrac{8RT}{4M}}$. $\quad(16)$

For air at room temperature 20°C, Eqs. (15) and (16) are expressed as follows,

$$C \text{ [l/s]} = 12.1d^3/L,\tag{15'}$$
$$C \text{ [l/s]} = (137/L)a^2b^2/\{(a^2+b^2)/2\}^{1/2},\tag{16'}$$

where unit of length is cm.

Effective conductance C_e in serial connection; $\quad \dfrac{1}{C_e} = \sum_i \dfrac{1}{C_i}$. $\qquad(17)$

Using these relations among conductances, we try to obtain "effective pumping speed" S_e along the vacuum duct[6]. Vacuum ducts of electron storage rings have different cross sections, and pumps are arranged locally and/or distributed, so that distances between pumps vary. We define conductance from a point of interest to a vacuum pump as serially connected conductance, as shown in Eq. (17). We now define "unit conductance" for long-tube conductance, Eqs. (15) and (16), divided by number of unit, which is $n=L/L_0$, where L is tube length and L_0 is unit length. We rewrite Eq. (15) as $C=C^*/L$. Thus unit conductance C_0 is given as $C_0=C^*/L_0$, and $1/C=\Sigma(1/C_0) = n/C_0$.

Next, We assume that there is a pump at a local spot i and its pumping speed is $S_p(i)$, which includes the conductances of slit and pumping port. At the spot i, gas molecules find three channels, that is, to the pump at i, to downstream, and to upstream. Conductance at i to downstream is $C_d(i)$, and to upstream $C_u(i)$:

$$C_d(i) = \sum_{i=0}^{i-1} \frac{1}{C_d(i-1)} \tag{18}$$

$$C_u(i) = \sum_{i=k}^{i+1} \frac{1}{C_u(i+1)}. \tag{19}$$

The calculation of $C_d(i)$ should be done from i=0 to i=1, i=2 and to i=i-1. On the contrary, the calculation of $C_u(i)$ should be done i=k to i=k-1, i=k-2 and to i+1 where k indicates the end of the tube. The effective pumping speed at i is defined as

$$S_e(i) = S_p(i) + C_d(i) + C_u(i). \tag{20}$$

When there are distributed ion pumps and/or NEG pumps, the $S_e(i)$ values distributed. As explained above, we calculate the pumping speed and conductance of the tube to both the upstream and downstream directions at every position. These give us the distribution of the local effective pumping speeds along the vacuum duct, and indicate the pumping performance of the vacuum system. We can compare pumping performance in different electron storage rings by using the $S_e(i)$ values, and find which machine has the most powerful pumping system.

4   Pressure Distribution along the Vacuum Duct

Pressure p is determined by gas load Q and pumping speed S of the vacuum system, and defined as

$$p = Q/S. \tag{21}$$

In real electron storage rings, since both Q and S are distributed, calculating them is complicated. We show a procedure to obtain local pressure $p(i)$ using the effective pumping speed $S_e(i)$ obtained in Eq. (20) along the vacuum duct (i indicates the location of interest).

[A] Pressures with a single gas source
    If we assume that there is a *single gas source* at i, then $p(i)$ is simply defined as

$$p(i) = Q(i)/S_e(i). \tag{22}$$

The gas molecules in $Q(i)$ divide into three flows: (1) flow $Q_p(i)$ to pump i, (2) flow to the downstream side, $Q_d(i)$, and (3) flow to the upstream side, $Q_u(i)$, as shown in Fig. 7.

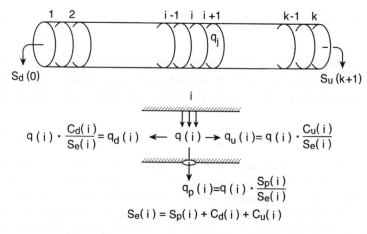

$$S_e(i) = S_p(i) + C_d(i) + C_u(i)$$

Fig. 7 Gas flows in a vacuum duct where there is outgassing, $Q(i)$, at point i.

The amount of dividing flow is proportional to the local pumping speed and two the two kinds of the conductance, C_d and C_u. The flows are defined as

$$
\begin{array}{lll}
Q_p(i)=Q(i)\,S_p(i)/S_e(i) & \text{flow to the pump i,} & (23) \\
Q_d(i)=Q(i)\,C_d(i)/S_e(i) & \text{flow to downstream (i -1),} & (24) \\
Q_u(i)=Q(i)\,C_u(i)/S_e(i) & \text{flow to upstream (i+1).} & (25)
\end{array}
$$

At the next place, that is i-1, we can define $p(i -1)$ as follows,

$$p(i -1)=Q_d(i)/S_e(i -1). \tag{26}$$

The flows at i-1 are expressed as

$$Q_p(i-1)=Q_d(i) \, S_p(i-1)/S_e(i-1) \qquad \text{flow to the pump i-1,} \qquad (27)$$
$$Q_d(i-1)= Q_d(i) \, C_d(i-1)/S_e(i-1) \qquad \text{flow to (i-2),} \qquad (28)$$
$$Q_u(i-1)= Q_d(i) \, C_u(i-1)/S_e(i-1) \qquad \text{back-flow to i.} \qquad (29)$$

At the next place, i-2, pressure and flows are defined as,

$$p(i-2)= Q_d(i-1)/S_e(i-2),$$
$$Q_p(i-2)= Q_d(i-1) \, S_p(i-2)/S_e(i-2) \qquad \text{flow to the pump i-2,} \qquad (30)$$
$$Q_d(i-2)= Q_d(i-1) \, C_d(i-2)/S_e(i-2) \qquad \text{flow to (i-3),} \qquad (31)$$
$$Q_u(i-2)= Q_d(i-1) \, C_u(i-2)/S_e(i-2) \qquad \text{back-flow to i-1.} \qquad (32)$$

Therefore, we can calculate flow rates Q_p, Q_d and Q_u, and the local pressures along the vacuum duct. The local pressures are expressed as $p(i, j)$, which means the pressure at i when there is a single gas source at j.

[B] Pressures with two gas sources

If there are *two gas sources* at i and j, we can obtain the effective pressures $p_e(i, j)$ along the vacuum duct by adding $p(i, j=i)$ and $p(i, j=j)$, i.e. $p_e(i, j)= p(i, j=i)+ p(i, j=j)$. Pressures are additive in molecular flow. In Fig. 8, pressures $p(i, j=i)$ and $p(i, j=j)$ are shown by dashed lines, and total pressures by a solid line.

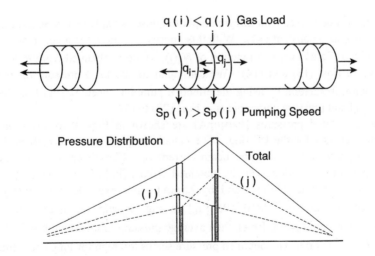

Fig. 8 Pressures in a tube with two gas sources, at i and j.

[C] Pressure with distributed gas sources

If *many gas sources* are distributed in the vacuum duct of the electron storage ring, the above procedure can be extended and applied to each gas source along the duct.

To obtain the local effective pressure at i $p_e(i, j)$, we first calculate pressures $p(i, j=j)$ with a single gas source j, and then continue to calculate $p(i, j)$ for different j. Each $p(i, j)$ can be added at every point i because *components of pressure are additive in a molecular flow region.* The component pressures $p(i, j)$ contributed from each gas source, j, have to be summed up from j=0 to j=k, that is,

$$p_e(i) = p(i, j=0) + p(i, j=1) + p(i, j=2) + p(i, j=3) + \cdots$$
$$+ p(i,j=i-1) + p(i, j=i) + p(i, j=i+1) + p(i, j=i+2) + \cdots$$
$$+ p(i, j=k-1) + p(i, j=k) = \sum_{j=0}^{k} p(i,j) \tag{33}$$

[D] Pressure distribution under photon irradiation

Gas load $Q(j)$ at each position j is expressed as follows:

$$Q(j) \text{ [Pam}^3/\text{s/m]} = k[\text{Pam}^3/\text{molecule}]N_p(j)[\text{photons/s/cm}^2]$$
$$\times \eta[\text{molecules/photon}]s[\text{cm}]dL[\text{cm}], \tag{34}$$

In Eq. (34), s is the perimeter of a cross section of the duct. It is convenient to define j as a unit number, given by L/L_0, where L is the duct length and L_0 is a unit length.

Though η decreases with integrated photon dose as explained in Section 2, here we assume η as a constant (=1). With this approximation, $k\eta$ is a constant, and the gas load $Q(j)$ [Pam^3/s/m] is a function of $N_p(j)$, which is obtained from F_p/s. F_p is explained in Eqs. (13) and (14). We can calculate the local pressure distribution under photon irradiation by using the effective pumping speed $S_e(j)$ in Eq. (20) and the gas load and pressure components in Eqs. (22) to (34)[6].

The calculated pressures [Pa/η/mA] are shown in Figs. 9 to 11, where the parameters are set for the PF ring. The calculated pressure is normalized by the photodesorption yield η and the beam current I_B. Consequently, real pressures under operating conditions can be obtained from $p[\text{Pa}]/\eta[\text{molec/photon}]/I_B[\text{mA}]$ multiplied by $\eta[\text{molec/photon}]$ and by $I_B[\text{mA}]$. For example, the value of the normalized pressure is 7.2×10^{-4}[Pa/η/I_B] in Fig. 9, and we assume $\eta = 1 \times 10^{-5}$ [molec/photon] and I_B=300 [mA], then average pressure in this section is 2.16×10^{-6} [Pa] $= 1.6 \times 10^{-8}$ [Torr]. The obtained pressure seems too high for long beam lifetime.

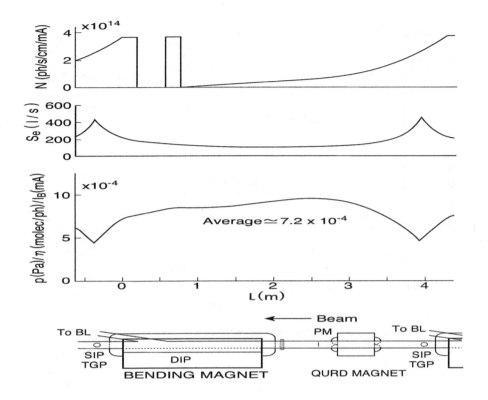

Fig. 9 Pressure distribution, effective pumping speed and photon flux in a normal cell of the PF ring. The magnet and pump arrangement is shown at the bottom. The electron beam comes from the right.

If the main residual gas is hydrogen, beam lifetime is expected to be over 20 hours, from Fig. 1. But since carbon monoxide is one of the main residual gases in real electron storage rings, the expected beam lifetime will be restricted to 4 to 5 hours. As the value of η in the PF is now estimated to be 10^{-6} or below, beam lifetime by gas scattering is sufficiently long and it is affected by the Touschek effect in the high brilliance lattice of the PF.

In Fig. 10, pumping speed effects are shown, where the parameters are getter activities. That "getter activity is 1" means fully active state of getter just after sublimation of titanium. That "getter activity is 0.3" means that getter activity is only 30% of the initial value.

Figure 11 shows photon scraper effect, and also the pumping effect on the beam line. The photon scraper can protect components made of stainless steel.

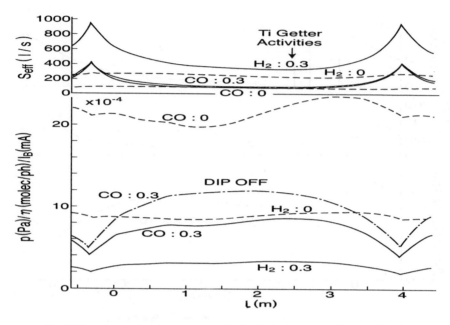

Fig. 10 Pumping speed effect on pressure distribution in the same section as in Fig. 9.

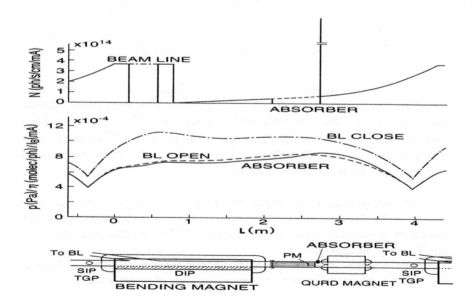

Fig. 11 Photon scraper (photon absorber) effect on the pressure.

The pressure normalized by photodesorption yield and by beam current is useful for understanding and comparing potential performance of vacuum systems in various electron storage rings under the same conditions, i.e. independently from photodesorption yield and beam current.

5   Practical System Design of the Electron Storage Ring

[A] Vacuum duct designing

In the design process, we first need to discuss 1) the geometrical conditions around the magnets. Then we estimate 2) the gas load distribution caused by photon irradiation along the vacuum ducts, 3) the required effective pumping speed at the pumping ports, and 4) the arrangement of evacuation pumps. Next we carefully design 5) the installation and/or the alignment standard for each vacuum duct. After these conditions are discussed, we need to discuss 6) the BPM structure and its mounting on the magnets, and 7) the bellows arrangement, i.e. stroke, locations and number. We must not forget mechanical stress by thermal expansion and distortion of the vacuum ducts when the ducts and pumps are baked in situ, i.e. on the magnets. To protect bellows and flanges from intense photon irradiation, we have to arrange photon scrapers upstream of them.

After we design the vacuum ducts, including evacuation pumps, BPMs and bellows, we have to discuss how to maintain their performance in long-term operation of the storage ring.

Analysis of thermal load becomes more and more important because the beam current is higher and the radiation power intensively localized in low emittance machines.

[B] Pressure measurement[12]

Vacuum gauges are necessary for monitoring vacuum pressures not only during operation but also during shutdown of the machine. A vacuum gauge with digital readout is useful; minimum detection range should be around 10^{-12} Torr (10^{-10} Pa). With the vacuum gauges arranged along the ring about every 4 meters, we can detect tiny vacuum leaks from small pressure changes by using acetone or helium. Pressure records are necessary. The history and trend analysis of pressures shows unexpected slow vacuum leaks and loss of pumping speed of getter pumps (TSP and NEG).

In the early stage of commissioning the ring, the pressure is so high that a hot cathode gauge risks having its cathode burn out. A cold cathode gauge (CCG) fits the required pressure range. After long-term operation the pressure goes down to the ultrahigh vacuum (UHV) region, 10^{-8} Pa or less, where the cold cathode gauge is not sufficient for reliable measurement. The ion current I_i of the cold cathode gauge is easily affected by surface contamination of electrodes, so that its $\delta I_i / \delta p$

characteristics change after long-term operation. The cold cathode gauge is inadequate for correct measurement of pressures. The hot cathode gauge, especially the Bayard-Alpert type gauge (B-AG) is suitable in the UHV region. For example, we have used CCG and B-AG, with a B-AG placed about every 4 meters, a total of 55.

Current output, not voltage output, gives reliable signals at I/O boards, which are arranged far from the B-AG controllers. Pressures from 48 B-AG's are readout about every 0.1 sec. The values are recorded in a personal computer, and the data are transferred to a data-channel in a workstation.

[C] Interlock for the vacuum system

An interlock system is necessary for stable operation of the vacuum system, especially of the electron storage ring. Usually we watch vacuum pressure and operate vacuum valves and switch power supplies. The threshold pressure to drive the interlock system varies from case by case. Accidental leaks from beam lines are larger than that in the ring itself. But when the magnet power supply misses and the beam orbit is distorted, strong irradiation may hit a component, and such radiation is not anticipated in the design process. Such accidents can occur on the vacuum components arranged downstream at a multipole wiggler (MPW). A sudden pressure rise has to be observed, that is, the time constant of the B-AG circuit is important. Consequently, when we design an interlock system, we must consider the response time of the vacuum system. The time constant τ of pressure rise in the vacuum system is expressed as $Q_a/V - pS_e/V$, where Q_a is accidental gas evolution, V is the volume of the chamber interest, and S_e is the effective pumping speed of the chamber. Therefore high pumping speed is helpful to protect the vacuum system.

6   Example of the Electron Storage Ring; Antechamber Structure

As explained above, photon irradiation on the vacuum chamber causes photodesorption, and it is the main gas load for the electron storage ring. To minimize this effect, the ALS group at LBL presented a new structure of the vacuum duct[13, 14]. It is shown in Fig. 12.

Figure 13 shows another example designed at APS, and also an ordinary type of duct[15]. The latter one is the duct of a bending magnet section designed 21 years ago at PF[16].

The new duct has a beam space and photon extraction space called an antechamber. It is expected that most photons can not hit the inner wall of the beam chamber and they can escape through the thin throat to the antechamber. Photons in the antechamber are disposed of on a photon absorber and by rump pumps with large pumping speed. The antechamber has a large effect on magnet structure, but it is relatively easy to mount a large size vacuum pump.

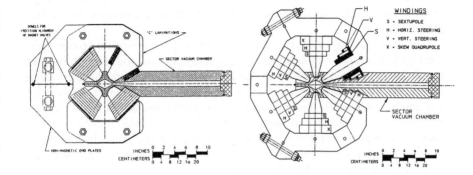

Fig. 12 Antechamber structure in quadrupole and sextupole magnet[13].

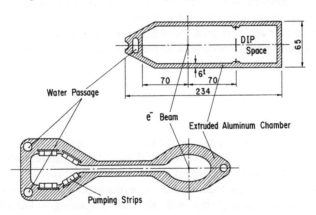

Fig. 13 An ordinary single cell beam duct[16] and an antechamber[15].

When the antechamber was proposed, pressure in the beam space was considered to be defined by thermal desorption and by duct conductance, that is, $p_{beam} = Q_{thermal}/C$. Usually the desorption rate in a thermal process is about 1/10,000 of photodesorption. Pressure in the beam space was expected to drop to the 10^{-9} Pa region in operation. This is much too optimistic. The pressures in the antechamber affect the pressures in the beam chamber through the extraction throat. Photodesorption occurs intensively on the photon absorber in the antechamber. We can design an absorber on which the photon flux is more intense than on the beam chamber wall. Photodesorption decreases according to the accumulation of photons, as mentioned in Section 2. Thus the beam-cleaning process can progress quickly on the absorber, though massive outgassing occurs in the initial stage of photon irradiation. Moreover, we can arrange a pump with a high pumping speed, so that the pressure in the antechamber does not get so high. Therefore pressures in the antechamber are in the acceptable range, and pressures in the beam space are kept sufficiently low.

Another outgassing source is in the antechamber system. Vertical divergence of low energy photons is larger than that of high energy photons, and some of them can not pass through the extraction throat and hit the wall of the beam chamber. If we increased the throat height in order to increase number of escaping photons, the throat conductance would become larger. Gas molecules would have a chance return from the antechamber to the beam space, so such a wide throat can not be adopted. The beam cleaning process in the beam space is determined by low energy photons, which hit the inside of the beam space. This cleaning process occurs similarly in the ordinary single cell chamber.

It would be interesting to compare the pressure decrement in the antechamber with that in the ordinary single cell duct, but reported experimental results are insufficient to discuss the beam-cleaning process in the antechamber system.

The antechamber is wider than the ordinary single cell duct, and mechanical deformation by atmospheric pressure is severe. This is a disadvantage in designing of the beam position monitor (BPM).

Summary
In this lecture we stressed the importance of the vacuum for obtaining long beam lifetime. The vacuum system should be designed quantitatively with regard to pressure, and qualitatively in with regard to atomic number Z.

Photon spectrum and photodesorption are explained. Photodesorption has an intense outgassing rate comparing with thermal outgassing. Photodesorption depends on materials and surface treatment of vacuum ducts, and it decreases with integrated photon dose. The irradiation intensity along the vacuum duct causes a distributed gas load.

We explained the method calculating pressure distribution. Effective pumping speed is defined, and three flows are defined: to a pump and downstream and upstream of the vacuum duct. From additive pressures in the molecular flow region, pressure component is calculated for every gas source, and summed up. The summed pressure gives the pressure distribution along the vacuum duct.

Characteristics of the vacuum system of an electron storage ring are discussed using the PF ring as an example.

Some comments are given on the process of designing the vacuum system of an electron storage ring. The antechamber is introduced and its advantages and disadvantages are discussed.

Reference

1. J.M.Lafferty, Ed., Foundations of Vacuum Science and Technology, Wiley (1998).
2. A.Roth, Vacuum technology, North-Holland (1983).
3. S.Kamada, KEK-Report **79-20**, (August 1979) 1.
4. J.Schwinger, Phys. Rev. **75**, (1949) 1912.
5. A.A.Sokolov and I.M.Ternov, Synchrotron Radiation, Academic-Verlag (1968).
6. M.Kobayashi, American Vacuum Society Series 5, Vacuum Design of Advanced and Compact Synchrotron Light Sources, AIP conference Proceedings No. 171, (Upton, NY, 1988) 155.
7. O. Groebner, A.G.Mathewson, H.Stoeri, P.Strubin, Vacuum **33** (7) (1983) 397.
8. Y.Hori, M.Kobayashi and Y.Takiyama, J.Vac.Sci.Technol. **A12** (4) (1994) 1644.
9. M.Kobayashi, American Vacuum Society Series 12, Vacuum Design of Synchrotron Light Sources, AIP conference Proceedings No. 236, (Argonne, Il, 1990) 332.
10. M.Kobayashi, Y.Hori, Vacuum **44** (5-7) (1993) 493.
11. S.Ueda, M.Matsumoto, T.Kobari, T.Ikeguchi, M.Kobayashi and Y.Hori, Vacuum **41** (1990) 1928.
12. PF Activity Report
13. ALS Conceptual Design Report (July 1986), PUB-5172 Rev.
14. K.Kennedy, American Vacuum Society Series 5, Vacuum Design of Advanced and Compact Synchrotron Light Sources, AIP conference Proceedings No. 171, (Upton, NY, 1988) 52.
15. R.B.Wehrle and R.W.Nielsen, American Vacuum Society Series 5, Vacuum Design of Advanced and Compact Synchrotron Light Sources, AIP conference Proceedings No. 171, (Upton, NY, 1988) 60.
16. M.Kobayashi, G.Horikoshi and H.Mizuno, Nuclear Instrum. Methods **177** (1980) 111.

BEAM INSTRUMENTATION AND FEEDBACK

JOHN D. FOX

Stanford Linear Accelerator Center, Menlo Park, CA, 94025 USA
E-mail: jdfox@slac.stanford.edu

This tutorial is offered to students interested in understanding some of the beam instrumentation in common use at accelerators and light sources. This note is a summary of material presented in November 1999 at the Asian Accelerator School, and is more limited in scope than the original lectures. I have chosen to highlight several essential and basic concepts that underlie the core of beam instrumentation, and hope that these core concepts help the reader navigate the references.

1 Introduction

Particle accelerators and light sources are complex machines, and understanding their performance and operation requires equally complicated instrumentation and measurement instruments. Common control room instrument needs might include:

• Total Intensity (total number of particles - a current)
• Relative intensity of bunches
• Distribution of charge within a bunch (or moments of charge*position, for both transverse and longitudinal coordinates)
• Positions of bunches in a magnetic guide field or vacuum chamber (a trajectory or orbit)
• Transverse oscillatory motion about a central orbit (betatron motion) - measurement of oscillation frequencies (tunes), amplitudes and modal patterns
• Longitudinal oscillatory motion about a mean synchronous arrival time (synchrotron motion) - measurement of frequencies (tunes), amplitudes and modal patterns

All of these measurements require some means to extract information from the moving particles, and some processing technique to separate the desired information from other signals or noise. There are limitless numbers of approaches to these problems - this note concentrates on the most fundamental issues, and refers the reader for specifics and details to the excellent tutorials and articles found in the references.

2 Organization

Before attempting to examine accelerator specific instruments which draw on electronic technology a brief overview of some essential engineering and signal processing ideas is a helpful starting point. This overview covers signals in the time and frequency domains, and the properties of linear time-invariant systems. We will look at the coupling between bunched beams and pick-up electrodes, and try to understand the signals beams generate in the time and frequency domains. Several special cases of uniformly filled rings with simple betatron and synchrotron motion are shown to introduce the student to the unfortunately-complex cases he or she will be confronted with in the control room.

The second portion of this note will present more engineering-oriented material on common control room instrumentation, and continues the time domain - frequency domain formalism by presenting material on fast oscilloscopes and spectrum analyzers. The universal measurement need of beam position monitors is examined as a specialized accelerator instrument. Example position monitor systems from several labs are presented to show the technical choices various designers made, and the unique ways they addressed common problems. Finally, a very brief introduction to feedback is offered, as a means to introduce the various feedback-control systems often used in accelerators.

3 Signal Processing Concepts

Some diagnostics and instrument techniques make direct use of induced signals and extract the desired information in the time domain. Other approaches represent information in the frequency domain, and many accelerator physics problems require information to be represented in both domains. Fourier and Laplace transform techniques are central mathematical tools for transforming measurements from one representation to another. These are tools of linear system theory, and are so universally used that many students may not realize the (surprisingly common) accelerator situations where these tools cannot be applied.

A function f(x) may be Fourier transformed into a function F(s),

$$F(s) = \int_{-\infty}^{\infty} f(x)e^{-i2\pi xs}dx \tag{1}$$

and likewise a function F(s) can be transformed into a function f(x).

$$f(x) = \int_{-\infty}^{\infty} F(s)e^{i2\pi xs}ds \tag{2}$$

The Laplace transform is related to the Fourier Transform but involves an integral from 0 to infinity.

The Fourier transform is a decomposition of an arbitrary function into sines and cosines of various frequencies. For systems involving discrete samples of data, such as from sampling circuits or from samples taken from circulating bunches, the discrete-time Fourier transform is similar[1].

$$F(\nu) = \frac{1}{N} \sum_{\tau=0}^{N-1} f(\tau)e^{-i2\pi(\nu/N)\tau} \tag{3}$$

$$f(\tau) = \sum_{\nu=0}^{N-1} F(\nu)e^{i2\pi(\nu/N)\tau} \tag{4}$$

There is a related transform, the Z transform, which is the discrete-time equivalent of the Laplace transform [2]. There are numerous interesting purposes and properties of these transform techniques - the references contain several suggested texts to learn more.

Figure 1 is a graphical table of some important functions and their transforms[3]. One important feature is readily seen - a function that is limited in extent in one domain (such as an impulse in time) has a corresponding transform which is of wide extent (the frequency components to make up the impulse require all frequencies from DC to infinity). Similarly, a single frequency in the frequency domain has a time transform which extends from negative infinity to positive infinity. An important transform pair, of the rectangular gate signal rect(x) (sometimes written as II(x)) which has as a transform the sinc(x) function $(\sin(x)/x)$, occurs often in systems with a gating function used to select some portion of a waveform.

3.1 Convolution of two functions

The convolution of two functions f(x) and g(x) is defined as $f(x) * g(x)$

$$f(x) * g(x) = \int_{-\infty}^{\infty} f(u)g(x-u)du \tag{5}$$

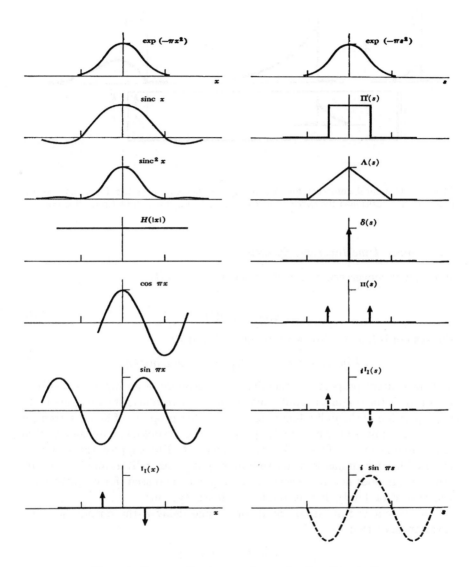

Figure 1. Common Fourier transform pairs, from Bracewell.

In a pictorial form, the convolution can be seen as the overlap integral of the two functions, one plotted backwards, as the origin offset of the two functions is varied[3].

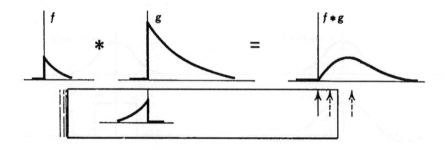

Figure 2. Graphical construct for visualizing the convolution of two functions - as the functions are positioned the overlap integral is sketched at the output (from Bracewell).

3.2 Linear Time Invariant Systems

If a system converts an input $u(t)$ into an output $y(t)$

$$y(t) = L\left[u(t)\right] \qquad (6)$$

the system is linear if for two constants a1 and a2

$$L\left[a_1 u_1 + a_2 u_2\right] = a_1 L\left[u_1(t)\right] + a_2 L\left[u_2(t)\right]. \qquad (7)$$

This is a direct property of linearity - the response of an input, scaled by a factor a1, is the original output scaled by a1. The response of two inputs is the superposition of the individual outputs. Another property of linear systems is that if an input is only a single frequency ω, the output can only contain that single frequency ω. This looks straightforward. But suppose the system is an amplifier which has some maximum output it can deliver, and it saturates (the output does not get larger, as the input is increased, for outputs above the saturation level). Is this amplifier a linear system?

A system is time invariant if for a time delay δ the output has shift invariance, or that

$$L\left[u(t)\right] = y(t) \qquad (8)$$

$$L\left[u(t - \delta)\right] = y(t - \delta) \qquad (9)$$

This statement of shift invariance looks very straightforward - and it is. But suppose our system is an amplifier, and the gain of the amplifier is a function of the power supply voltage on the amplifier. If the power supply voltage has some variation with time or temperature, is this a time invariant system?

3.3 System Responses

The output signal for a linear time-invariant system , from any arbitrary input, can be found as a convolution of an invariant impulse response with the specific input signal. If we can determine this impulse response, the signals from a complicated excitation may be computed using the many techniques and theorems of Fourier analysis and linear systems theory. The impulse response $I(t)$ of a system is found by exciting the system with a δ-function in the time domain.

Figure 3. Impulse response of an LTI system.

In the laboratory these responses are found by driving the input with a fast impulse (or sometimes a step signal) and observing the response with an oscilloscope.

Equivalently, the frequency response $H(s)$ is the transfer function in the frequency domain. In the laboratory these transfer functions are measured with a network analyzer, which measures the gain and phase of the output signal $H(s)$ relative to the input signal $A(s)$ as a function of frequency s.

The frequency domain response, and the time domain response are directly related for LTI systems, for the impulse response is the inverse Fourier transform of the frequency response. The general output in the time domain is the convolution of the time-domain input $u(t)$ with the impulse response $I(t)$. The output can also be found as the inverse transform of the product of the Fourier transform of the input $u(t)$ and the frequency response $H(s)$

$$y(t) = u(t) * I(t) \tag{10}$$

$$y(t) = IFT\left[FT(u(t))H(s)\right] \tag{11}$$

In this form we see that the impulse response of an LTI system is exactly the inverse Fourier transform of the frequency response of the system. A product

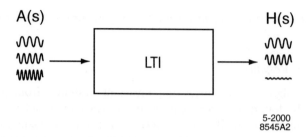

Figure 4. Frequency response of an LTI system found by measurements of magnitude and phase vs. frequency.

in one domain is a convolution in the other domain. For example, the output of a sensor in the time domain can be found from a convolution of the sensor impulse response with an input signal in the time domain. Equivalently, it can be found as an inverse transform of the product of the sensor frequency response and the input frequency spectrum.

As another example, suppose that we have a signal, $f(t)$, which is modulated (multiplied) by a signal $g(t)$. Such a case might occur if we have a train of circulating bunches, each with a characteristic structure (the $f(t)$ function) , but with a variation in current between the bunches (the $g(t)$ current distribution). Such a case, where the time domain signal $f(t)g(t)$ is a product, can be understood in the frequency domain as a convolution of the two function transforms $F(s) * G(s)$.

3.4 Quiz on LTI circuits

Consider the electrical circuit of figure 5- it is a simple voltage divider with one input and one output. Is it a LTI system?

Consider the circuit of figure 6- it is a simple high-pass filter, which removes any DC on a signal, yet pass without phase shift all signals well above its' cut-off frequency $\omega_{3dB} = 1/(2\pi RC)$. Is it a LTI system?

Finally, consider the simple circuit of figure 7 - it is a bidirectional limiter, which passes without change all inputs below a clipping level, and limits the output magnitudes if the inputs exceed the clipping level. Such a circuit might be found at the input of a sensitive amplifier to protect it from excessive input voltages. Is this a LTI circuit?

Answers to these questions are found in Appendix A.

The reader may be wondering why this material is relevant to beam instrumentation - there are several important reasons. The first is that short

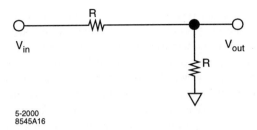

Figure 5. Resistive voltage divider.

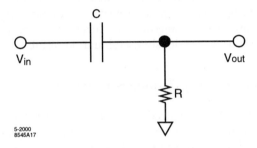

Figure 6. RC high-pass filter.

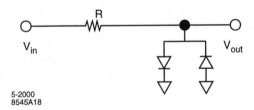

Figure 7. Symmetric diode clamp.

bunched beams excite detection systems much like delta functions, and understanding this formalism helps explain the nature of the instrument response. Understanding the complicated frequency domain spectrum of orbiting bunches may be understood as the result of multiplying an envelope and modulation function with a series of delta-function beam bunches. Another, more subtle and important reason, is that many common accelerator devices and technical components are not linear, time invariant devices. Common circuit devices, or circuit techniques that use non-linear devices, like diodes, or use magnetic coupling, such as transformers, often are NOT linear time invariant devices. Active circuitry in amplifiers is another common non-LTI system. The time response of a beam signal, as processed by an real physical amplifier, may not in any way be well represented by the inverse transform of product of the Fourier transforms of the amplifier frequency response and input signal spectrum. Because so often electronic functions or systems are described in the frequency domain (such as microwave amplifiers or processing components), it is easy to pretend one can predict the time response of such an amplifier to a beam signal by referring to a data sheet displaying a frequency response. Be forewarned, and remember to think before assuming Fourier transform techniques are appropriate to a beam measurement situation. A little knowledge is sometimes a dangerous thing. But the concepts, properly applied, are tremendously useful and powerful.

4   Signals From Beams, Couplings to Beams

The charged particles in an accelerator can be sensed from their electro-magnetic fields, using various devices and structures generally referred to as "pickups". Essentially all of these devices measure the image charges in the electrically-conducting vacuum chamber[a].

A charge in a perfectly conducting chamber produces an image charge, of equal magnitude and opposite sign. The exact structure of the fields of a moving charge, and the charge density of the image charge, depend on the velocity and trajectory of the particles and the geometry of the duct. For simple geometries it is often possible to analytically calculate the field distribution and image charge density, while complex geometries may be best solved numerically.

Figure 8 shows an example for a relativistic bunched beam (where the electric field distribution is largely radial) and figure 9 the field distribution for a rectangular cross-section duct (based on the PEP-II chamber cross-section)[4]. The references detail the calculation of these fields[5,9,6].

The most direct sort of pick-up is the gap monitor (figure 10), which is a finite-impedance resistive structure placed in a gap of the conducting chamber[7]. For good high-frequency response the resistive elements are typically spread uniformly in azimuth around the beam pipe. The image charges flow through this resistance, producing a voltage proportional to the beam

---

[a]There are alternative techniques, using synchrotron radiation, for example, but for this tutorial I am restricting this discussion to the most common techniques in widest use.

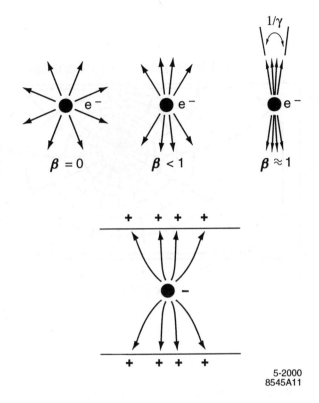

5-2000
8545A11

Figure 8. Field of a moving charge, and image charges in a perfectly conducting beam pipe.

charge. The capacitance of the gap, in conjunction with the resistance, form a low-pass filter with true DC response. These monitors can be constructed with frequency response up to 5 GHz, which can allow information about total current, and longitudinal current distribution, to be measured.

Button or electrode pick-ups make up another important type of beam monitor. These pickups are typically small capacitively-coupled structures[7,12]. Ring-style and multiple tapered electrodes are shown in figures 11 and 12. The equivalent circuit is a high-pass filter, so these monitors do not have any DC response. The frequency response of these pickups can extend to 10 GHz, though care is needed in the design of the button, vacuum feedthrough and coaxial output transition to avoid creating spurious resonances in the structure

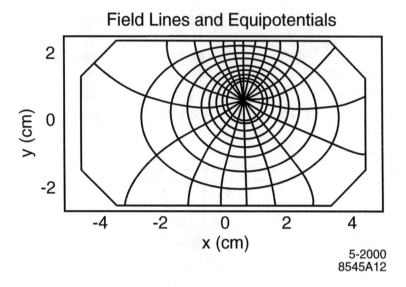

Figure 9. Electric fields for an off-center beam in a rectangular duct, as solved with conformal mapping techniques (from Smith).

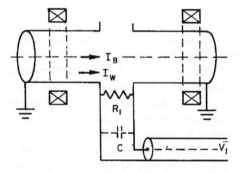

Figure 10. A gap monitor using a finite-conductivity resistance to sense the image current.

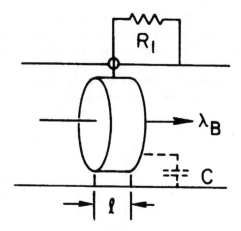

Figure 11. A capacitively-coupled ring electrode. Such a monitor has a high-pass filter response.

at high frequencies.

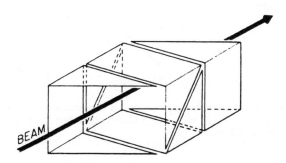

Figure 12. Four capacitively-coupled shaped electrodes. The trajectory of the beam is revealed in the ratio of the four electrode responses.

Stripline electrodes comprise a third general class of pickup (figure 13). These function in a manner somewhat like a directional coupler. The stripline pick-up is a transmission line located within the vacuum chamber. Each line has a characteristic impedance Z_0 and a length l. The lines are coupled to the beam electromagnetic field (capacitively to the beam electric field, inductively through the beam's magnetic field looping between the stripline and conducting wall). This monitor has a periodic frequency response, and for frequencies where the length $l = (2n+1)\frac{\lambda}{4}$ long the response is a maxima. while the response is zero for lengths $n\lambda/2$. In the time domain the impulse response of these monitors has a pair of opposite polarity impulses, spaced in time by $2l/c$.

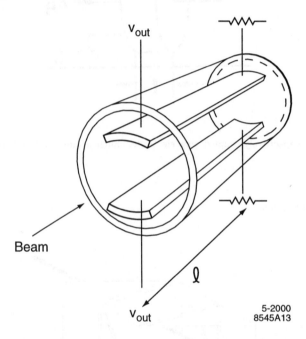

Figure 13. Stripline electrodes comprise transmission line structures coupled to the beam. Such structures can act both as pickups and as kickers, to deflect the beam from an externally applied current excitation.

A pair of striplines can also act as a "kicker" to apply external fields to the beam[5]. If an opposing pair is driven differentially, the transverse electric field, and magnetic field from the two differential currents, act to deflect the beam at an angle. Such a excitation is useful to drive the beam to measure transverse beam transfer functions, or as a power stage to apply correction signals to counteract beam motion. The directional nature of the device is easily seen if the driving currents are applied at the upstream port rather than the downstream - in this case the electric and magnetic kicks cancel out.

5   Common Control Room Instrumentation

Most control rooms contain a mix of commercial, general purpose instruments and lab-designed, specialized instruments.

5.1   Oscilloscopes

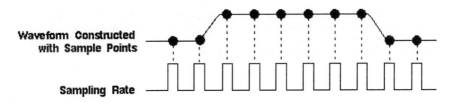

Figure 14. Continous time sampling in an oscilloscope, allowing a single transient to be measured. Such techniques are very complex and expensive for sampling rates above 1 - 5 Gs/sec., allowing real-time bandwidths of the order of a GHz (from Tektronix).

In the time domain the observation of many signals is done via commercial oscilloscopes. There are numerous technology choices, based on analog CRT vs. digital sampling technologies. A distinction needs to be made between instruments which can sample a unique transient in real time (figure 14), as opposed to equivalent time or sampling oscilloscopes. These equivalent time approaches are used for the highest bandwidth instruments, but are unable to record a transient in a single continous sequence(figure 15). The triggering of these instruments requires care, because for a 20 GHz signal to be faithfully reconstructed from numerous triggered samples, the triggering signal must have time stability consistent with the required sweep rate, and this implies very fast signals with sub-nanosecond rise times - which is not something a low-frequency source, like a pulse generator with TTL outputs, is capable of providing. Several oscilloscope manufacturers, particularly Agilent (Hewlett-Packard), Tektronix, and LeCroy provide very helpful application notes and guides to understand the variety and appropriate selection of oscilloscopes.

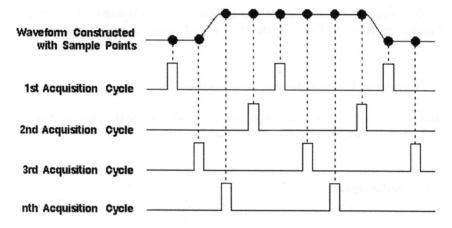

Figure 15. Equivalent-time sampling in an oscilloscope, in which a periodic signal is successively sampled over many repetitions. The highest-speed sampling oscilloscopes achieve 50 GHz of equivalent-time bandwidth. A repetitive signal and high-stability trigger are necessary to use these techniques (from Tektronix).

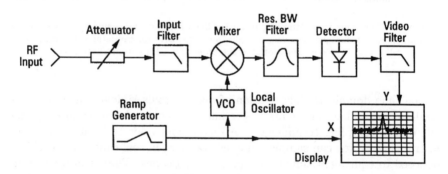

Figure 16. A swept spectrum analyzer uses a tuned receiver, swept across the band of interest. A detection circuit measures the magnitude of the signal at each frequency (from Agilent).

5.2 Spectrum Analyzer

The central frequency domain instrument, the spectrum analyzer, falls into two general classes. The traditional analyzer is a heterodyning instrument, which functions as a swept receiver, sweeping across the band of interest and recording power vs. frequency within a resolution bandwidth (figure 16). High-end spectrum analyzers can process signals up to 50 GHz or more, and many instruments include some amount of digital processing for the video processing, such as averaging and peak detection functions. In contrast, there exist real-time spectrum analyzers, which sample the input and through numeric computation compute an FFT of the input sequence. Such instruments are limited in processing bandwidth to tens of MHz, which might be heterodyned down from some higher frequency carrier. Usually these FFT based instruments provide a faster measurement of a given bandwidth than a swept receiver, for they are effectively computing all the frequency bins in parallel, and generating the spectrum all at once, rather than building it up from sucessive narrowband measurements. Again, most manufacturers, particularly Stanford Research Systems, Agilent (Hewlett-Packard) and Tektronix provide excellent application notes to understand the trade-offs in different instruments and architectures.

6 Common Control Room Measurements

Among the most common control measurement needs are:

6.1 Particle Current (total)

A total current is often measured via integrating current transformers (DCCT - direct current current transformers). These instruments use a magnetic core, which is a toroid encircling the moving beam charges. A sense winding, and a feedback winding, are used to develop a current in the feedback winding which cancels the magnetic field in the toroid from the beam. These instruments can be designed to allow microampere resolution with ampere dynamic range, though care is required to design the magnetic core and associated electromagnetic shielding to provide good frequency response. A simple gap monitor, consisting of a resistive gap in the beam pipe can provide signals proportional to individual bunch charge, and subsequent processing can integrate the responses to get a total current.

6.2 Particle current distributions (bunch by bunch distributions)

Measuring the relative current in each bunch requires a wideband (fast) pickup and time domain processing, such as sampling circuits, to measure a signal proportional to the charge on each bunch. Fast synchrotron light signal processing, using wideband photodiodes or phtomultipliers, followed by sampling or gating circuits, can also be used.

6.3 Orbits

Particle trajectories can be measured from position monitors located around the ring. Such a measurement of trajectory is often called an orbit. The changes in the orbit, due to modifications of magnet strengths, or changes in RF frequency, can be processed to reveal lattice functions (β and η) as a function of position in the ring. There are special techniques to use peturbations in the orbits to identify errors, and quantify strengths on the magnetic elements[16].

6.4 Longitudinal Bunch Profile

Streak cameras (from optical or electrical singnals) provide high-bandwidth measurements of longitudinal charge density. Wideband gap monitors and fast pick-ups can be directly observed in both time and frequency domains. The frequency domain distribution functions from such wideband pickups reveal a wealth of information about the bunch structure[8].

6.5 Transverse Bunch profiles

Transverse density information can be found via synchrotron light images, which can offer a TV rate real-time image. Such TV pictures are often useful for observing transverse bunch profiles or the presence of instabilities, though depending on the optics and system the vertical profile can be difficult to quantify due to diffraction-limited effects from very small bunched profiles. Alternative methods include intercepting techniques such as scraper measurements, in which an aperture edge is inserted into the vacuum chamber, and lifetime (loss rate of particles) is measured as the scraper beings to intercept the far edges of the bunch distribution. If some gaussian functional form is assumed the sigma of the transverse distribution can be estimated. Another intercepting technique uses wire scanners, in which a thin filament is passed through the beam, and during the transverse passage either intercepted current or scattered radiation is measured to estimate the transverse profile. An interesting optical interferometer technique has been developed[17], which uses a pair of interfering optical fields to generate an interference pattern in the vacuum chamber. As this pattern has periodic maxima and minima, as determined by the optical wavelength, a scan of the optical interference pattern over the beam, while measuring the compton scattering process, can measure very small transverse profiles.

In colliders, it is also possible to infer the transverse profiles by carefully scanning one beam across an opposing colliding bunch, for as the beams collide at some impact parameter the resulting beam-beam deflection changes the bunch trajectories, which can be measured via beam position monitors. Such beam-beam techniques are useful for measuring the beam profiles, as well as extremely useful in measuring the colinearity and alignment of the colliding bunches.

6.6 Tunes

The stored beam in a circular accelerator or storage ring is confined by various restoring potentials, resulting in natural oscillation frequencies for the confined particles. The transverse motion of particles, as they are focussed by the magnetic guide fields, exhibit betatron oscillations. Machines are most often designed with independent vertical and horizontal lattice functions, so that there are usually separate horizontal and vertical oscillatory frequencies (often described as a tune, measured as the ratio of the natural oscillation frequency to the revolution frequency). The longitudinal motion of a stored beam is also an interplay of the RF accelerating voltage waveform, and the magnetic guide field, and the bunches exhibit synchrotron oscillations as well.

The knowledge and manipulation of these oscillatory frequencies is among the most important of control room instrumentation. The beam signals can usually be directly sensed via a pick-up electrode, and processed by a spectrum analyzer to reveal the frequencies of various motions. As the beam is sensed at a particular point in the ring, there is an inherent sampling of the natural motion at the revolution frequency, and a concomitant aliasing of frequencies higher than $1/2$ the revolution frequency (the Nyquist sampling limit). For synchrotron motion the natural oscillation frequency is typically much lower than the revolution frequency, (a tune much less than $1/2$). Betatron tunes are typically much greater than $1/2$, so that what is measured on a control room spectrum analyzer is the aliased frequency, or the fractional part of the tune evaluated modulo $1/2$. The integer portion of the tune is implicit in the design of the lattice.

Usually the noise sources acting on the beam are sufficient to excite this natural motion to an easily measured level. Figure 17 is such a spectrum from the PEP-II accelerator, showing naturally-excited betatron motion.

Sometimes it is helpful to excite the beam with a coherent excitation from a special "kicker" magnet or transverse stripline kicker, and measure the

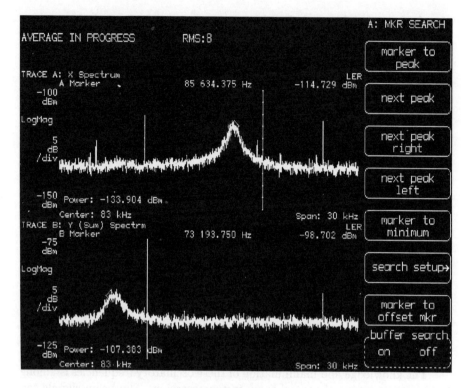

Figure 17. A spectrum from PEP-II, showing naturally excited betatron motion.

driven motion of the beam, as shown in figure 18. In doing such a measurement, it is important to consider and control the magnitude of the excitation applied, as it is possible to drive the beam to such an extent non-linear responses, and amplitude-dependent tune shifts, are observed (it is also possible to kick the beam out of the accelerator). Such driven measurements can be extended using a network analyzer to measure beam transfer functions. A network analyzer measures the complex response (magnitude and phase) with respect to an excitation. Sensitive measurements of the lattice functions and RF system are made by observing changes in the oscillation frequencies and oscillation phases as machine parameters are varied.[8]

In applications where it is desired to regulate the tune, or to compensate for some unregulated element which causes the tune to deviate from a desired

I700393-008

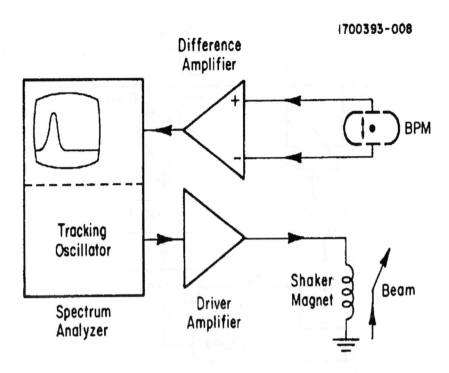

Figure 18. A driven tune measurement, using the tracking oscillator of a spectrum analyzer to excite the beam while observing the resulting response on transverse electrodes (from Billing).

value, a tracking tune measurement using a phased-lock-loop detection scheme is often implemented. Figure 19 shows the topology (the phase-locked-loop is a type of feedback system), and in this approach the voltage controlled oscillator (VCO) tracks the varying tune, either as a frequency measured via a frequency counter, or via the DC control voltage of the VCO. What is measured is the phase of the beam response with respect to the oscillation, which in a high-Q system is a sensitive discriminant. The tune can then be regulated by a feedback loop to some desired value.

7 Signatures of Bunch Motion

The signals from circulating bunches are sometimes overwhelming in structure to a first-time observer. It is easier to understand the frequency-domain

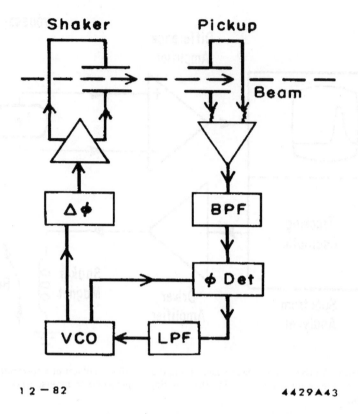

Shaker **Pickup**

Beam

Δφ **BPF**

φ Det

VCO **LPF**

1 2 − 82 4429A43

Figure 19. Lock-In tune measurement, using a phase detector to adjust the VCO control voltage to keep the phase of the beam response at a fixed value (from Littauer).

signals from a pick-up by first understanding the spectra from the simplest case of a single stored bunch, and then try understand the spectra of more complex fills, with transverse or longitudinal motion of the bunches.

7.1 Single Bunch Spectrum

If a single bunch is orbiting in a storage ring, the signal from a pick-up monitor may be understood as a series of impulses - each with a characteristic shape related to the bunch structure and the pick-up response.

$$f(t) = \sum_{k=-\infty}^{\infty} \delta(t - kT_0), \tag{12}$$

where T_0 is the revolution period of the ring.

In the frequency domain there is a line spectrum, with spacing $\Delta\omega_0 = 1/2\pi T_0$

$$f(\omega) = \int_{-\infty}^{\infty} f(t) \exp(-i\omega t) dt \tag{13}$$

$$f(\omega) = \omega_0 \sum_{m=-\infty}^{\infty} \delta(\omega - m\omega_0) \tag{14}$$

The envelope of the line spectrum falls off with frequency, as determined by the Fourier Transform of the bunch shape (or the frequency response of the pickup). For bunches in a typical electron storage ring, the spectrum extends to 20 GHz and beyond (figure 20).

7.2 Multi-bunch Spectrum

If we fill a ring with M identical, uniformly spaced bunches the signals in the time domain become

$$f(t) = \sum_{k=-\infty}^{\infty} \sum_{n=0}^{M-1} \delta(t - kT_0 - nT_0/M) \tag{15}$$

and in the frequency domain

$$F(\omega) = M\omega_0 \sum_{m=-\infty}^{\infty} \delta(\omega - mM\omega_0). \tag{16}$$

For this case of identical, uniformly spaced bunches, the ring spectrum simply looks like a scaled version of the single bunch case - the ring looks as if it is $1/M$ smaller.

If the bunches are not uniformly spaced, or there are current variations among the bunches, the time domain signal is multiplied by a modulating function $g(t)$

$$f(t) = g(t) \sum_{k=-\infty}^{\infty} \sum_{n=0}^{M-1} \delta(t - kT_0 - nT_0/M) \tag{17}$$

In the frequency domain the resulting spectrum $H(\omega)$ is a convolution

$$H(\omega) = G(\omega) * F(\omega) \tag{18}$$

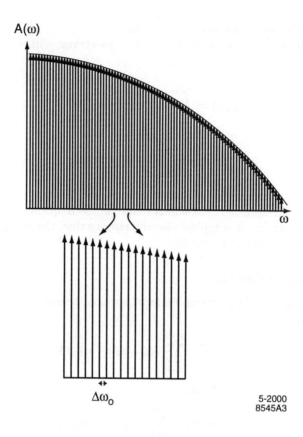

Figure 20. A single-bunch spectrum is a forest of spectral lines, each a multiple of the revolution frequency ω_0. The envelope of the lines contains information about the longitudinal bunch distribution.

As an example, consider a uniformly filled ring of identical bunches, but with a 10% gap (10% of the circumference has no filled bunches). This rectangular gap is a multiplication with a modulating function, and the transform of the rectangular modulating function will be of $sin(x)/x$ form. As a result, a $sin(x)/x$ envelope will be convolved with the uniform line spectra. A variation in the uniformity of bunch current produces similar effects, with the line spectra of the uniform fill convolved with the transform of the current variation. [b]

7.3 Spectra of Oscillating bunches, Bunch Motion, and Signatures of Instabilities

The transverse (betatron) motion and longitudinal (synchrotron) motion of the particles is often observed via signals generated from pick-up monitors. In the frequency domain the circulating bunches generate a complex spectrum, which contains frequency information which can identify the type of particle motion.

Consider a single bunch undergoing transverse (betatron) oscillations. A position sensitive pickup will produce an output signal which contains the transverse motion as an amplitude modulation (AM) at the betatron frequency (ω_β)

$$f(t) = A_\beta \cos(\omega_\beta t) \sum_{k=-\infty}^{\infty} \delta(t - kT_0), \qquad (19)$$

In the frequency domain the resulting spectrum is found as a convolution of the cosine modulating function with the line spectrum of the orbiting bunch

$$f(\omega) = \frac{A_\beta \omega_0}{2} \sum_{m=-\infty}^{\infty} (\delta(\omega - m\omega_0 + \omega_\beta) + \delta(\omega - m\omega_0 \omega_\beta)). \qquad (20)$$

Every revolution harmonic will have betatron sidebands from the amplitude modulation.

Motion in the longitudinal plane (synchrotron oscillations, or energy oscillations) produce similar spectra. A single bunch, executing simple harmonic motion about the synchronous phase

$$f(t) = \sum_{k=-\infty}^{\infty} \delta(t + \tau sin(\omega_s t + \phi) - kT_0) \qquad (21)$$

---

[b]In actual accelerator operations, it is very difficult to store bunches with exactly the same number of particles per bunch. Usually every peak corresponding to the revolution harmonics is seen. The current distribution function multiplies the perfect uniform impulse train in the time domain, so that in the frequency domain the transform of the current distribution is convolved with the perfect uniform beam spectrum, placing information around every revolution harmonic.

194

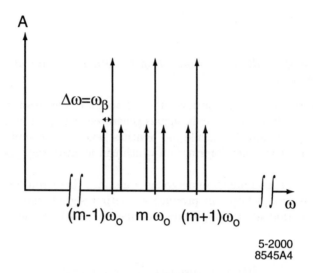

Figure 21. Single-bunch betatron spectrum. Every revolution harmonic has betatron sidebands at ω_β away from the revolution harmonics.

is a signal containing phase modulation of magnitude τ at the synchrotron frequency ω_s. The resulting spectrum is

$$F(\omega) = \omega_0 \sum_{l=-\infty}^{\infty} e^{il\varphi} J_l(\omega_\tau) \sum_{m=-\infty}^{\infty} \delta(\omega - l\omega_s - m\omega_0) \tag{22}$$

For small oscillation amplitudes we can just consider the lowest order terms of the Bessel functions (the l=-1,0,1 terms) and the spectrum will have the form of figure 22.

The ratio of the amplitudes of the l=0 and l=+1/-1 sidebands can be used to measure the magnitude of the phase oscillation.

8 Multi-Bunch Motion and Modal Decomposition

Consider M identical bunches, weakly coupled to each other (such couplings arise from resonators in the ring vacuum structure, particularly in the RF cavities. A passing bunch can excite such a resonator, and a subsequent bunch experiences a perturbation in trajectory or energy from these transients). A simple picture of weakly coupled pendula is helpful to picture the motion of the bunches (though slightly misleading, in that the coupling impedances

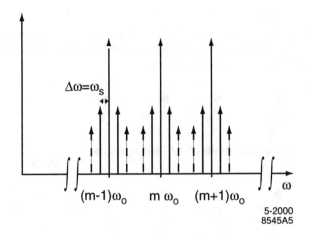

$$\Delta\omega = \omega_s$$

$$(m-1)\omega_0 \qquad m\,\omega_0 \qquad (m+1)\omega_0$$

5-2000
8545A5

Figure 22. Single bunch synchrotron motion is revealed in synchrotron sidebands, each spaced $l\,\omega_s$ away from the revolution harmonics. Only $l = -1, 0, 1$ sidebands are shown, as for small amplitudes the higher terms fall off in amplitude.

are complex, so that the "springs" connecting the bunches should really be complex. A general connecting impedance might have a dashpot, of positive or negative coefficient relating force to velocity, and possible mass-like properties).

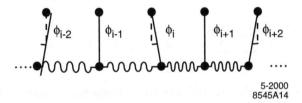

$$\phi_{i-2} \qquad \phi_{i-1} \qquad \phi_i \qquad \phi_{i+1} \qquad \phi_{i+2}$$

5-2000
8545A14

Figure 23. Coupled oscillator analogy to multi-bunch coupled motion.

For M bunches the arbitrary motion can be decomposed into M normal modes. Each normal mode has a single eigenfrequency, and the phase relationship between bunches is specified for each eigenmode M as

$$\Phi_i = \Phi_0 - 2\pi \frac{im}{M} \tag{23}$$

196

where the index i runs $i = 0, 1 ... M - 1$

The lowest mode (zero mode) has all bunches oscillating in phase, while the highest frequency mode (sometimes called the π mode) has a 180 degree (π phase shift) between adjacent bunches).

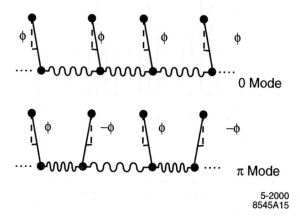

0 Mode

π Mode

5-2000
8545A15

Figure 24. The lowest frequency mode (zero) has all bunches oscillating in phase, while the highest frequency mode (π) has the largest phase shift between the bunches.

A pickup signal observing multi-bunch betatron motion for a ring of M identical bunches will show amplitude modulation of the form

$$f(t) = \sum_{k=-\infty}^{\infty} \sum_{n=0}^{M-1} \cos(\omega_\beta t + \varphi_n)\delta(t - (kT_0 + nT_0/M)) \qquad (24)$$

and in the frequency domain the sideband information for each mode is present as an upper sideband, and a lower sideband. Across M revolution harmonics are found the spectral information for the M normal modes. The spectrum repeats this information every M revolution harmonics, so that to fully characterize a beam with M uniformly spaced bunches require M revolution harmonics worth of spectral bandwidth.

A similar spectrum is generated by multi-bunch synchrotron motion, where M revolution harmonics span the range of M normal modes (figure 26. Each mode is represented by an upper sideband (at the mth revolution harmonic in the band), and a lower sideband (at the M-mth revolution harmonic).

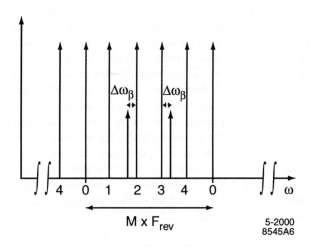

Figure 25. Multi-bunch betatron spectrum, for the M=5 case. Motion of mode 2 is seen as a lower sideband around the second revolution harmonic, and an upper sideband of the M-2 revolution harmonic.

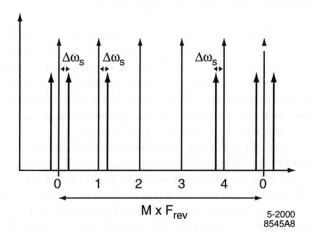

Figure 26. Simplified spectrum for multi-bunch synchrotron motion, showing only the first-order lines. The case M=5 is shown, with motion at mode 0 and mode 1.

8.1 Signals from non-uniform bunches, or uneven fills

The spectral information from these uniform multi-bunch fills is simply represented as normal modes which project exactly onto the Fourier eigenmodes of the symmetric, uniformly filled ring. But in a real accelerator, bunches cannot be filled exactly uniformly, and often there are gaps in the filling pattern. In these cases, the true eigenmodes (normal modes) of the system are different from the even-fill eigenmodes. The uneven-fill eigenmodes will be some linear combination of the even-fill eigenmodes, though calculating or deriving the exact linear combination is very difficult unless one has exact knowledge of the coupling matrix of the system. If we did have exact knowledge of the couplings in the system, one could calculate the eigenvalues and eigenmodes of the non-uniform fill case. However, such coupling information is not generally known. Instead, for practical reasons most accelerator measurements project the motion unto the uniform fill eigenmodes, with the knowledge this is an incorrect basis to represent the problem. This projection unto the uniform-fill eigenmodes leads to couplings of "modes" in the resulting spectrum, and confusion, in that a single true beam normal mode may appear as several uniform-fill eigenmodes. If the fill is almost-uniform, the errors are usually forgiven but for fills with complicated structures understanding the measured spectrum can be very difficult.

9 Beam Position Monitors and Signal Processing

The basic goal of position monitoring is to measure the beam trajectory through an electrode or cavity structure. Many beam diagnostic techniques use the image charge, and the azimuthal variation of charge with beam trajectory to measure the position of the charge centroid with respect to some fixed co-ordinates in the duct. These Beam Position measurements are central to measurements of beam orbits in an accelerator or storage ring, and for diagnostic measurements of tunes, β and η functions.

There are numerous pick-up geometries based on button-type monitors, stripline monitors and shaped electrode monitors, all of which encode the bunch trajectory information in the relative amplitudes of the signals on the various electrodes. Cavity position monitors also are used, in which the bunch trajectory excites a resonant cavity, and the relative strength of various modes in the cavity encodes the trajectory. The important parameters for a position monitor system are:

• Accuracy - a system design will have an absolute accuracy, as well as a relative accuracy.

- Resolution
- SNR of measurement
- Dynamic Range - both for variation in particle current, and variation in trajectory (essentially the allowed range of trajectory compared with the system resolution)
- Linearity of detection - many electrode monitors have non-linear outputs vs. trajectory, requiring post-processing to achieve good accuracy
- Location of zero position (electrical center vs., mechanical center)
- Mechanical zero position and alignment
- Calibration technique

General purpose position monitors sense the steady-state orbit, but there are special purposes that emphasise certain system features. Systems intended for use as reference position information for a light source (a "photon source BPM") require excellent long-term repeatability but have very low bandwidth requirements. Single-turn position monitors are intended for instrumentation functions, and must implement the sensitivity and bandwidth to perform a single-shot measurement (though often at reduced resolution from a multi-turn measurement). Single-pass systems, coupled with appropriate memory systems, can provide time history and frequency domain information (which can show the machine tunes or unstable motion). Finally, there are systems intended to detect the oscillatory motion around the mean orbit, while rejecting the DC mean orbit. Such systems are useful in oscillation control or feedback systems, where the feedback system is damping the oscillatory motion, but not trying to change the equilibrium mean orbit in any way.

9.1 Common Processing Issues

Any BPM processing design reflects the choices and trade-offs of the system designer. There are numerous choices facing the system designers.

The information from the beam is represented across a tremendous width of spectrum (the bunch signals can contain components limited by the bunch length, or possibly 20 - 40 GHz). Some BPM designers choose to heterodyne information from a particular frequency band down to a processing channel. Such a heterodyned approach is like a radio receiver, in that it selects some particular frequency, translates it to an intermediate processing frequency, and detects the signal magnitude and sometimes phase. Alternatively, some designers prefer to work directly with the electrode or button signal, do some operation on it (filter? peak detect? sample?) and then process the resulting information at baseband. Both approaches often include a final synchronous demodulator, followed by a bandwidth narrowing low pass filter to improve the measurement SNR.

All approaches need to allow operation over variations in bunch current of several orders of magnitude. This requirement can be met using various AGC techniques, or the division of the operating current range into sub-ranges, with adjustable gains or attenuation applied to the channel as the beam current changes. With this approach there are often difficulties maintaining system accuracy at the overlap of the operating ranges.

In most designs the position information is found by computing the difference/sum of some set of pickups - for a straightforward geometry the basic idea is to use the difference/sum of signals from 4 monitors to compute the centroid of the beam

$$X = \frac{(A + C) - (B + D)}{A + B + C + D}$$

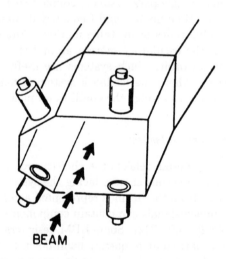

Figure 27. A vacuum chamber with 4 button electrodes designed to sense the trajectory of the beam.

This difference/sum processing can be implemented before the amplitude detection process, or after. RF techniques, based on hybrid junctions, allow the sums/differences to be done directly on the RF signals. The matching and

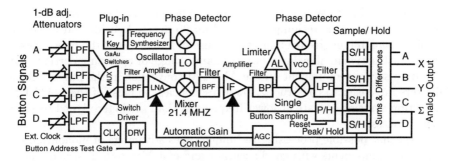

Figure 28. Four channel heterodyned BPM processing architecture.

isolation of the hybrids then sets a limit on the dynamic range and accuracy of the measurement. Alternatively, the difference/sum processing can be done as digital (or analog) computation, on the 4 detected signals. With this approach the matching and calibration of the 4 channels is important, as the trajectory measurement is derived from the difference of large values.

Some example designs from the various labs illustrate these many possibilities. Figure 28 presents a heterodyned BPM processing system, based on a multiplexed homodyne amplitude detection technique[10]. The basic idea is to sequentially measure the amplitude of each of the 4 electrode signals, though a single down-converter which translates the BPM information at the operating frequency to a 21 MHz IF frequency (essentially a tuned heterodyned radio receiver with an AGC loop). The limiter and VCO oscillator are used to create a phase-coherent constant amplitude signal to multiply the BPM signal, forming a homodyne magnitude detector. Four sample/hold circuits are used as the multiplexer cycles, providing a magnitude signal for each button. The four detected amplitudes are processed by difference/sum circuitry to generate the X and Y co-ordinates. A sum signal (A+B+C+D) is also computed and used to adjust the system gain to allow operation over a wide range of beam currents. This basic idea has been implemented at many labs, and is available in commercial BPM processor products. The multiplexing of a single channel relaxes the absolute gain requirements and matching of a multi-channel processor (since all 4 signals come through the same processing path an absolute gain is not required). A disadvantage of this approach is that wideband measurements are not possible.

An alternate processing technique, based on amplitude to phase modulation techniques, is shown in figure 29. The heart of this approach is the conversion of amplitude ratio information into relative phase differences of constant amplitude signals[10]. The phase difference is then measured in a phase detector. This technique is based on 90 degree hybrids and a pair of in-phase and 180 degree combiners. The use of limiters, and the phase detection, is attractive in allowing a large dynamic range of operating current, but it puts a severe requirement on the AM- PM conversion in the limiter. Such limiters are easier to construct a lower frequencies, so most of these AM-PM detection systems operate at lower frequencies than the heterodyned receiver type processors.

Another possible processing technique takes advantage of monolithic log amplifier components with good accuracy and matching over a large dynamic range. The basic idea in these processing channels (figure 30)is to replace the ratio measurement of two signals with the difference of their logarithms. These techniques, like the AM/PM technique, are more popular at detection frequencies in the 10 - 100 MHz range, rather than higher frequencies.

A baseband processing scheme, intended for linac instrumentation, is sketched in figure 31. This approach is intended to work with beams at low repitition rates, and is self-triggering on a sum signal[12]. Any of these schemes is only useful in a BPM instrument if it can be consistently calibrated, operated and tested to verify proper operation. Internal calibration signals, and post-processing correction software are important system features to allow consistent and useful position monitor signals.

10   Feedback Basics

Feedback techniques are central tools of control theory and electronic systems. In this short tutorial only the most superficial insight into the techniques and applications of feedback can be presented. I hope this short introduction

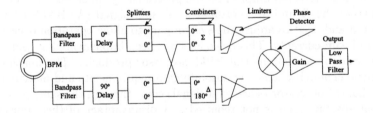

Figure 29. Amplitude modulation to phase modulation processing scheme (AM to PM). This approach requires a accurate 90 and 180 degree hybrids and good amplitude limiting without uncontrolled phase shifts.

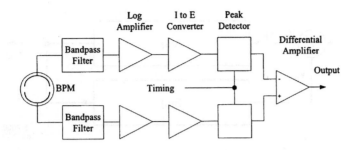

Figure 30. Log-Ratio BPM processing architecture, based on differences of logarithmic amplifier signals.

is helpful, in that it may help readers to navigate the numerous text and reference books.

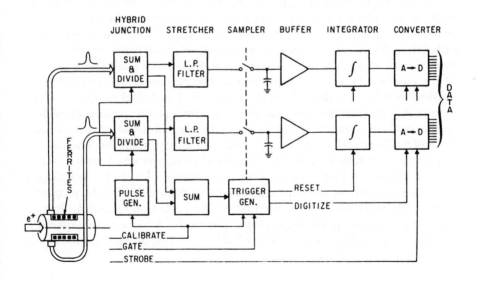

Figure 31. Baseband processing , for linac applications, with hybrid sum/difference circuitry before detection, and self-triggering capability based on a sum signal.

204

The central idea of a feedback system is illustrated in figure 32.

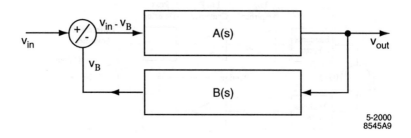

5-2000
8545A9

Figure 32. Every feedback system has an error signal, computed from a summing node, a plant system A, and output signal and a feedback network B.

Any block diagram of such a feedback-stabilized system has a summing node, in which an error signal is computed as the difference between an input or reference variable and some function of the output, a system A (sometimes called the plant in control terminology) which generates some output variable from the error signal, and a feedback network B, which generates a function of the output. This basic block diagram can be applied to essentially any feedback-regulated system. Literally hundreds of technical components in a storage ring or accelerator may use this formalism - power supplies, amplifiers, beam oscillation damping systems, etc. The original invention and applications of feedback techniques were to audio amplifiers, where feedback is used to regulate and make uniform the response of amplifiers which were not identical or uniform in frequency response. The central idea of this technique is that the properties of the fed-back system are determined by the properties of the B feedback network, rather than by the open-loop system A. To have this work, one usually requires greater gain in the A block than required in the final system. The response of the fed-back system is found from the loop topology as

$$\frac{A(s)}{1 + A(s)B(s)}. \tag{25}$$

If the original raw amplifier has gain which varies by a factor of 2, say 1000 to 2000 - but in a feedback circuit, with B = 1/10 the fed-back amplifier gain only changes from 9.95 to 9.90 - or the factor of 2 variation is reduced to 0.005. The larger the loop gain AB, the less sensitive the output is to variations in the system A.

10.1 Stability of Feedback systems

This argument suggests that a large loop gain is helpful in a feedback system - it is, but not if the phase shifts or time delays around the loop get large enough, because a fed-back system can become an oscillator. This is easy to see - consider the system shown, and break the loop at point X.

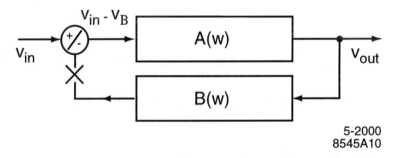

5-2000
8545A10

Figure 33. Loop stability can be understood by breaking the loop at X and examining the signal magnitude and phase around the loop.

If the magnitude of the loop gain $A(\omega)B(\omega) = 1$ and the phase around the loop path $AB = n2\pi$, the system will oscillate, with the signal around the loop exactly replicating itself. The stability of the fed-back system is important in understanding the loop behavior. (There are many analysis techniques to understand loop stability - among them Bode plots, Nyquist plots, root-locus plots, etc. - the reader will find in the references entire textbooks on stability analysis). What is important for our tutorial is an understanding that the feedback network B, and the open loop system A determine the properties of the combined system, and that incorrectly applied, a feedback signal can cause oscillation of a system.

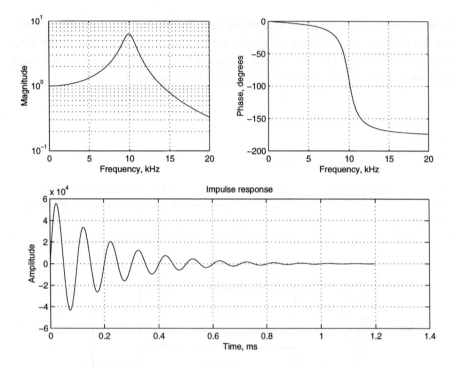

Figure 34. Transfer function and impulse response for the simple harmonic oscillator.

10.2 Feedback control of Dynamic systems

If one applies feedback around a harmonic oscillator, the behavior of the system can be dramatically changed. A simple harmonic oscillator is a second-order differential equation of the form

$$\ddot{x} + 2\gamma\dot{x} + \omega_0^2 x = U_{ext}(t) \tag{27}$$

where $U_{ext}(t)$ represents a driving term. Such an equation of motion represents a mass/spring/dashpot system (where $\omega = \sqrt{k/m}$), or an electrical RLC circuit ($\omega = \sqrt{1/LC}$). Many physical systems display this exact form of response - including bunches in storage rings. The natural frequency ω for a bunch executing synchrotron oscillations is determined by a "spring constant" of the slope of the RF voltage at the synchronous phase (a return force per unit displacement) and the momentum compaction factor of the lattice (a description of the variation in revolution time vs. changes in bunch energy, which is analogous to an inertial mass-like term). Betatron oscillations are also well-described as simple harmonic oscillators. Each normal mode in a multiple bunch system is described by the same equation of motion.

In the frequency domain the transfer function describing the response to an excitation is

$$X(s) = \frac{1}{s^2 + 2\gamma s + \omega_0^2} \tag{28}$$

which is shown in figure 34 along with the impulse response of the system. Note that in this second order system the phase shifts π radians through the resonance, and that the Q of the system is shown in the phase slope through resonance, and the width of the resonant peak at ω_0.

The damping in the system determines the Q factor of the system - a strongly damped system has a well-behaved impulse response which decays quickly - a weakly damped (high Q) system has an impulse response which oscillates for many cycles before decaying away. The envelopes of the impulse response are $e^{-\gamma t}$. If the damping is negative the impulse response grows in time. Such a system with negative damping is unstable, in that a perturbation grows exponentially in time.

We can change this response with feedback around the harmonic oscillator. The simplest sort of feedback is proportional - a term proportional to the output is added back at the input. The equation of motion of this system is

$$\ddot{x} + 2\gamma\dot{x} + \omega_0^2 x = -\alpha x + U_{ext}(t) \tag{29}$$

with frequency response

$$X(s) = \frac{1}{s^2 + 2\gamma s + (\omega_0^2 + \alpha)} \tag{30}$$

Figure 35 shows the resulting frequency response and impulse response. The proportional feedback shifts the resonant frequency higher. This increase of frequency response may be useful in some applications, such as to extend the bandwidth of a transducer. The proportional feedback has also reduced the overall transfer gain from input to output, which could be desirable, or the original transfer gain could be restored, with this increased system bandwidth, with a modest application of electronic gain to the system.

If a feedback signal proportional to the integral of the output is fed-back, a different change in the system dynamics is made. The new equation of motion, incorporating the integral feedback is

$$\ddot{x} + 2\gamma\dot{x} + \omega_0^2 x = -\alpha \int x(t)dt + U_{ext}(t) \tag{31}$$

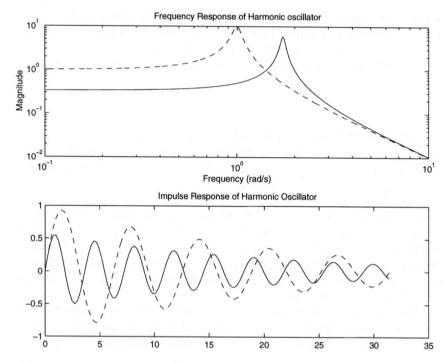

Figure 35. Proportional feedback around the SHO acts to increase the system bandwidth (and reduce the overall gain).

with frequency response

$$X(s) = \frac{s}{s(s^2 + 2\gamma s + \omega^2) + \alpha} \qquad (32)$$

Figure 36 shows the action of integral feedback on the original harmonic oscillator. Integral feedback has reduced the response of the system to DC and low frequency disturbances, but has left the oscillatory dynamics near the resonance unchanged. This sort of response resists changes to the DC or equilibrium position.

If a correction term proportional to the derivative of the motion is applied, the new equation of motion is

$$\ddot{x} + 2\gamma\dot{x} + \omega_0^2 x = -\alpha\dot{x} + U_{ext}(t) \qquad (33)$$

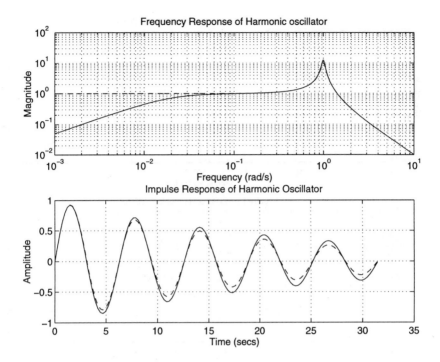

Figure 36. Integral feedback around the SHO acts to reduce the system DC response, while leaving the behavior near the resonance unchanged.

with frequency response

$$X(s) = \frac{1}{s^2 + (\alpha + 2\gamma)s + \omega_0^2} \tag{34}$$

As shown in figure 37, derivative feedback acts to change the Q of the system, and changes the transient response but not the natural frequency or the DC behavior. The figure shows the response for two signs of derivative correction - note that by applying the opposite sign of correction the Q can be increased, and the impulse response made more oscillatory. This sort of feedback acts to damp or antidamp the motion - it is exactly analogous to adding a dashpot or shock absorber to the mechanical system (or a resistance to the LCR oscillator).

As a practical matter, it is sometimes difficult to implement a true dif-

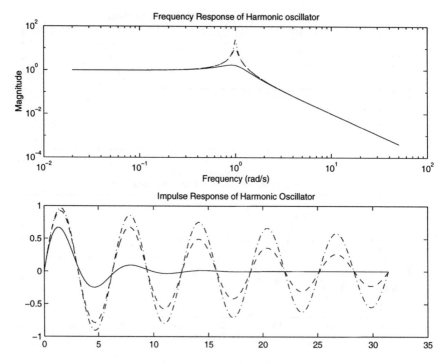

Figure 37. Derivative feedback around the SHO changes the damping in the system. The damping can be increased or decreased, depending on the sign of the derivative feedback. The system can be stabilized or destabilized by this technique.

ferentiator for this sort of application, as a perfect differentiator has gain increasing with frequency, and acts to inject lots of high-frequency noise into the system. Often a band-pass characteristic is used to approximate the differentiator over some frequency band of interest.

These three simple feedback strategies are often combined into one controller, which implements some amount of each, and is called a PID (proportional-integral-derivative) controller. In general, choosing the right amount of each term is a trade-off of improvement in the dynamics of the uncontrolled system, and sensitivity to accidentally driving the controlled system into oscillation.

There are many formal methods to study the stability of these fed-back systems, and there are numerous controllers and controller design techniques. The references suggest several good texts which provide an introduction to control theory and applications[13,14,15].

11 Summary

This brief tutorial covered a wide range of topics, and in a superficial manner. I hope the material presented helps the students of the course to understand the more detailed reference material. This note has attempted to stress the importance of signal processing concepts, particularly the application of transform techniques and the properties of linear time-invariant systems. These are the essential and central tools for any understanding of the beam signals. I also have provided a warning about "too much confidence" in assuming time-domain behavior from frequency domain data.

The signals from the beams can be sensed capacitively, inductively, or through both means. The time responses of various detectors can be understood as convolutions of beam functions with the impulse responses of the pick-up electrodes. The signatures of bunch motion are very complex, especially for real-world cases in the control room. Motion of single-bunch systems can be readily understood using the transform techniques, and more complex cases can be understood as convolutions of bunch current or amplitude modulation functions with the single-bunch response.

More significantly, this note has made no treatment of low-energy or unbunched beams, and has made no attempt to give a quantitative method to calculate the interaction of the beams with pickups and kickers. The concept of the impedance of an electrode or vacuum structure is critical, and readers are urged to study the references for more complete discussions[5,9]

Common control room instruments and measurements have been described, in the hope that a student will feel better qualified to participate in the control room, and have a basic understanding of the purposes of beam measurements, and the rationale for selecting instruments to a measurement need. A special class of beam position monitor instruments provides a hint of the techniques and choices that the engineering of beam instruments can offer.

Finally, the simple harmonic oscillator, and the manner in which simple feedback techniques can change the dynamics of a system, were presented to give some insight into the purposes of feedback control techniques as applied to beam systems. Such systems, intended to control unstable beam motion, are becoming more common at light sources and accelerators as currents are pushed upwards.

Agsin, the superficial treatment means that much important material was not discussed. Readers will find conference papers and beam instrumentation workshop proceedings helpful in understanding the principles of synchrotron light monitors and the kinds of information they provide[7,10,11]. A better treatment of BPM systems should include quantitative examples, with estimates of BPM errors, sensitivity and limits to the system resolution[4,12,6]. The discussion of feedback techniques, which had a very limited presentation on choices of control techniques and stability analysis techniques, will hopefully motivate a reader to consult a reference on control theory to better understand these important topics.

12 Acknowledgments

In preparing this tutorial I have freely used figures from Jim Hinkson (LBL), Jean-Louis Pellegrin (SLAC) and R. Littauer (Cornell). I thank them for their generosity. I thank E. Kikutani of KEK for his recent collaboration on the Montreaux accelerator school, which provided some of the slides and material for the AAS. D. Teytelman (SLAC), S. Prabhakar (SLAC), H. Hindi (Stanford), M. Tobiyama (KEK), M. Serio (LNF-INFN) and F. Pedersen (CERN) also have been important collaborators and contributors to my thinking on these topics. I also want to thank Steve Smith (SLAC) and Greg Stover (LBL) for reference material, and acknowledge the excellent tutorials on accelerator instrumentation which have been prepared and presented at various accelerator schools and the Beam Instrumentation Workshop. I enjoyed looking over these tutorials and hope the original authors enjoy seeing their ideas represented one more time. I also thank the organizers of the AAS for arranging the school and for the opportunity for me to participate. Terry Anderson and Michael Hyde (SLAC) skillfully transformed my crude sketches into the figures of this paper, and I thank them for their expertise. Jim Lewandowski (SLAC) kindly provided the PEP-II tune measurement example. Work supported by U.S. Department of Energy contract DE-AC03-76SF0051

Appendix

Answers to the LTI Quiz

Question #1 - The resistive divider is an excellent example of an LTI system, at least in the limit of applied voltages which do not damage the resistors due to excessive power dissipation, or arc over due to excessive voltage.

Question #2 - The high-pass filter has an attenuation which is a function of frequency, and a frequency-dependent phase shift. It is still an LTI circuit, at least considering the voltage and power limits as for example #1.

Question #3 - the symmetric clamp is clearly non-linear, in that for large amplitude inputs it saturates. It is therefore not an LTI circuit - and it produces harmonics of the applied clamped signal at the output, another important clue as to it not being an LTI system. Note that if the input signals were restricted in amplitude, to a level below the clipping level, the behavior of the circuit would be linear and time invarariant.

References for more information

In preparing this tutorial have drawn extensively on the figures, ideas and examples from previous papers covering the broad field of beam instrumentation. This list is not inclusive - it is just a place to get started.

1. E. Oran Brigham, The Fast Fourier Transform, Prentice-Hall, New Jersey, 1974
2. Alan V. Oppenheim, Ronald W. Schafer, Discrete-Time Signal Processing, Prentice-Hall, New Jersey, 1989
3. Ronald N. Bracewell, The Fourier Transform and its Applications, McGraw-Hill, New York, 1978
4. S. R. Smith,"Beam Position Monitor Engineering", Proc. Beam Instrumentation Workshop, Argonne IL 1996
5. D. A. Goldberg and G. R. Lambertson, "Dynamic Devices - A Primer on Pickups and Kickers", Physics of High-Energy Particle Accelerators, AIP conference proceedings 249
6. R. C. Webber, "Charged Particle Beam Current Monitoring Tutorial", Proc Beam Instr. Workshop, Vancouver, 1994
7. R. Littauer, "Beam Instrumentation", Physics of High-Energy Particle Accelerators, AIP conference proceedings 105, SLAC summer School 1982
8. R. H. Siemann, "Spectral Analysis of Relativistic Bunched Beams", Proc Beam Instrumentation Workshop, Argonne, IL 1996
9. J. N. Corelett,"Impedance of Accelerator Components", Proc. Beam Instrumentation Workshop, Argonne IL 1996
10. C. H. Kim and James Hinkson, "Advanced Light Source Instrumentation Overview", Proc. Beam Instrumentation Workshop, Berkeley CA 1992
11. M. Billing,"Introduction to Beam Diagnostics and Instrumentation for Circular Accelerators", Proc. Beam Instrumentation Workshop, Berkeley CA 1992

12. J-L Pellegrin, "A Review of Accelerator Instrumentation", Proc. 11th International Conf. on High-Energy Accelerators, Geneva, 1980
13. Gene F. Franklin, J. David Powell, Abbas Emani-Naeini, Feedback Control of Dynamic Systems, Addison-Wesley, Massachusetts, 1991
14. Gene F. Franklin, J. David Powell, Michael L. Workman, Digital Control of Dynamic Systems, Addison-Wesley, Massachusetts
15. Robert D. Strum, Donald E. Kirk, Discrete Systems and Digital Signal Processing, Adison-Wesley, Massachusetts, 1989
16. James Safranek, "Beam-Based Lattice Diagnostics", Proceedings of the Beam Measurement US-CERN-Japan-Russia School on Particle Accelerators, Montreaux, Switzerland, May 1998
17. Peter Tenenbaum and Tsumoru Shintake, "Measurement of Small Electron Beam Spots", Ann.Rev.Nucl.Part.Sci.49:125-162,1999

BEAM INSTABILITIES*

ALEX CHAO

Stanford Linear Accelerator Center, Stanford University, CA 94309, USA
E-mail: achao@slac.stanford.edu

These lectures treat some of the common collective beam instability effects encountered in accelerators. In choosing the material for these lectures, it is attempted to introduce this subject with a more practical approach, instead of a more theoretical approach starting with first principles. After introducing the terminologies, emphasis will be placed on how to apply the lecture material to perform calculations and to make estimates of various instability effects.

In the first half of the lectures, after briefly introducing the concepts of impedance and wake field, we will discuss a selected list of formulas for the impedances of various accelerator components. Detailed derivations are omitted, allowing time for the students to think through the process of how to apply the knowledge learned. The list of impedances to be covered include: space charge, resistive wall, resonator, wall roughness, and small perturbation on the vacuum chamber wall.

Assuming impedances are known, the second half of the lectures addresses the question of how to calculate the power of beam heating, the growth rates, and the thresholds for a list of selected beam instability effects. Again with minimal detailed derivations, our aim is to introduce a collection of formulas, and apply them to linear as well as circular accelerators. The list of beam instability effects to be covered include: loss factor, beam break-up, BNS damping, bunch lengthening, resistive wall instability, head-tail instability, longitudinal head-tail instability, Landau damping, microwave instability, and mode coupling instability.

1 Wakefields

When a charged particle beam traverses a discontinuity in the conducting vacuum chamber, an electromagnetic *wakefield* is generated, as sketched in Fig.1. An intense beam will generate a strong wakefield. When the wakefield is strong enough, the beam becomes unstable.

Let the drive beam have charge q and travels down the beam pipe along its axis. Consider a test charge e following the drive beam. Let both the drive beam and the test charge move with speed $v \approx c$, holding the spacing z between them fixed. Let $\vec{E}$ and $\vec{B}$ be the electric and magnetic parts of the wakefield seen by the test charge. The test charge experiences the wake

---

*WORK SUPPORTED BY DEPARTMENT OF ENERGY CONTRACT DE–AC03–76SF00515.

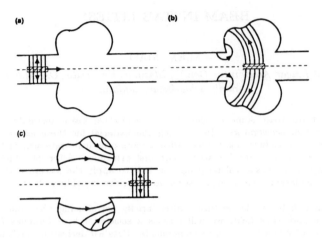

Figure 1. Wakefield driven by a beam when there is a discontinuity in the vacuum chamber. (a) before the drive beam traverses the vacuum chamber discontinuity, (b) during the traversal, (c) after the traversal.

force $\vec{F} = e(\vec{E} + \vec{v} \times \vec{B})$. Integrating $\vec{F}$ over the traversal results in the *wake potential*

$$\overline{\vec{F}} = \int_{-\infty}^{\infty} ds \vec{F} \tag{1}$$

$\overline{\vec{F}}$ is a function of the spacing between the test charge and the drive beam z. It also depends on (r, θ), the transverse coordinates of the test charge. Furthermore, it satisfies the *Panofsky-Wenzel theorem*,

$$\nabla_{\perp} \overline{F}_{\parallel} = \frac{\partial}{\partial z} \overline{F}_{\perp} \tag{2}$$

Here $\parallel$ denotes longitudinal and $\perp$ denotes transverse components.

One can also consider a drive beam which travels down the axis of the beam pipe with an mth moment I_m. ($m = 0, 1, 2$ correspond to monopole, dipole and quadrupole moments. When $m = 0$, one has $I_0 = q$.) In case the discontinuity is axially symmetric, one can write

$$\overline{\vec{F}}_{\perp}(r, \theta, z) = -eI_m W_m(z)\, mr^{m-1}(\hat{r}\cos m\theta - \hat{\theta}\sin m\theta)$$
$$\overline{F}_{\parallel}(r, \theta, z) = -eI_m W'_m(z)\, r^m \cos m\theta \tag{3}$$

Here a prime denotes d/dz. $W_m(z)$ is called the *transverse wake function* and $W'_m(z)$ the *longitudinal wake function*.

We omit its derivation, but Eq.(3) is quite amazing. An efficient application of the Maxwell equations has yielded a wealth of information built-in in it. The fact that $\vec{F}$ is proportional to e and I_m is straightforward. On the other hand, without even specifying the geometry of the vacuum chamber discontinuity, or the chamber wall's resistivity, one sees that the m, r, and θ dependences of the wake potentials have been explicitly solved. The only unknown in Eq.(3) is the wake function $W_m(z)$, which depends only on z. Furthermore, the transverse and the longitudinal wake potentials involve the *same* funtion $W_m(z)$.

Homework

One consequence of the Maxwell equations is the Panofsky-Wenzel theorem. Show that the Panofsky-Wenzel theorem is built-in in Eq.(3).

In Cartesian coordinates, for each of the mth moment, the drive beam has two components—one normal and another skewed. Table below lists the two moments (first the normal moment and then the skewed moment) and the associated transverse and longitudinal wake potentials seen by a test charge e with transverse coordinates x, y that follows, at a distance $|z|$ behind, a beam which possesses an mth moment. The question being asked is what kick is being received by the test charge e as the beam and the test charge complete the traversal of the chamber discontinuity. We have the convention that $z < 0$ if the test charge trails the drive beam. A bracket $\langle\ \rangle$ means averaging over the transverse distribution of the drive beam; $\hat{x}$ and $\hat{y}$ are the unit vectors in the x- and y-directions.

m	Distribution Moments of Beam	Longitudinal Wake Potential	Transverse Wake Potential
0	q	$-eq\,W_0'(z)$	0
1	$\begin{cases} q\langle x\rangle \\ q\langle y\rangle \end{cases}$	$-eq\langle x\rangle x W_1'(z)$ $-eq\langle y\rangle y W_1'(z)$	$-eq\langle x\rangle W_1(z)\hat{x}$ $-eq\langle y\rangle W_1(z)\hat{y}$
2	$\begin{cases} q\langle x^2-y^2\rangle \\ q\langle 2xy\rangle \end{cases}$	$-eq\langle x^2-y^2\rangle(x^2-y^2)W_2'(z)$ $-eq\langle 2xy\rangle 2xy\,W_2'(z)$	$-2eq\langle x^2-y^2\rangle W_2(z)(x\hat{x}-y\hat{y})$ $-2eq\langle 2xy\rangle W_2(z)(y\hat{x}+x\hat{y})$
3	$\begin{cases} q\langle x^3-3xy^2\rangle \\ q\langle 3x^2y-y^3\rangle \end{cases}$	$-eq\langle x^3-3xy^2\rangle$ $\times(x^3-3xy^2)W_3'(z)$ $-eq\langle 3x^2y-y^3\rangle$ $\times(3x^2y-y^3)W_3'(z)$	$-3eq\langle x^3-3xy^2\rangle W_3(z)$ $\times[(x^2-y^2)\hat{x}-2xy\hat{y}]$ $-3eq\langle 3x^2y-y^3\rangle W_3(z)$ $\times[2xy\hat{x}+(x^2-y^2)\hat{y}]$

$$(4)$$

Eq.(4) contains rich information about the wake effects and should be studied with attention. Note that the leading longitudinal wake potential is driven by the monopole ($m = 0$) moment of the drive beam, while the leading transverse wake potential is driven by the dipole moment ($m = 1$) of the beam. However, when the geometry is not axially symmetric, one must not forget that a monopole beam moment can also drive transverse wake potential.

Dimensionalities of the wake functions are $[W_m] = [\Omega s^{-1} m^{-2m+1}]$, $[W'_m] = [\Omega s^{-1} m^{-2m}]$. The most important transverse wake function is that for $m = 1$, $[W_1] = [\Omega s^{-1} m^{-1}]$. The most important longitudinal wake function has $m = 0$, $[W'_0] = [\Omega s^{-1}]$.

Properties of wake functions

- $W_m(z) = 0$, $W'_m(z) = 0$ for $z > 0$ (causality).

  - $W_m(z) \leq 0$, $W'_m(z) \geq 0$ for $z \to 0^-$.
  - $W_m(0) = 0$ (in most cases, except space charge).
  - $W'_m(0) = \frac{1}{2} W'_m(0^-)$ (fundamental theorem of beam loading).
  - $W'_m(0^-) \geq |W'_m(z)|$ for all z.
  - $\int_{-\infty}^{0} W'_m(z) dz \geq 0$.

In general, $W_m(z)$ is a sine-like function, while $W'_m(z)$ is a cosine-like function, as sketched below.

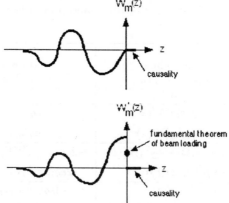

2 Impedances

Impedances are just Fourier transforms of wake functions:

$$Z_m^{\parallel}(\omega) = \int_{-\infty}^{\infty} \frac{dz}{v} e^{-i\omega z/v} W'_m(z)$$

$$Z_m^\perp(\omega) = \frac{i}{v/c} \int \frac{dz}{v} e^{-i\omega z/v} W_m(z) \tag{5}$$

Time dependence of $e^{-i\omega t}$ is assumed.

Dimensionalities are $[Z_m^\parallel] = [\Omega \mathrm{m}^{-2m}]$, $[Z_m^\perp] = [\Omega \mathrm{m}^{-2m+1}]$. The most important transverse impedance is that for $m = 1$, $[Z_1^\perp] = [\Omega \mathrm{m}^{-1}]$. The most important longitudinal impedance has $m = 0$, $[Z_0^\parallel] = [\Omega]$.

Properties of impedances

- $Z_m^\parallel(\omega) = \dfrac{\omega}{c} Z_m^\perp(\omega)$ (Panofsky-Wenzel theorem in frequency domain).

- $\begin{cases} Z_m^{\parallel\,*}(\omega) = Z_m^\parallel(-\omega) \\ Z_m^{\perp\,*}(\omega) = -Z_m^\perp(-\omega) \end{cases}$ (reality of wake functions).

- $\begin{cases} \int_0^\infty d\omega\, \mathrm{Im} Z_m^\perp(\omega) = 0 \\ \int_0^\infty d\omega\, \dfrac{\mathrm{Im} Z_m^\parallel(\omega)}{\omega} = 0 \quad (W_m(0) = 0, \text{ in most cases}). \\ \mathrm{Re} Z_m^\parallel(0) = 0 \end{cases}$

- $\begin{cases} \mathrm{Re} Z_m^\parallel(\omega) = \dfrac{1}{\pi} \mathrm{PV} \displaystyle\int_{-\infty}^\infty d\omega' \dfrac{\mathrm{Im} Z_m^\parallel(\omega')}{\omega' - \omega} \\ \mathrm{Im} Z_m^\parallel(\omega) = -\dfrac{1}{\pi} \mathrm{PV} \displaystyle\int_{-\infty}^\infty d\omega' \dfrac{\mathrm{Re} Z_m^\parallel(\omega')}{\omega' - \omega} \end{cases}$ (causality, Hilbert trans-

form) The same expressions apply to $Z_m^\perp$. PV means taking the principal value of the integral.

- $\begin{cases} \mathrm{Re} Z_m^\parallel(\omega) \geq 0 \text{ for all } \omega \\ \mathrm{Re} Z_m^\perp(\omega) \geq 0 \text{ if } \omega > 0, \quad \leq 0 \text{ if } \omega < 0 \end{cases}$

- $Z_1^\perp \approx \dfrac{2c}{b^2\omega} Z_0^\parallel,\ Z_m^\perp \approx \dfrac{2c}{b^{2m}\omega} Z_0^\parallel,\ Z_m^\parallel \approx \dfrac{2}{b^{2m}} Z_0^\parallel$ These are approximate expressions relating transverse and longitudinal impedances, $b =$ pipe radius. They are exact for resistive round pipe.

3 Calculation of Impedances

To calculate the impedance for a given vacuum chamber discontinuity, one needs to solve for the electromagnetic fields produced in the vacuum chamber by a given beam current. Over the years, a large arsenal of techniques have been developed to calculate the impedances.

The first method is to solve Maxwell equations analytically with appropriate boundary conditions. This method applies only to the simplest cases.

We omit the derivations and give only the results for some examples.

Space charge

See Fig.2 for the wakefield patterns in the transverse plane. The z-dependence is a δ-function $\delta(z)$. With a beam of radius a in a perfectly conducting round pipe of radius b and length L,

Impedances	Wake functions
$Z_0^\parallel = i\dfrac{Z_0 L\omega}{4\pi c\gamma^2}\left(1 + 2\ln\dfrac{b}{a}\right)$	$W_0' = \dfrac{Z_0 cL}{4\pi\gamma^2}\left(1 + 2\ln\dfrac{b}{a}\right)\delta'(z)$
$Z_{m\neq 0}^\perp = i\dfrac{Z_0 L}{2\pi\gamma^2 m}\left(\dfrac{1}{a^{2m}} - \dfrac{1}{b^{2m}}\right)$	$W_{m\neq 0} = \dfrac{Z_0 cL}{2\pi\gamma^2 m}\left(\dfrac{1}{a^{2m}} - \dfrac{1}{b^{2m}}\right)\delta(z)$

$$\tag{6}$$

where $Z_0 = \sqrt{\mu_0/\epsilon_0} \approx 377\ \Omega$ is the free-space impedance, ϵ_0 and μ_0 are the free-space dielectric constant and magnetic permeability. Because of the factor $1/\gamma^2$, space charge effects are most significant for low-to-medium energy proton or heavy ion accelerators.

The space charge impedance is purely imaginary, and is proportional to $i\omega$. Its ω-dependence is as if it is a pure inductance. However, its sign is as if it is a capacitance. By convention, we call the space charge impedance "capacitive".

Resistive wall

Another case solvable analytically is for a round resistive pipe with radius b, conductivity σ_c, and length L. Defining the skin depth (change 0.066 to 0.086 for aluminum, and to 0.43 for stainless steel)

$$\delta_{\text{skin}} = \sqrt{\frac{2c}{|\omega| Z_0 \sigma_c}}, \qquad \delta_{\text{skin}}\,[\text{mm}] = \frac{0.066}{\sqrt{f\,[\text{MHz}]}}\ \text{for copper, room temp.} \tag{7}$$

one finds

Impedances		Wake functions
$Z_m^\parallel = \dfrac{\omega}{c}Z_m^\perp$		$W_m = -\dfrac{c}{\pi b^{m+1}(1+\delta_{m0})}\sqrt{\dfrac{Z_0}{\pi\sigma_c}}\dfrac{L}{|z|^{1/2}}$
$Z_m^\parallel = \dfrac{1 - \text{sgn}(\omega)i}{1+\delta_{0m}}$	$\dfrac{L}{\pi\sigma_c\delta_{\text{skin}}b^{2m+1}}$	$W_m' = -\dfrac{c}{2\pi b^{m+1}(1+\delta_{m0})}\sqrt{\dfrac{Z_0}{\pi\sigma_c}}\dfrac{L}{|z|^{3/2}}$

$$\tag{8}$$

The impedance is proportional to $(1 - i)$, i.e. one might say that it is half resistive and half inductive.

Displaced beam in a resistive pipe

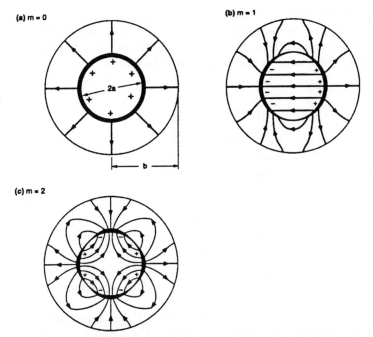

Figure 2. Space charge wakefields in the x-y plane which contains the ring-shaped, infinitely thin, $\cos m\theta$-distribution drive beam.

If the beam is not centered, but is displaced a distance x_0 in the beam pipe, the impedances can still be found analytically,

$$Z_0^{\parallel} = \frac{[1 - \mathrm{sgn}(\omega)i]L}{2\pi\sigma_c\delta_{\mathrm{skin}}b} \times \frac{b^2 + x_0^2}{b^2 - x_0^2}$$

$$Z_1^{\{x,y\}} = \frac{[1 - \mathrm{sgn}(\omega)i]cL}{\pi\omega\sigma_c\delta_{\mathrm{skin}}b} \times \frac{b^4}{(b^2 - x_0^2)^3}\{b^2 + 3x_0^2, b^2 - x_0^2\} \qquad (9)$$

Eq.(9) is useful when the beam is close to the chamber wall, such as in the case of a metallic collimator. On the other hand, in those cases, one should also include contribution from the wall surface roughness (see later).

Field matching method

For more complicated cases, solutions are often to be found numerically. For example, in structures which can be subdivided into a few simple subregions (e.g., Fig.3 for the case of a pill-box cavity), a field matching method can be applied. Solution in each subregion is expanded in terms of the eigen functions of the respective subregion with yet-to-be-determined coefficients. Field matching on the subregion interfaces and boundary conditions on the vacuum chamber walls leads to an infinite system of linear equations for these coefficients, which usually can be truncated to a finite size.

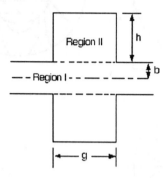

Figure 3. Pill-box cavity and field matching method.

Pill-box cavity

The pill-box cavity is a complicated object. Its accurate impedances are to be found numerically. However, approximate impedances of a pill-box cavity can be found by combining the following:

(i) <u>Resonant impedances</u>: A pill-box cavity can trap some modes, each giving rise to a sharp, narrow-band impedance. The resonant frequencies are determined approximately ($d = h + b$) by (ignoring pipe radius b, up to frequency $\omega \sim c/b$)

$$\frac{\omega_{mnp}^2}{c^2} = \frac{x_{mn}^2}{d^2} + \frac{p^2 \pi^2}{g^2} \tag{10}$$

where n, p, m are the radial, longitudinal, and azimuthal mode indices, x_{mn} is the nth zero of Bessel function J_m. The shunt impedance R_s and the Q-value of the modes satisy

$$\left[\frac{R_s}{Q}\right]_{0np} = \frac{Z_0}{x_{0n}^2 J_0'^2(x_{0n})} \frac{8c}{\pi g \omega_{0np}} \begin{cases} \sin^2 \frac{g \omega_{0np}}{2\beta c} \times (1+\delta_{0p}) & p \text{ even} \\ \cos^2 \frac{g \omega_{0np}}{2\beta c} & p \text{ odd} \end{cases}$$

$$\left[\frac{R_s}{Q}\right]_{1np} = \frac{Z_0}{J_1'^2(x_{1n})} \frac{2c^2}{\pi g d^2 \omega_{1np}^2} \begin{cases} \sin^2 \frac{g \omega_{1np}}{2\beta c} \times (1+\delta_{0p}) & p \text{ even} \\ \cos^2 \frac{g \omega_{1np}}{2\beta c} & p \text{ odd} \end{cases} \tag{11}$$

(ii) <u>Low-frequency broad-band impedance</u>: In addition to trapped modes, a pill-box cavity also gives rise to a broad-band impedance. At low frequencies

$\omega \ll c/b$, and for a small pill-box $(g, h \ll b)$ this part of impedance behaves as an inductance,

Impedances		Wake functions	
$Z_0^{\parallel} =$	$\begin{cases} -i\frac{\omega Z_0}{2\pi cb}\left(gh - \frac{g^2}{2\pi}\right) & g \leq h \\ -i\frac{\omega Z_0 h^2}{\pi^2 cb}\left(\ln\frac{2\pi g}{h} + \frac{1}{2}\right) & h \ll g \end{cases}$	$W_0' =$	$\begin{cases} -\frac{Z_0 c}{2\pi b}\left(gh - \frac{g^2}{2\pi}\right)\delta'(z) & g \leq h \\ -\frac{Z_0 ch^2}{\pi^2 b}\left(\ln\frac{2\pi g}{h} + \frac{1}{2}\right)\delta'(z) & h \ll g \end{cases}$
$Z_1^{\perp} =$	$\begin{cases} -i\frac{Z_0}{\pi b^3}\left(gh - \frac{g^2}{2\pi}\right) & g \leq h \\ -i\frac{2Z_0 h^2}{\pi^2 b^3}\left(\ln\frac{2\pi g}{h} + \frac{1}{2}\right) & h \ll g \end{cases}$	$W_1 =$	$\begin{cases} -\frac{Z_0 c}{\pi b^3}\left(gh - \frac{g^2}{2\pi}\right)\delta(z) & g \leq h \\ -\frac{2Z_0 ch^2}{\pi^2 b^3}\left(\ln\frac{2\pi g}{h} + \frac{1}{2}\right)\delta(z) & h \ll g \end{cases}$

$$(12)$$

(iii) High-frequency broad-band impedance: Diffraction theory can be used to calculate the impedances at high frequencies, $\omega \gg c/b$, yielding

Impedances	Wake functions
$Z_m^{\parallel} = \frac{[1 + \text{sgn}(\omega)i]Z_0}{(1 + \delta_{m0})\pi^{3/2}b^{2m+1}}\sqrt{\frac{cg}{|\omega|}}$	$W_m = -\frac{2Z_0 c\sqrt{2g}}{(1+\delta_{m0})\pi^2 b^{2m+1}}|z|^{1/2}$
$Z_m^{\parallel} = \frac{\omega}{c}Z_m^{\perp}$	$W_m' = \frac{Z_0 c\sqrt{2g}}{(1+\delta_{m0})\pi^2 b^{2m+1}}|z|^{-1/2}$

$$(13)$$

The impedance $Z_0^{\parallel} \propto 1 + i$, i.e. half resistive and half capacitive.

Broad band resonator model

If one is interested only in the short range wake, a cavity whose dimensions are comparable to the pipe radius b can be approximated by a broad band resonator (see Eq.(18) for resonator model in general). The impedances *per cavity* are approximated by

$$R_S^{(0)} \approx 60\Omega, \quad Q \approx 1, \quad \omega_r \approx \frac{c}{b} \quad \text{for } Z_0^{\parallel}$$

$$R_S^{(1)} \approx 60\Omega \times \frac{1}{b^2}, \quad Q \approx 1, \quad \omega_r \approx \frac{c}{b} \quad \text{for } Z_1^{\perp} \qquad (14)$$

If one further contends by finding only an order-of-magnitude estimate, one can take the impedance values at $\omega \sim \omega_r$ (or equivalently the wake function values at $|z| \sim b$), and obtains, very roughly, per cavity,

Impedances	Wake functions
$Z_m^{\parallel} \sim \frac{Z_0}{4\pi}\frac{1}{b^{2m}}$	$W_m \sim \frac{Z_0 c}{4\pi}\frac{1}{b^{2m}}$
$Z_m^{\perp} \sim \frac{Z_0}{4\pi}\frac{1}{b^{2m-1}}$	$W_m' \sim \frac{Z_0 c}{4\pi}\frac{1}{b^{2m+1}}$

$$(15)$$

In particular, $Z_0^\parallel \sim 30\,\Omega$ and $Z_1^\perp \sim 30\,\Omega/b$. Note that $Z_0^\parallel$ per cavity is independent of the cavity size or the pipe radius b as long as they are comparable. The longitudinal wake function can be rewritten in another convenient form as W_0' [V/pC] $\sim 0.9/b$ [cm]. Eq.(15) is useful for linacs with its accelerating cavities of dimensions $\sim b$.

Periodic pill-box array

For an array of pill-boxes, each with gap length g, box spacing L, and beam pipe radius b, the high-frequency impedance per box is ($k = \omega/c$)

$$Z_0^\parallel = \frac{iZ_0L}{\pi k b^2}\left[1 + (1 + i\,\mathrm{sgn}(k))\frac{\alpha L}{b}\sqrt{\frac{\pi}{|k|g}}\right]^{-1} \tag{16}$$

$$\alpha = \begin{cases} 1 & \text{when } g/L \ll 1 \\ 0.4648 & \text{when } g/L = 1, \text{infinitely thin irises} \end{cases}$$

The real part of $Z_0^\parallel$ goes like $\sim k^{-3/2}$ for large k, while the imaginary part goes like $\sim k^{-1}$. This is in contrast with a single pill-box, whose impedance (both real and imaginary parts, see Eq.(13)) at high frequencies behave like $\sim \omega^{-1/2}$.

In addition to Eq.(16) and the trapped modes, there is a resonator-type impedance generated by the pill-box array. If the boxes are small, they contribute to a single-frequency impedance mode at $k_0 = \omega_0/c = \sqrt{2L/bg\delta}$ (δ is cavity depth),

Impedances		Wake functions
$\dfrac{Z_0^\parallel}{L} = \dfrac{Z_0 c}{2\pi b^2}\left[\pi\delta(\omega-\omega_0)+\pi\delta(\omega+\omega_0)+\dfrac{i}{\omega-\omega_0}+\dfrac{i}{\omega+\omega_0}\right]$		$\dfrac{W_1(z)}{L} = \dfrac{2Z_0 c}{\pi b^4 k_0}\sin k_0 z$
$\dfrac{Z_1^\perp}{L} = \dfrac{2c}{b^2\omega}\dfrac{Z_0^\parallel}{L}$		$\dfrac{W_0'(z)}{L} = \dfrac{Z_0 c}{\pi b^2}\cos k_0 z$

$$\tag{17}$$

The corresponding resonator has $(\frac{1}{L})(R_s^{(0)}/Q) = Z_0/(\pi b^2 k_0)$.

Time domain calculations

Another way to calculate the impedances numerically is to do it in the time domain. By evolving the Maxwell equations on a mesh, wakefields driven by a rigid ultra-short gaussian beam are calculated as functions of time. Integrating the wakefields seen by a test charge that trails the drive beam produces the wake functions. Fourier transforming the wake functions then gives the impedances. Fig.4 is an example of such a calculation using the program MAFIA.

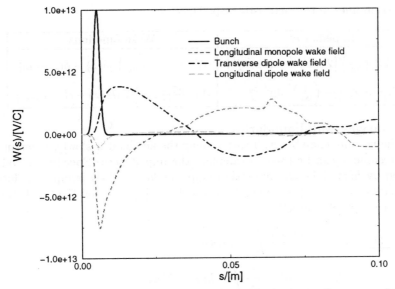

Figure 4. The longitudinal and transverse wake potentials of a 3-cell structure obtained by a time domain calculation. The drive beam has $\sigma_z = 1\text{mm}$. The offset of the drive beam and the test charge is 1 mm.

Resonator model

Sometimes it is useful to model an impedance by an equivalent circuit. In this approach, a complicated object is modeled as a transmission line or as an RLC-circuit. The most notable example is the resonator model ($R_s^{(m)}$ is the shunt impedance, Q is the quality factor, ω_r is the resonant frequency),

Impedances	Wake functions
$Z_m^{\parallel} = \dfrac{R_s^{(m)}}{1 + iQ\left(\omega_r/\omega - \omega/\omega_r\right)}$	$W_m(z < 0) = \dfrac{R_s^{(m)} c\,\omega_r}{Q\bar{\omega}_r}\,e^{\alpha z/c}\sin\dfrac{\bar{\omega}_r z}{c}$
$Z_m^{\perp} = \dfrac{c}{\omega}\,Z_m^{\parallel}$	$\alpha = \omega_r/(2Q),\ \bar{\omega}_r = \sqrt{|\omega_r^2 - \alpha^2|}$

$$(18)$$

Stripline beam-position monitor

Another example of an equivalent circuit impedance is that for strip-line BPMs (two strip lines of length L, subtending angle per strip ϕ_0, forming transmission lines of characteristic impedance Z_c).

Impedances	Wake functions
$Z_0^{\parallel} = 2Z_c \left(\frac{\phi_0}{2\pi}\right)^2 [2\sin^2\frac{\omega L}{c} - i\sin\frac{2\omega L}{c}]$	$W_0' = 2Z_c c \left(\frac{\phi_0}{2\pi}\right)^2 [\delta(z) - \delta(z+2L)]$
$Z_1^{\perp} = Z_0^{\parallel} \frac{c}{b^2\omega} \left(\frac{4}{\phi_0}\right)^2 \sin^2\frac{\phi_0}{2}$	$W_1 = \frac{8Z_c c}{\pi^2 b^2} \sin^2\frac{\phi_0}{2} [H(z) - H(z+2L)]$

$$(19)$$

Slowly varying wall boundaries

If the vacuum chamber wall varies along the accelerator slowly, a perturbation technique can be used to calculate the impedances. Specify the wall variation by $h(z)$ (1-D axisymmetric bump), or $h(z,\theta)$ (2-D bump). At low frequencies $k = \omega/c \ll$ (bump length or width)$^{-1}$, $|h| \ll b$, and $|\nabla h| \ll 1$, the impedance is purely inductive,

$$\underline{1-D}: \quad Z_0^{\parallel} = -\frac{2ikZ_0}{b} \int_0^{\infty} \kappa |\tilde{h}(\kappa)|^2 d\kappa$$

$$\text{where} \quad \tilde{h}(k) = \frac{1}{2\pi} \int_{-\infty}^{\infty} h(z)e^{-ikz} dz$$

$$\underline{2-D}: \quad Z_0^{\parallel} = -\frac{4ikZ_0}{b} \sum_{m=-\infty}^{\infty} \int_{-\infty}^{\infty} \frac{\kappa^2}{\sqrt{\kappa^2 + m^2/b^2}} |\tilde{h}_m(\kappa)|^2 d\kappa$$

$$\text{where} \quad \tilde{h}_m(k) = \frac{1}{(2\pi)^2} \int_0^{2\pi} d\theta \int_{-\infty}^{\infty} dz\, h(z,\theta) e^{-ikz - im\theta} \qquad (20)$$

Eq.(20) can be used to calculate the impedance of a tapered discontinuity. Note that the impedance is quadratic in the height of the small obstacle. The impedance is the same whether the wall bulges into or out of the pipe region.

Homework

Calculate the broad-band low-frequency impedance $Z_0^{\parallel}(\omega)$ for the beam pipe discontinuity as shown below.

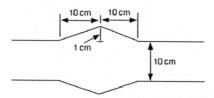

Small obstacles

A special class of calculation applies when there is a small obstacle on the vacuum chamber wall. Using a theory developed by Bethe, one obtains, for low frequencies (ϕ is the azimuthal angle of the obstable relative to the direction of the wake force being considered), a purely inductive impedance

Impedances	Wake functions
$Z_0^{\parallel} = -i\dfrac{\omega Z_0}{c}\dfrac{\alpha}{4\pi^2 b^2}$	$W_0' = -Z_0 c\dfrac{\alpha}{4\pi^2 b^2}\,\delta'(z)$
$Z_1^{\perp} = -i\dfrac{Z_0\alpha}{\pi^2 b^4}\cos\phi$	$W_1 = -Z_0 c\dfrac{\alpha}{\pi^2 b^4}\cos\phi\,\delta(z)$

$$(21)$$

The parameter α is related to the electric polarizability and magnetic susceptibility of the obstacle in the Bethe theory, and is a parameter (dimensionality $= L^3$) determined by the geometry of the obstacle. For examples,

$$
\alpha = \begin{cases}
\dfrac{2a^3}{3} & \text{circular hole of radius } a \\[2mm]
\dfrac{\pi d^4 [\ln(4a/d) - 1]}{3a} & \text{elliptical hole with major radius } a \text{ along} \\
 & \text{the pipe and minor radius } d \ll a \\[2mm]
w^3(0.1814 - 0.0344\dfrac{w}{L}) & \text{rectangular slot of length } L \text{ and width } w \ll L \\[2mm]
w^3(0.1334 - 0.0500\dfrac{w}{L}) & \text{rounded-end slot of length } L \text{ and width } w \ll L \\[2mm]
\pi a^3 & \text{half spherical protrusion of radius } a \\[2mm]
\dfrac{2\pi h^3}{3[\ln(2h/a) - 1]} & \text{circular protrusion with height } a \ll h
\end{cases}
$$

$$(22)$$

The first four results apply when the pipe wall thickness is much smaller than the size of the small obstacle. If that is not the case, the value of α would be reduced by a multiplicative factor of about 0.6.

Trapped modes

A small indent of the vacuum chamber wall out of the beam pipe region can contribute a sharp impedance, corresponding to a weakly trapped mode whose field pattern extends over a large distance over the indent. Consider a ring-shaped indent of cross-section area A in an axi-symmetric beam pipe. The impedance resonant frequency is located slightly below the TM$_{01}$ pipe cut-off frequency $\omega_{01} = x_{01}c/b$ by an amount

$$\Delta\omega_{01} = \omega_{01}\frac{x_{01}^2}{2}\left(\frac{A}{b^2}\right)^2 \tag{23}$$

where $x_{01} = 2.405$ is the first root of Bessel function $J_0(x)$. When the wall has a slight resistivity, the mode damps with damping rate $\gamma_{01} = \omega_{01}\delta_{\text{skin}}/(2b)$, where δ_{skin} is the skin depth. The trapped mode disappears when $\gamma_{01} > \Delta\omega_{01}$. Otherwise, it leads to a sharp longitudinal impedance with the shunt impedance

$$R_{01} = \frac{4Z_0 x_{01} A^3}{\pi\delta_{\text{skin}} b^5 J_1^2(x_{01})}, \qquad J_1(x_{01}) = 0.519 \tag{24}$$

<u>Homework</u>
Does the ring-shaped indentation in the following figure allow a trapped mode? Consider a copper beam pipe.

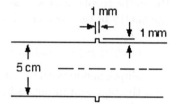

Figure 5. A possibility of trapped mode.

Rough metallic surface

The result for slowly varying surface and the result for small obstacles can also be used to estimate the low-frequency inductive impedance of a rough metallic surface provided the surface characteristics are known. A long pipe with rough wall surface behaves like a periodic array of small pill-boxes, and a resonator-type contribution like (17) will also be present.

4 Parasitic Heating

The energy change (parasitic loss) of a bunch of charge q and normalized line density $\lambda(t)$, traversing a structure with longitudinal impedance $Z_0^{\|}$ is given by

$$\Delta\mathcal{E} = -\kappa^{\|} q^2 \tag{25}$$

where $\kappa^{\|}$ is the *loss factor*, in units of V/pC,

$$\kappa^{\|}(\sigma) = \frac{1}{\pi} \int_0^\infty d\omega \, \text{Re} Z_0^{\|}(\omega) \, |\tilde{\lambda}(\omega)|^2 \tag{26}$$

where $\sigma = \sigma_z/\beta c$ is the rms beam bunch length in sec. For a gaussian bunch, $\lambda = e^{-t^2/2\sigma^2}/(\sqrt{2\pi}\sigma)$, $\tilde{\lambda}(\omega) = e^{-\omega^2\sigma^2/2}$.

Only the real part of the impedance contributes to the parasitic loss. Inductive impedances and the space charge impedance do not introduce a net power loss to the beam. Energy loss by particles at the head of the bunch is recovered by particles in the tail of the bunch (see e.g., Eq.(35) later).

One can also write the loss factor in terms of the wake function. For a gaussian bunch,

$$\kappa^{\parallel} = \frac{1}{2\sqrt{\pi}\,\sigma} \int_{-\infty}^{0} \frac{dz}{c}\, W_0'(z)\, e^{-z^2/4\sigma^2 c^2} \tag{27}$$

Resistive wall A resistive wall impedance gives

$$\frac{\kappa^{\parallel}(\sigma)}{L} = \frac{\Gamma(\frac{3}{4})c}{4\pi^2 b\sigma_z^{3/2}} \left(\frac{Z_0}{2\sigma_c}\right)^{1/2}, \qquad \Gamma(\frac{3}{4}) = 1.225 \tag{28}$$

Pill-box cavity For a bunch traversing a pill-box cavity, $\kappa^{\parallel}$ is given by a sum over cavity modes up to the cut-off frequency, plus a contribution from the diffraction model impedance. Each of the cavity modes contributes a resonator impedance, and each resonator impedance contributes

$$\kappa^{\parallel} \approx \begin{cases} \dfrac{\omega_r R_s}{2Q_r} e^{-\omega_r^2 \sigma^2} & \text{high-}Q \text{ resonator} \\[2mm] \dfrac{\omega_r R_s}{2Q_r} & \text{low-}Q \text{ resonator, short bunch } \omega_r \sigma \ll 1 \\[2mm] \dfrac{R_s}{4\sqrt{\pi}Q_r^2 \omega_r^2 \sigma^3} & \text{low-}Q \text{ resonator, long bunch } \omega_r \sigma \gg 1 \end{cases} \tag{29}$$

The contribution from the high frequency diffraction impedance is

$$\kappa^{\parallel} \approx \frac{\Gamma(\frac{1}{4})Z_0}{4\pi^{5/2}b} \sqrt{\frac{cg}{\sigma}} \tag{30}$$

For a single bunch in a circular accelerator, the integral in Eq.(26) is replaced by an infinite sum,

$$\kappa^{\parallel}(\sigma) = \frac{\omega_0}{2\pi} \sum_{p=-\infty}^{\infty} Z_0^{\parallel}(p\omega_0)\, |\tilde{\lambda}(p\omega_0)|^2 \tag{31}$$

For short bunches in large machines ($\omega_0 \ll 1/\sigma$), the sum can be replaced by an integral, and the difference between single passes and multiple passes disappears as it should.

Homework

Consider a storage ring with smooth resistive pipe. Let circumference $C = 100$ m, $b = 5$ cm, $N = 10^{11}$ protons (ignore synchrotron radiation), $\sigma_z = 1$ cm, $\sigma_c = 3.5 \times 10^7 \Omega^{-1} m^{-1}$ (aluminum). Let there be one single cavity structure in the ring with $g = 10$ cm, which has 6 trapped HOMs with $R/Q = 60\,\Omega$ each. (1) Estimate the parasitic heating power due to resistive wall. (2) Estimate the parasitic heating power due to the cavity HOMs below cut-off. (3) Estimate the parasitic heating power due to the cavity above cut-off. (4) How are the above heating powers distributed around the accelerator? What if the pipe is made of stainless steel? (5) Do we need water cooling for the pipe? Do we need water cooling for the rf cavity due to the parasitic heating?

Hints

(2) Since $\omega_{cut-off} \approx c/b$, the trapped modes most likely will have $\omega_R \ll c/\sigma_z$. This means we should use the short bunch formula in Eq.(29). Let the 6 trapped modes have $\omega_R = (1, 5/6, 4/6, 3/6, 2/6, 1/6)c/b$, add up the heating due to the 6 modes. This heating is trapped by the cavity.

(3) Use diffraction model. This power propagates down the two directions from the cavity.

(4) First estimate the attenuation length of the untrapped modes in (3). Let this power dissipation be given by $e^{-2\beta z}$, then very roughly, $\beta \approx c/(4\pi b \sigma_c \delta_{skin})$. Is the untrapped wave absorbed near the cavity, or does it propagate around the ring more or less evenly?

(5) Without water cooling, the heated area will have to cool by black body radiation. The black body radiation per meter is given by $dQ/dt = -(2\pi b)\pi^2 k_B^4 T^4/(15\hbar^3 c^2)$. Therefore, in equilibrium, $dQ_{parasitic\ heating}/dt = (2\pi b)\pi^2 k_B^4 [T^4 - T_0^4]/(15\hbar^3 c^2)$ where T_0 is the room temperature. Estimate the equilibrium temperature T at the cavity and on the beam pipe around the ring to see if water cooling is needed.

5   Collective Effects in High Energy Linacs

Energy variation along the bunch length

As a beam bunch travels down the linac, the longitudinal wake induces an energy variation along the length of the bunch (consult Eq.(4) for the wake force),

$$-\Delta E(z) = Ne^2 \int_z^\infty dz' \lambda(z') W_0'(z - z')$$

$$= \frac{Ne^2}{2\pi} \int_{-\infty}^\infty d\omega \, e^{i\omega z/c} Z_0^\parallel(\omega) \tilde{\lambda}(\omega) \tag{32}$$

This variation can be compensated for by properly phasing the bunch center relative to crest of the acceleration rf. The total energy change of the beam is given by

$$\Delta \mathcal{E} = N \int_{-\infty}^\infty dz \, \Delta E(z) \lambda(z) \tag{33}$$

which is just the $\Delta \mathcal{E}$ in Eq.(25). A rough order-of-magnitude estimate of the head-tail energy split is

$$-\Delta E \sim \frac{Ne^2}{2} W_0' \tag{34}$$

provided one has an estimate of the magnitude of W_0'.

For longitudinally gaussian bunch of charge Ne, rms length σ_z, and transverse radius a, travelling in cylindrical, perfectly conducting beam pipe of radius b, the space charge wake-induced energy variation along the bunch is

$$\frac{\Delta E(z)}{L} = \frac{1}{4\pi\epsilon_0} \sqrt{\frac{2}{\pi}} \frac{Ne^2}{\gamma^2 \sigma_z^2} \left(\ln \frac{b}{a} + \frac{1}{2} \right) \frac{z}{\sigma_z} \exp(-z^2/2\sigma_z^2) \tag{35}$$

The total loss of the beam is zero, as mentioned before, because the space charge impedance is purely imaginary.

For a resistive wall,

$$\frac{\Delta E(z)}{L} = \frac{1}{4\pi\epsilon_0} \frac{Ne^2}{4b\sigma_z^{3/2}} \sqrt{\frac{2}{Z_0 \sigma_c}} f\left(\frac{z}{\sigma_z} \right) \tag{36}$$

$$f(u) = -|u|^{3/2} e^{-u^2/4} \left[(I_{-1/4} - I_{3/4}) \mathrm{sgn}(u) - I_{1/4} + I_{-3/4} \right]$$

where $I_{\pm 1/4}$ and $I_{\pm 3/4}$ are modified Bessel functions evaluated at $u^2/4$. See Fig.6.

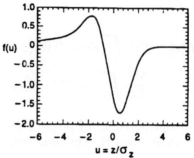

Figure 6. Function $f(u)$ in Eq.(36).

Homework

Estimate the wake-induced energy variation for the SLAC linac. Let $N = 5 \times 10^{10}$, and $L = 3000$ m. Model the SLAC linac cavities as an array of pill-boxes whose dimensions are comparable to the pipe radius $b = 1$ cm. What is the head-tail $\Delta E/E$-split of the beam when the beam reaches 50 GeV at the end of the linac if we do not compensate for this energy variation by properly phasing the rf?

Hints

Use Eq.(34), with a rough estimate of W_0' per cavity given by Eq.(15). The total number of cavities is about L/b.

Transverse beam break-up

The equation of motion is (consult Eq.(4) for the wake force)

$$\frac{d}{ds}\left(E(s)\frac{d}{ds}x(z,s)\right) + E(s)k^2(s)x(z,s) = \frac{Ne^2}{L}\int_z^\infty \lambda(z')W_1(z'-z)x(z',s)dz'$$

(37)

Given the wake function $W_1(z)$ of the entire linac (of length L), the longitudinal bunch distribution $\lambda(z)$, the energy acceleration $E(s)$ and betatron focussing $k(s)$, Eq.(37) is to be solved for the betatron motion $x(z,s)$ as a function of s for a particle located at position z relative to the bunch center.

For uniform $\lambda = 1/\ell$ (ℓ is total length of bunch), $W_1(z) = W_1'z$, and when the variations of E and k are adiabatic, the asymptotic solution is

$$\frac{\text{final amplitude}}{\text{initial amplitude}} \approx \sqrt{\frac{E(0)k(0)}{E(L)k(L)}}\frac{\eta^{-1/6}}{\sqrt{6\pi}}\exp\left(\frac{3\sqrt{3}}{4}\eta^{1/3}\right) \qquad (\eta \gg 1) \qquad (38)$$

where

$$\eta(z) = \frac{Ne^2W_1'}{L\ell}\left(z - \frac{\ell}{2}\right)^2 \int_0^L \frac{ds}{E(s)k(s)}$$

The bunch head has $z = \ell/2$ and thus $\eta = 0$. The exponential growth in Eq.(38) illustrates the *beam break-up instability*. This problem is usually cured by the *BNS damping*.

Homework
Estimate how potentially serious is the beam break-up instability problem for the SLAC linac (no BNS damping). Let $E(s)[\text{GeV}] = 1.2 + 0.017s[\text{m}]$, $k(s) = 0.06/\text{m}$, $N = 5 \times 10^{10}$, $\ell = 2$ mm, $b = 1$ cm, and $L = 3000$ m.

Hint
Use Eq.(15) for W_1 per cavity. Our model then corresponds to taking $W_1' \sim (W_1/b) \times (L/b)$. Consider the particle at the tail of the bunch by taking $z = -\ell/2$.

BNS damping
BNS damping is accomplished by having the particles at the tail of the bunch being focussed more strongly than particles at the head of the bunch. The stronger focusing balances out the defocusing effect of the transverse wake. The net result is that the wake-induced emittance growth is minimized. The condition for this to occur is called the *autophasing condition*. In case of uniform acceleration and uniform betatron focusing, it reads

$$\frac{\Delta k(z)}{k} = -\frac{Ne^2}{2k^2LE_f}\ln\frac{E_f}{E_i}\int_z^\infty \lambda(z')W_1(z - z')dz' \qquad (39)$$

An order-of-magnitude estimate of the needed BNS focusing is given by

$$\frac{\Delta k}{k} \approx -\frac{Ne^2W_1(\ell)}{4k^2LE_f}\ln\frac{E_f}{E_i} \qquad (40)$$

where $W_1(\ell)/L$ is the transverse wake function per unit length seen by a particle at the tail of the bunch.

Homework
Estimate the BNS focusing needed for the SLAC linac.

Hint
Using Eq.(15) to obtain $W_1(\ell) \sim (1/b^2) \times (L/b) \times (\ell/b)$, we obtain $\Delta k/k \sim (Ne^2\ell/4k^2E_fb^4)\ln(E_f/E_i)$.

Multibunch transverse dynamics

Consider a beam with n_B equally-popu-lated, equally-spaced bunches, each with charge Ne. Motion of each bunch is affected by the transverse wake left behind by the previous bunches (or the same bunch in previous turns). The analysis is similar to that of the BBU effect within a single bunch. The difference is that here W_1 is dominated by one or a few resonators having large shunt impedances. In case of a single isolated resonator, the amplitude blowup factor of the last bunch takes a form $\sim e^{\sqrt{\eta}}$ where

$$\eta = n_B Ne^2 \frac{R_s \omega_r}{Q} \int_0^L \frac{ds}{E(s)k(s)} \gg 1 \tag{41}$$

The condition for tolerable emittance growth due to multi-bunch transverse wake is roughly given by $\eta \lesssim 1$.

6 Robinson Instability

Instability mechanisms in a linac are comparatively simple because of the absence of synchrotron oscillation. In a circular accelerator, synchrotron oscillation plays a critical role. (Fortunately, the role is often a stabilizing one.) The most basic collective instability in a circular accelerator is the Robinson instability.

Single-bunch, point-charge beam

Consider a beam with a single bunch executing synchrotron motion in a circular accelerator. First let the bunch be represented as a point charge Ne without any internal structure. Given the impedance $Z_0^{\parallel}(\omega)$ of the accelerator, the stability of this beam is analyzed by assuming the beam is executing a collective motion as $z(t) \sim e^{-i\Omega t}$. The key quantity to be calculated is the *collective mode frequency* Ω. It is a complex quantity directly related to the impedance. The real part of Ω is the perturbed synchrotron oscillation frequency of the collective beam motion, while the imaginary part gives its growth rate (or damping rate if negative). The result of the growth rate is

$$\tau^{-1} = \text{Im}(\Omega - \omega_s) = \frac{Ne^2\eta}{2ET_0^2\omega_s} \sum_{p=-\infty}^{\infty} (p\omega_0 + \omega_s)\text{Re}Z_0^{\parallel}(p\omega_0 + \omega_s) \tag{42}$$

where η is the momentum slippage factor, ω_0 is the revolution angular frequency, $T_0 = 2\pi/\omega_0$ is the revolution period, ω_s is the unperturbed synchrotron oscillation frequency. Note that it is the real part of the impedance that contributes to the instability growth rate. A space charge impedance causes a mode frequncy shift, but not an instability.

Eq.(42) applies to any impedance. The largest impedance to be considered is the resonator impedance of the fundamental cavity mode, which has $\omega_R \approx$

$h\omega_0$ (h = integer is the *harmonic number*). The only significant contributions to the growth rate come from two terms in the summation, namely $p = \pm h$,

$$\tau^{-1} \approx \frac{Ne^2\eta h\omega_0}{2ET_0^2\omega_s}\left[\mathrm{Re}Z_0^{\parallel}(h\omega_0 + \omega_s) - \mathrm{Re}Z_0^{\parallel}(h\omega_0 - \omega_s)\right] \tag{43}$$

Beam stability requires $\tau^{-1} \leq 0$. That is, the real part of the impedance must be lower at frequency $h\omega_0 + \omega_s$ than at frequency $h\omega_0 - \omega_s$ if $\eta > 0$ (above transition), and the other way around if $\eta < 0$ (below transition). This condition implies the *Robinson criterion* that, above transition, the resonant frequency ω_R of the fundamental cavity mode should be slightly detuned downwards from an exact integral multiple of ω_0. Below transition, stability requires ω_R be slightly higher than $h\omega_0$. In the figure below, (a) is stable above transition, while (b) is stable below transition.

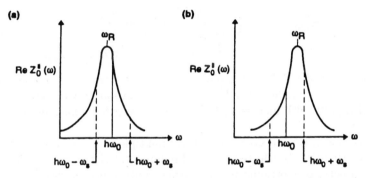

After properly tuning the fundamental rf cavity mode, one should consider the contributions from the higher rf modes, paying attention to accidentally landing the frequencies $p\omega_0$ for some integer p on the wrong side of some higher order impedance peak. Sometimes it becomes necessary to damp the higher order modes of the rf to avoid excessive instability growth rates.

Multi-bunch beam

Eq.(42) is restricted because (i) it applies to a beam with only one bunch, and (ii) the beam bunch is a point-charge. Consider a beam with n_B evenly-spaced, equally-populated bunches. This time also consider bunches with finite length to allow internal structure of the mode patterns within each bunch. There are now a large number of collective modes in our problem, and the mode frequency Ω depends on the mode being considered. A mode is now specified by two mode indices μ and ℓ, where $\mu = 0, 1, 2, ..., (n_B - 1)$ is the multi-bunch mode number, and ℓ is the internal bunch structure mode number ($\ell = 1$ dipole, $\ell = 2$ quadrupole, etc.). The growth rate is found to be

$$\frac{1}{\tau^{(\ell,\mu)}} = \text{Im}(\Omega^{(\ell,\mu)} - \ell\omega_s) = \frac{2n_B N e^2 \eta c^2}{ET_0^2 \omega_s \hat{z}^2} \ell \sum_{p=-\infty}^{\infty} \frac{\text{Re}Z_0^{\|}(\omega')}{\omega'} J_\ell^2\left(\frac{\omega'\hat{z}}{c}\right) \quad (44)$$

where N is the number of particles per bunch, $\omega' = pn_B\omega_0 + \mu\omega_0 + \ell\omega_s$. In deriving Eq.(44), it has been assumed that the unperturbed bunch distribution is described by a "waterbag model". The full extent of the bunch length is $2\hat{z}$.

The lowest bunch structure mode has $\ell = 1$ (dipole). For short bunches with $\omega'\hat{z}/c \ll 1$, and for $n_B = 1$, Eq.(44) reduces to Eq.(42). On the other hand, Eq.(44) also applies to higher order modes $\ell > 1$ and arbitrary bunch length $\hat{z}$. In particular, the fundamental rf mode contributes to a Robinson growth rate for the ℓth beam mode,

$$\frac{1}{\tau^{(\ell,0)}} = \frac{\ell}{(\ell!)^2}\left(\frac{h\omega_0\hat{z}}{2c}\right)^{2\ell-2}\frac{n_B N e^2 \eta h \omega_0}{2ET_0^2 \omega_s}$$
$$\times [\text{Re}\,Z_0^{\|}(h\omega_0 + \ell\omega_s) - \text{Re}Z_0^{\|}(h\omega_0 - \ell\omega_s)] \quad (45)$$

It follows that the Robinson stability criterion for the $\ell = 1$ mode also stabilizes the $\ell > 1$ modes. The higher order Robinson growth rates, however, drop off rapidly with increasing ℓ if the bunch length is much less than the rf wavelength.

Note that h is necessarily an integral multiple of n_B, thus only the $\mu = 0$ multi-bunch mode is affected by the fundamental rf mode, and the growth rate is directly proportional to n_B. Other multi-bunch modes ($\mu \neq 0$) must be considered when studying the effects of the higher order rf modes.

7 Transverse Robinson Instability

Robinson instability has a transverse counterpart. Basically the same physical mechanism applies when we consider the transverse beam motion driven by the transverse impedance. Letting $y \propto \exp(-i\Omega t)$, the instability growth rate is found to be

$$\tau^{-1} = \text{Im}(\Omega - \omega_\beta) \approx -\frac{Ne^2 c}{2ET_0^2 \omega_\beta}\sum_{p=-\infty}^{\infty}\text{Re}Z_1^{\perp}(p\omega_0 + \omega_\beta) \quad (46)$$

where $Z_1^{\perp}$ is the total impedance around the accelerator circumference. Only the real part of the impedance contributes to the instability growth rate. The space charge impedance does not cause instability of the Robinson type.

A transverse Robinson instability occurs when $\mathrm{Re}Z_1^\perp(\omega)$ contains sharp resonant peaks. If a resonant frequency ω_R is close to $h\omega_0$ for some integer h, then

$$\tau^{-1} \approx -\frac{Ne^2c}{2ET_0^2\omega_\beta}\left[\mathrm{Re}Z_1^\perp(h\omega_0 + \Delta_\beta\omega_0) - \mathrm{Re}Z_1^\perp(h\omega_0 - \Delta_\beta\omega_0)\right] \qquad (47)$$

where Δ_β is the non-integer part of the betatron tune $\nu_\beta = \omega_\beta/\omega_0$ and we have chosen $-1/2 < \Delta_\beta < 1/2$. A positive Δ_β means ν_β is above an integer; a negative Δ_β means ν_β is below an integer. For stability, ω_R should be slightly above $h\omega_0$ if $\Delta_\beta > 0$ and below $h\omega_0$ if $\Delta_\beta < 0$. The stability criterion of the transverse Robinson instability does not depend on whether the accelerator is operated above or below transition. Instead, it depends on whether the betatron tune is above or below an integer.

Homework
Find the transverse Robinson instability growth rate for the case $N = 10^{11}$, $R_S = 40$ MΩ/m^2, $Q = 2000$, $E = 1$ GeV (electron beam), $\omega_0 = 9.4 \times 10^6$ s^{-1}, $\nu_\beta = 6.05$, $h = 518$, and $(\omega_R - h\omega_0)/\omega_0 = \pm h/2\sqrt{3}Q$ (the worst case).

Like its longitudinal counterpart, to control the transverse Robinson instability, it is often necessary to de-Q the transverse higher order modes of the rf cavities by either passively or actively removing their field energies. Unlike its longitudinal counterpart, the transverse Robinson instability does not have the strong damping provided by the fundamental rf cavity mode (assumed properly tuned), which makes it more of a serious concern.

In case of a broad-band impedance (short-ranged wake field), the summation over p can be approximated by an integral over p. It then follows from Eq.(46) and the fact that $\mathrm{Re}Z_1^\perp$ is an odd function of ω that $\tau^{-1} = 0$. Broad-band impedances therefore do not cause transverse Robinson instability, a situation similar to the longitudinal case. However, as will be explained later, a broad-band impedance does cause an instability when the betatron frequency of a particle is not a constant, as assumed so far, but depends on its relative energy deviation $\delta = \Delta E/E$.

Resistive wall instability

Eq.(46) also gives the instability growth rate for an accelerator with a resistive vacuum chamber. Substituting Eq.(8) into Eq.(46), we obtain

$$\tau^{-1} \approx -\frac{Ne^2c^2}{b^3 E\omega_\beta T_0\sqrt{\pi\sigma_c\omega_0}} f(\Delta_\beta) \qquad (48)$$

$$f(\Delta_\beta) = \frac{1}{\sqrt{2}} \sum_{p=-\infty}^{\infty} \frac{\mathrm{sgn}(p + \Delta_\beta)}{|p + \Delta_\beta|^{1/2}}$$

when $|\Delta_\beta| \ll 1$, the $p = 0$ term dominates in the summation for $f(\Delta_\beta)$.

The function $f(\Delta_\beta)$ is positive (so that $\tau^{-1} < 0$ and the beam is stable) if $0 < \Delta_\beta < 1/2$, and negative if $-1/2 < \Delta_\beta < 0$. This means one should choose the betatron tune above an integer to assure stability against the resistive wall instability. However, because the resistive wall instability is usually rather weak (Very large storage rings are an exception because they have small ω_0, and according to Eq.(48), can potentially have large growth rates.), a small spread of betatron tune of the particles in the beam will stabilize the beam even if the betatron tune is below an integer (see Landau damping later).

Homework
Estimate the transverse resistive wall instability growth rate for the case $N = 10^{11}$, $b = 5$ cm, $E = 1$ GeV (electron beam), $\nu_\beta = 5.9$ (below an integer, therefore the beam is unstable), $\omega_0 = 9.4 \times 10^6$ s^{-1}, and $\sigma = 3 \times 10^{17}$ s^{-1}.

8 Strong Head-Tail Instability

The strong head-tail instability (also called *transverse mode coupling* instability or *transverse turbulence* instability) is the circular accelerator's counterpart of the dipole beam breakup in a linac. The difference from the linac case is that now the beam particles execute synchrotron oscillations, thus constantly changing their relative longitudinal positions with a slow synchrotron frequency ω_s.

In the discussion of transverse Robinson instability, it was stated that a broad-band impedance (short ranged wake) does not cause an instability. However, this is only true when one ignores coupling between the "azimuthal modes" of the beam motion. The dimensionless parameter that describes the strength of the mode coupling is given by

$$\Upsilon = -\frac{\pi N e^2 W_1 c^2}{4 E C \omega_\beta \omega_s} \tag{49}$$

where $W_1 < 0$ is the short-range wake function (integrated over the accelerator circumference C), assumed to be constant over the bunch length.

For low beam intensities, $\Upsilon \ll 1$, mode coupling is negligible and adjacent azimuthal modes have their frequencies well separated by ω_s. The beam is stable for short-ranged wake, as we discussed in the context of Robinson instability. As beam intensity increases, however, mode coupling becomes significant. As the mode frequencies shift by amounts comparable to ω_s, adjacent azimuthal mode frequencies may merge into each other, and the beam becomes unstable even with a short-range wake. This is called the strong head-tail instability. The threshold of this instability occurs at

$$\Upsilon_{\text{th}} = 2 \tag{50}$$

The reason it is called *strong* head-tail instability is that once the threshold is exceeded, the instability tends to grow very fast.

Homework
The strong head-tail instability is one of the cleanest instabilities to observe in electron storage rings. The instability threshold observed at PEP occurred when $N_{\text{th}} = 6.4 \times 10^{11}$, $\omega_\beta/\omega_0 = 18.19$, $\omega_s/\omega_0 = 0.044$, $E = 14.5$ GeV, and $\omega_0 = 0.86 \times 10^6$ s^{-1}. By relating these parameters to $\Upsilon_{\text{th}} = 2$, estimate the wake function W_1 and the transverse impedance $Z_1^\perp$ for PEP. Take a beam pipe radius $b = 5$ cm. Once $Z_1^\perp$ is found, relate it to give an estmate of $Z_0^\parallel/n$.

Hint
To estimate $Z_1^\perp$ once W_1 is found, use $Z_1^\perp \approx -bW_1/c$.

9   Head-Tail Instability

In our analysis of the strong head-tail instability, we assumed that the betatron and the synchrotron motions are decoupled. In doing so, we have ignored an important instability known as the head-tail instability.

The betatron oscillation frequency of a particle depends on $\delta = \Delta E/E$ of the particle through the chromaticity parameter ξ,

$$\omega_\beta(\delta) = \omega_\beta(1 + \xi\delta) \tag{51}$$

As we will see, in order to avoid head-tail instability, ξ must have a definite sign. The main reason for introducing sextupoles in circular accelerators is in fact to control ξ.

Because of a nonzero chromaticity, the betatron phase of a particle is modulated by the longitudinal position z of the particle. The modulation is slow and weak, but is sufficient to drive the head-tail instability. The magnitude of this phase modulation is specified by the *head-tail phase*,

$$\chi = \frac{\xi\omega_\beta\hat{z}}{c\eta} \tag{52}$$

with $\pm\hat{z}$ the extent of the bunch length. For example, an electron accelerator with $\eta = 0.003$, $\xi = 0.2$, $\hat{z} = 3$ cm, and $\omega_\beta = 1.4 \times 10^7$ s^{-1}, would have $\chi \approx 2\pi \times 0.016$.

The strong head-tail instability is a threshold effect. It occurs when the beam current exceeds a certain critical value. Once the threshold is exceeded, the instabilty is very strong. In contrast, the head-tail instability is not a threshold effect. It is unstable even for weak beam currents, although the growth rate is slow. Note however that the *same* transverse impedance (typically broad-band) drives both the head-tail and the strong head-tail instabilities.

Taking into account of the chromaticity effect, the transverse instability growth rate for the ℓth mode ($\ell = 0$ for monopole, $\ell = 1$ for dipole, etc.) is found to be

$$\frac{1}{\tau^{(\ell)}} = \text{Im}(\Omega - \omega_\beta - \ell\omega_s) \approx -\frac{Ne^2c}{2ET_0^2\omega_\beta} \sum_{p=-\infty}^{\infty} \text{Re}Z_1^\perp(\omega') J_\ell^2\left(\frac{\omega'\hat{z}}{c} - \chi\right) \quad (53)$$

where $\omega' = p\omega_0 + \omega_\beta + \ell\omega_s$. Note how the head-tail phase appears in shifting the beam spectrum which is then convoluted with the impedance when summing over p. Eq.(53) is a rather general result. For example, When $\ell = 0$, $\chi = 0$, and $\hat{z} = 0$, it reduces to Eq.(46). However, it did assume an "airbag model" for the unperturbed longitudinal beam distribution, although the impedance is left aribitrary.

When $\chi \ll 1$, Eq.(53) gives the head-tail instability growth rate, to first order in χ,

$$\frac{1}{\tau^{(\ell)}} \approx \frac{Ne^2c}{\pi ET_0\omega_\beta} \chi \int_0^\infty d\omega \ \text{Re}Z_1^\perp(\omega) J_\ell\left(\frac{\omega\hat{z}}{c}\right) J_\ell'\left(\frac{\omega\hat{z}}{c}\right) \quad (54)$$

where we have considered a broad-band impedance, so that the summation in p can be replaced by an integral.

For the case of a constant wake function ($W_1(z)$ is independent of z, $W_1 < 0$), Eq.(54) gives

$$\frac{1}{\tau^{(\ell)}} = -\frac{Ne^2cW_1}{ET_0\omega_\beta} \chi \frac{2}{\pi^2(4\ell^2 - 1)} \quad (55)$$

According to Eq.(55), the $\ell = 0$ mode is unstable when $\xi/\eta < 0$, while the $\ell \geq 1$ modes are unstable when $\xi/\eta > 0$. We conclude from this that the only value of ξ that assures a stable beam is $\xi = 0$. However, the $\ell = 0$ growth rate is the strongest. The presence of some stabilizing mechanisms (such as Landau damping, or radiation damping in the case of circular electron accelerators) therefore leads us to choose slightly positive values for ξ for operation above transition $(\eta > 0)$, and slightly negative ξ below transition $(\eta < 0)$.

The head-tail growth rate provides another way to measure the transverse impedance of an accelerator. To do so, ξ is made slightly positive (above transition), a beam centroid motion is excited by a kicker, and its subsequent damped motion is observed. Before applying Eqs.(54) or (55), make sure that $\xi \ll 1$. Usually, radiation damping is much weaker than the head-tail damping. Otherwise the contribution from radiation damping has to be subtracted out from the measured damping rate.

Homework

It was observed in the electron storage ring SPEAR I that the head-tail damping time is 1 ms under the conditions $I = 20$ mA, $E = 1.5$ GeV, $\xi = 0.67$, $\hat{z} = 13$ cm, $\eta = 0.037$, $C = 240$ m, $b = 5$ cm. Estimate the magnitudes of W_1, $Z_1^\perp$, and $Z_0^\parallel/n$.

10 Landau Damping

So far, we have ignored the important effect of Landau damping. Landau damping occurs when there is a sufficient spread in particles' natural oscillation frequencies. Table below gives the natural frequencies needed by the Landau damping mechanism in order to damp the respective instabilities.

	Bunched beam	Unbunched beam
Longitudnal	synchrotron frequency ω_s	revolution frequency ω_0
Transverse	betatron frequency ω_β	betatron frequency ω_β

Consider an unbunched beam executing longitudinal or transverse oscillation. Let the frequency spectrum $\rho(\omega)$ be normalized to unity (ω is the frequency relevant to the Landau damping mechanism, and is given in the table above), and be centered around $\bar{\omega}$ with spread $S \ll \bar{\omega}$. The strength of Landau damping depends on $\rho(\omega)$.

Transverse microwave instability

Consider an unbunched beam executing a transverse oscillation with dipole perturbation $\propto e^{-i\Omega t + in(s/R)}$ where n is the mode number. Whether the beam is stable is determined by the following steps:

(i) Given the impedance, one first calculates the complex mode frequency shift – its imaginary part is the instability growth rate – in the *absence* of Landau damping, designated with a subscript 0 (compare with Eq.(46)),

$$(\Delta\omega)_0 \equiv (\Omega - n\omega_0 - \bar{\omega})_0 = -\frac{Ne^2c}{2ET_0^2\omega_\beta}iZ_1^\perp(\bar{\omega} + n\omega_0) \qquad (56)$$

where $\bar{\omega}$ is the center of the betatron frequency spectrum.

(ii) Given the spectral distribution $\rho(\omega)$, one calculates the *unperturbed beam transfer function*, which is a dimensionless function of Ω,

$$\text{BTF} = f(u) + ig(u) \tag{57}$$

where (PV means taking principal value)

$$u = (\omega_\beta + n\omega_0 - \Omega)/S$$
$$f(u) = S \text{ PV} \int d\omega \frac{\rho(\omega)}{\omega - \Omega}$$
$$g(u) = \pi S \rho(\Omega) \tag{58}$$

(iii) With Landau damping, the complex mode frequency Ω, for a mode which is at the edge of stability, is determined by the *dispersion relation*

$$-\frac{(\Delta\omega)_0}{S} = \frac{1}{\text{BTF}} \tag{59}$$

To obtain boundary of stability: (a) trace the locus of the r.h.s. of Eq.(59) on a complex plane as u is scanned from $-\infty$ to ∞. This trace divides the complex plane into two regions, one contains the origin, the other doesn't. (b) Plot the l.h.s. of Eq.(59) as a single point on the same complex plane. (c) If this point lies in the region which contains the origin, the beam is stable; otherwise it is unstable. Fig.7 shows the stability region produced in step (b) for several spectral distributions. When the beam lies in the unstable region, the beam is said to have a *transverse microwave instability*.

A simplified stability boundary, as sketched in Fig.7(h), leads to the *Keil-Schnell criterion*,

$$|(\Delta\omega)_0| < \frac{1}{\sqrt{3}} S_{\frac{1}{2}} \tag{60}$$

where $S_{\frac{1}{2}}$ is the half width at half maximum of the frequency spectrum. Roughly speaking, when the instability growth rate or mode frequency shift (calculated in the absence of Landau damping) exceeds the beam frequency spread, one loses the protection of Landau damping, and the beam is most likely unstable.

Bunched beam

The simplified stability criterion Eq.(60) applies to unbunched beams, but it can be applied to a bunched beam if one replaces the beam density N/C by the peak local density of the bunched beam. This is called the *Boussard criterion*.

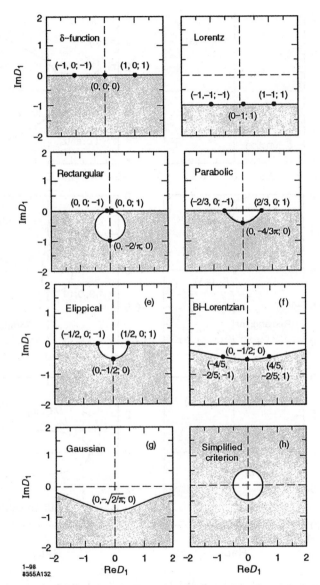

Figure 7. Stability diagram for transverse microwave instability for various beam spectra. Shaded regions are unstable.

244

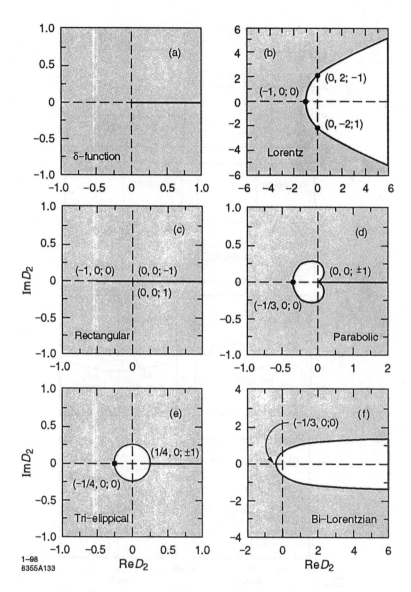

Figure 8. Stability diagram for longitudinal microwave instability for various beam spectra. Shaded regions are unstable.

For a bunched beam whose length is smaller than the pipe radius b, the simplified criterion against transverse microwave instability reads

$$|Z_1^\perp| < Z_0 \frac{\pi E |\eta| \omega_\beta \Delta \delta_{\frac{1}{2}} \Delta z_{\frac{1}{2}}}{3Ne^2 \omega_0 b} \tag{61}$$

Relating the longitudinal to the transverse impedance, it gives

$$\left|\frac{Z_0^\parallel}{n}\right| < Z_0 \frac{\pi E |\eta| \omega_\beta b \Delta \delta_{\frac{1}{2}} \Delta z_{\frac{1}{2}}}{6Ne^2 c} = Z_0 \frac{\pi E \omega_\beta \omega_s b \Delta z_{\frac{1}{2}}^2}{6Ne^2 c^2} \tag{62}$$

Longitudinal microwave instabilities

Similar steps apply to the longitudinal microwave instability of an unbunched beam. The complex mode frequency shift in the absence of Landau damping is now given by

$$(\Delta \omega)_0^2 \equiv (\Omega - n\omega_0)_0^2 = i \frac{2\pi Ne^2 n\eta}{ET_0^3} Z_0^\parallel (n\omega_0) \tag{63}$$

Landau damping comes from a spread in ω_0, spectrum $\rho(\omega_0)$ with spread S around center $\bar{\omega}_0$. The dispersion relation is

$$\frac{(\Delta \omega)_0^2}{n^2 S^2} = \frac{1}{F(u) + iG(u)} \tag{64}$$

$$u = (n\bar{\omega}_0 - \Omega)/(nS)$$

$$F(u) = nS^2 \, \text{PV} \int d\omega_0 \frac{\rho'(\omega_0)}{n\omega_0 - \Omega}$$

$$G(u) = \pi S^2 \rho'(\Omega/n)$$

Stability boundaries for various spectra are shown Fig.8. When the beam lies in the unstable region, the corresponding instability is called the *longitudinal microwave instability*. Note that Eqs.(63) and (64) are quadratic on the l.h.s. These are qualitatively different from Eqs.(56) and (59), and are the reason for the qualitative difference between Figs.7 and 8.

The Keil-Schell stability criterion is

$$|(\Delta \omega)_0^2| < 0.68 \, n^2 S_{\frac{1}{2}}^2 \tag{65}$$

If $S_{\frac{1}{2}}$ (the spread in revolution frequency ω_0) comes from energy spread, $S_{\frac{1}{2}} = \bar{\omega}_0 |\eta| \Delta \delta_{\frac{1}{2}}$, then the condition reads

$$\left|\frac{Z_0^\parallel (n\bar{\omega}_0)}{n}\right| < 0.68 \, Z_0 \frac{|\eta| EC}{2Ne^2} \Delta \delta_{\frac{1}{2}}^2 \tag{66}$$

bunched beam

For a bunched beam, applying the Boussard criterion, Eq.(66) becomes

$$\left|\frac{Z_0^{\parallel}(n\bar{\omega}_0)}{n}\right| < 0.66\, Z_0 \frac{|\eta|E}{Ne^2} \Delta\delta_{\frac{1}{2}}^2 \Delta z_{\frac{1}{2}} = 0.66\, Z_0 \frac{\omega_s^2 E}{|\eta|c^2 Ne^2} \Delta z_{\frac{1}{2}}^3 \qquad (67)$$

11 Potential-Well Distortion

The longitudinal wake field distorts the focusing field supplied by the rf, and thus distorts the equilibrium shape of a beam bunch. The mechanism is a static one; no part of the beam bunch is executing collective oscillation. The degree of the distortion increases with the beam intensity. This phenomenon is called the potential well distortion.

Without the wake field, the equilibrium distribution of an electron beam in a storage ring is bi-gaussian in the longitudinal phase space (z, δ), where $\delta = \Delta E/E$. It turns out that the wake field does not disturb the δ part of the stationary distribution, and only the z-distribution gets distorted, i.e.

$$\psi(z, \delta) = \frac{N}{\sqrt{2\pi}\sigma_\delta} \exp\left(-\frac{\delta^2}{2\sigma_\delta^2}\right) \lambda(z) \qquad (68)$$

where $\lambda(z)$ satisfies a transcendental equation, called the *Haissinski equation*,

$$\lambda(z) = \lambda(0)\exp\left[-\frac{z^2}{2\sigma_{z0}^2} + \frac{Ne^2}{\eta\sigma_\delta^2 EC}\int_0^z dz'' \int_{z''}^\infty dz'\, \lambda(z')W_0'(z'' - z')\right] \qquad (69)$$

where ω_s is the unperturbed synchrotron frequency, C is the storage ring circumference.

In the limit of zero beam intensity, the solution reduces to the bi-gaussian form, where the unperturbed rms bunch length is $\sigma_{z0} = \eta c\sigma_\delta/\omega_s$. For high beam intensities, $\lambda(z)$ deforms from gaussian. Once $W_0'(z)$ is known and σ_δ specified, the distorted distribution $\lambda(z)$ can be solved numerically using Eq.(69). Fig.9 shows the result of one such attempt for the electron damping ring for the SLAC Linear Collider. The calculated bunch shapes agree well with the measured results shown as open circles.

One feature of Fig.9 is that the distribution leans forward ($z > 0$) as the beam intensity increases. This effect comes from the parasitic loss of the beam bunch, and is a consequence of the real part of the impedance. Since the SLC damping ring is operated above transition, the bunch moves forward so that the parasitic energy loss can be compensated by the rf voltage.

Another feature of Fig.9 is that the bunch length increases as the beam intensity increases. The bunch shape distortion comes mainly from the imag-

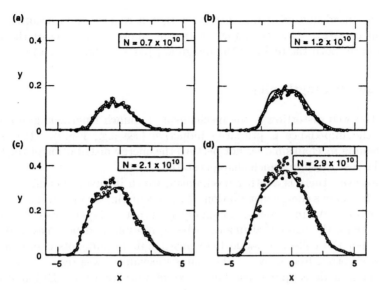

Figure 9. Potential-well distortion of bunch shape for various beam intensities for the SLC damping ring. The horizontal axis is $x = -z/\sigma_{z0}$. The vertical scale is $y = 4\pi Ne\lambda(z)/V'_{\mathrm{rf}}(0)\sigma_{z0}$.

inary part of the impedance. That the bunch *lengthens* is because the imaginary part of the impedance is mostly inductive.

Find impedance from bunch shape

It is possible to extract the impedance $Z_0^{\parallel}(\omega)$ from a detailed measurement of $\lambda(z)$ using a streak camera. Assuming $\lambda(z)$ is determined by potential well distortion alone. Any collective oscillations (e.g. due to microwave instability) are either not present, or averaged out in data averaging with multiple scans.

The procedure to extract $Z_0^{\parallel}(\omega)$ from $\lambda(z)$ is as follows. (i) Calculate the quantity

$$F(z) = \frac{EC\eta\sigma_\delta^2}{Ne^2}\left(\frac{\lambda'(z)}{\lambda(z)} + \frac{z}{\sigma_{z0}^2}\right)$$

(ii) The impedance is then given by

$$Z_0^{\parallel}(\omega) = \frac{1}{c}\frac{\tilde{F}(\omega)}{\tilde{\rho}(\omega)} \tag{70}$$

where $\tilde{F}(\omega) = \int_{-\infty}^{\infty} dz\, e^{-i\omega z/c} F(z)$ and $\tilde{\lambda}(\omega) = \int_{-\infty}^{\infty} dz\, e^{-i\omega z/c}\lambda(z)$.

One disadvantage of this method is that it requires very accurate information on $\lambda(z)$. One advantage is that it allows extraction of the entire impedance, both the real and the imaginary parts, as functions of ω.

12 Bunch Lengthening

Potential well distortion is one mechanism for bunch lengthening, especially when the impedance is primarily inductive (see e.g., Fig.9). This bunch lengthening mechanism does not perturb the energy distribution of the beam, and the distortion in bunch shape is static.

However, there is another mechanism, which involves dynamics and disturbs the beam's energy distribution. This is sometimes called *turbulent bunch lengthening*. Its mechanism nominally (but not always) involves the coupling among the azimuthal modes of beam motion, and becomes important only when the collective mode frequencies shift by amounts comparable to the synchrotron frequency ω_s. The same mechanism was responsible for the transverse mode coupling instability, except now we are in the longitudinal dimension.

The analysis of turbulent bunch lengthening is rather involved. If we ignore the potential-well distortion effect, we end up with solving the eigenvalue problem for the mode frequencies Ω,

$$\det \left(\frac{\Omega}{\omega_s} I - M \right) = 0 \tag{71}$$

where I is a unit matrix and M is a matrix with elements

$$M_{\ell\ell'} = \ell\delta_{\ell\ell'} + i \frac{Nr_0\eta c^2}{\pi\gamma T_0\omega_s^2\hat{z}^2} \ell i^{\ell-\ell'} \int_{-\infty}^{\infty} d\omega \frac{Z_0^{\|}(\omega)}{\omega} J_\ell\left(\frac{\omega\hat{z}}{c}\right) J_{\ell'}\left(\frac{\omega\hat{z}}{c}\right) \tag{72}$$

This result assumes the unperturbed beam has a water-bag distribution in the longitudinal phase space. It also assume a broad-band impedance.

Note that $\Omega = 0$ is always a solution; this is the mode that describes the static potential-well distortion.

As an illustration, we apply the result to the diffractive model, Eq.(13), rewritten as

$$Z_0^{\|}(\omega) = R_0 \left| \frac{\omega_0}{\omega} \right|^{1/2} [1 + \text{sgn}(\omega)i] \tag{73}$$

where R_0 is a real positive constant. Then,

$$M_{\ell\ell'} = \ell\delta_{\ell\ell'} - \ell\, C_{\ell\ell'}\Upsilon \tag{74}$$

$$C_{\ell\ell'} = \frac{\frac{1}{2}\Gamma\left(\frac{\ell+\ell'-\frac{1}{2}}{2}\right)}{\Gamma\left(\frac{\ell'-\ell+\frac{5}{2}}{2}\right)\Gamma\left(\frac{\ell+\ell'+\frac{5}{2}}{2}\right)\Gamma\left(\frac{\ell-\ell'+\frac{5}{2}}{2}\right)} \times \begin{cases} (-1)^{(\ell-\ell')/2} & \text{if } \ell-\ell' \text{ even} \\ (-1)^{(\ell-\ell'-1)/2} & \text{if } \ell-\ell' \text{ odd} \end{cases}$$

where we have defined a dimensionless parameter

$$\Upsilon = \frac{Ne^2\eta R_0}{E\omega_s^2}\left(\frac{c}{T_0\hat{z}}\right)^{3/2} \tag{75}$$

The eigenvalues Ω/ω_s calculated numerically as functions of Υ for the lowest few modes are shown in Fig.10. At $\Upsilon = 0$, the mode frequencies are simply multiples of ω_s. As Υ increases, the mode frequencies shift. As Υ reaches the critical value $\Upsilon_{\text{th}} \approx 1.45$, the $\ell = 1$ and $\ell = 2$ mode-frequency lines merge, and when $\Upsilon > \Upsilon_{\text{th}}$, they become imaginary and the beam is unstable. The parameter Υ_{th} thus defines the instability threshold of the beam.

Once a longitudinal mode-coupling instability occurs, one of its consequences is that the bunch lengthens. It lengthens just enough so that Υ stays at the instability threshold, i.e.,

$$\hat{z} = \frac{c}{T_0}\left(\frac{Nr_0\eta R_0}{\gamma\omega_s^2\Upsilon_{\text{th}}}\right)^{2/3} \tag{76}$$

The behavior of bunch length as a function of beam intensity looks like Fig.11. The curve above the bunch lengthening threshold has $\hat{z} \propto N^{2/3}$. Below the threshold, we have shown a slight potential-well bunch shortening. The change of bunch distribution due to potential-well distortion and that due to mode coupling instability are distinctly different. In the former case, the energy distribution of the beam (we consider an electron beam) is unaffected. In the latter case, the synchrotron oscillation translates the changes in $\hat{z}$ rapidly into proportional changes in $\hat{\delta}$.

Turbulence due to radial mode coupling

The above analysis ignores the important potential-well distortion effect. When included, instabilities can result from not only the coupling among azimuthal modes, but also from coupling among radial modes. This leads to a qualitatively different instability behavior. In particular, it might lower the instability threshold, and the mode frequencies do not have to shift as much as $\sim \omega_s$ to reach the threshold. The growth rate, however, tends to be slower in this mechanism because it is quadratic in the beam intensity.

250

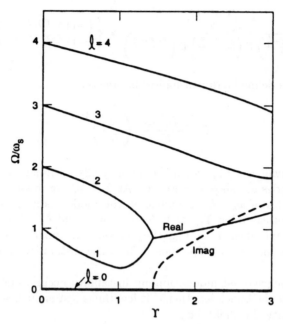

Figure 10. Longitudinal mode frequencies Ω/ω_s versus Υ for a water-bag beam with the diffraction model impedance. The solid curves give the real part of the mode frequencies; the dashed curve gives the imaginary part (magnitude only) of the $\ell = 1$ and $\ell = 2$ mode frequencies above threshold. There is always a static mode with $\Omega = 0$. The spectra for $\ell < 0$ are mirror images with respect to the $\Omega = 0$ line. Potential well distortion has been ignored.

13   Longitudinal Head-Tail Instability

We discussed a head-tail instability and a strong head-tail instability. Both are transverse effects. There is also a head-tail instability in the longitudinal dimension, first observed at CERN-SPS. The "longitudinal head-tail phase" comes from a dependence of the slippage factor η on δ.

Let $\eta = \eta_0 + \eta_1\delta$. Assuming a water-bag model, the longitidinal head-tail instability growth rate for the ℓth mode is found to be, to first order in η_1,

$$\frac{1}{\tau^{(\ell)}} = \frac{4\ell^2\eta_1 N e^2 c}{3\pi\eta_0 EC} \int_{-\infty}^{\infty} d\omega \, \frac{\mathrm{Re}Z_0^{\|}(\omega)}{\sigma^2} \left[\sigma J_\ell(\sigma)J_{\ell+1}(\sigma) + (1-\ell)J_\ell^2(\sigma)\right] \quad (77)$$

where $\sigma = \omega\hat{z}/2c$ and $\ell = 1, 2, 3\ldots$.

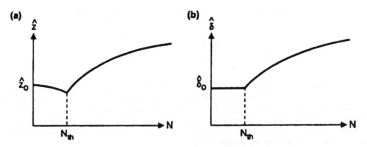

Figure 11. Bunch length $\hat{z}$ and energy spread $\hat{\delta}$ as functions of beam intensity N. Below threshold $N_{\rm th}$, $\hat{z}$ changes due to potential-well distortion, while $\hat{\delta}$ stays constant. Above $N_{\rm th}$, both $\hat{z}$ and $\hat{\delta}$ increase with N. If the impedance is given by the diffractive model, $\hat{z}$ and $\hat{\delta}$ are proportional to $N^{2/3}$ in the region $N > N_{\rm th}$.

Physically this instability occurs because of the following. Consider a bunch executing a longitudinal dipole oscillation ($\ell = 1$). Due to $\eta_1 \neq 0$, the bunch length $2\hat{z}$ is a little longer when the bunch has $\delta < 0$ and a little shorter when the bunch has $\delta > 0$ (assuming $\eta_1 \eta_0 < 0$). The bunch loses more energy when it is shorter. This means then that the beam loses more energy when it has $\delta > 0$ and loses less energy when $\delta < 0$. This leads to damping when $\eta_1 \eta_0 < 0$. Instability occurs when $\eta_1 \eta_0 > 0$.

This phenomenon may be important for the *isochronous rings* (considered for synchrotron radiation sources or muon colliders) when $\eta_0 \approx 0$.

14 Further Readings

These notes give a collection of formulas which might come handy in actual calculations. A good fraction of the notes can be found in the *Handbook of Accelerator Physics and Engineering*, ed. A. Chao and M. Tigner, World Scientific (1999), particularly from articles by W. Ng, T. Suzuki, B. Zotter, K. Thompson, K. Yokoya, T. Weiland, R. Gluckstern, S. Kurrenoy, P. Wilson, and A. Piwinski. For discussions emphasizing the physics principles, one may consult *Physics of Collective Beam Instabilities in High Energy Accelerators*, A. Chao, Wiley (1993). For a much more extensive discussion on impedances, one may consult *Impedances and Wakes in High-Energy Particle Accelerators*, B. Zotter and S. Kheifets, World Scientific (1997).

FUNDAMENTALS OF SUPERCONDUCTIVITY

MAURY TIGNER
Cornell University, Ithaca, NY 14853

Those fundamental physical aspects of superconductivity needed for engineering of accelerator magnets and radiofrequency cavities are presented. Beginning with the classical free electron theory of metals, the modifications of this theory required by quantum mechanics are discussed. The general electrical and magnetic properties of superconductors follow. Chapters dealing with specific features needed for accelerator magnet and rf cavity design complete the work. General references where much more detail may be found are given.

1 Free electron theory of metals

Figure 1 presents a schematic concept that has been used to develop a quantitative description of metals.

The classical theory which treats the current carrying valence (conduction) electrons in the metal like a gas of non interacting particles which obey Maxwell-Boltzmann, MB, statistics as in the kinetic theory of gasses. See Figure 2. They are pictured to move freely through the regular lattice of undisplaced ion cores. The principal scattering centers are the lattice vibrations which displace the ion cores and the interstitial impurity atoms. This theory successfully accounts for electrical and thermal conduction but quantum mechanics, QM, is needed to deal properly with specific heat and susceptibility. In the QM modified version, the conduction electrons are referred to as a free electron Fermi gas in which the conduction electrons are subject to the Pauli exclusion principle.

1.1 Electrical Conductivity in the classical free electron model

In between scatterings, which randomize the direction of motion, the conduction electrons acquire a velocity change Δv if subject to an electric field. In all practical cases, Δv is small compared to the thermal velocity of the electrons. The current density is then

$$\vec{j} = ne\Delta\vec{v}$$

where n is the conduction electron density and e the electronic charge. If τ is the mean time between collisions, E is the applied electric field and m is the electron mass, then

$$\Delta\vec{v} = e\vec{E}\tau / m$$

from Newton's law of non-relativistic motion. After each scattering the directed velocity gain is lost due to random thermal motion so that we can write

$$\vec{j} = \frac{ne^2\tau}{m}\vec{E} \quad \text{or} \quad \vec{j} = \sigma\vec{E} \quad \text{(Ohm' s Law)}$$

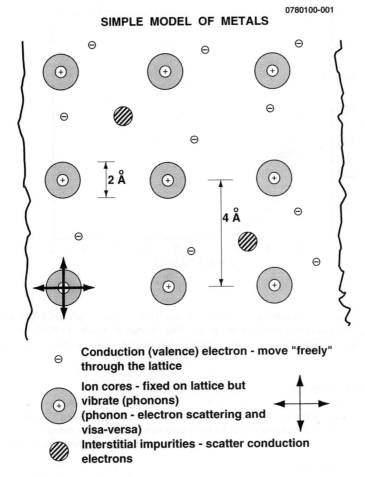

SIMPLE MODEL OF METALS

⊖   Conduction (valence) electron - move "freely" through the lattice

⊕   Ion cores - fixed on lattice but vibrate (phonons) (phonon - electron scattering and visa-versa)

⊘   Interstitial impurities - scatter conduction electrons

Figure 1. Schematic concept of a metal

where σ is the dc electrical conductivity $ne^2\tau/m$. By measuring σ at room temperature, knowing that $n \sim 10^{22}$ cm^{-3}, we can find $\tau \sim 10^{-14}$s. From the equipartition theorem we can find the typical electron thermal velocity

$$\frac{1}{2}mv^2 = \frac{3}{2}k_B T$$

from which $v \sim 10^7$ cm/s so that the mean free path, m.f.p., $\ell = vt = 10^{-7}$ cm or about 5 lattice spacings at room temperature. As $T \to 0$ the conductivity increases because the lattice vibrations, which are the primary scattering centers at high temperature, vanish at vanishing temperature leaving only impurity scattering. The ratio of the (normal state)

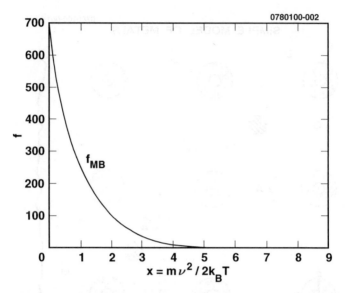

Figure 2. The Maxwell-Boltzmannn, MB, distribution

conductivity at low temperature to that at room temperature is called RRR (residual resistivity ratio) is thus a measure of purity. The theoretical maximum RRR for Nb, is 35,000. The best industrial material has RRR > 500. Figure 3 shows σ vs. T for various RRR.

1.2 Thermal Conductivity in the classical free electron model

Figure 4 shows a schematic model that can be used to compute the thermal conductivity. The arrows indicate the velocities of electrons contributing to the heat current driven by a thermal gradient. From this picture we can derive the differential equation of heat current.

In thermodynamic terms

$$j_q = -\kappa \frac{dT}{dx}$$

In microscopic terms the heat current density is the net number density, N, of electrons flowing across an area multiplied by their velocity, v_x and the energy, Q, carried by each electron

$$j_q = Nv_xQ$$

In terms of the model (Figure 4) we can rewrite the quantities (if τ is the same as in electrical conduction) as

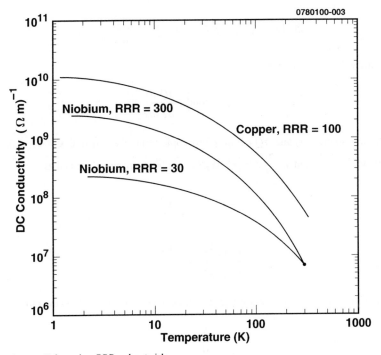

Figure 3. σ vs. T for various RRR and materials

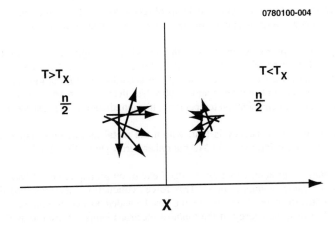

Figure 4. Schematic showing velocities of electrons contributing to heat current a position x

$$j_q = \frac{n}{2}v_x Q\big[T(x - v_x \tau)\big] - \frac{n}{2}v_x Q\big[T(x + v_x \tau)\big]$$

For the net energy transport per carrier we can write

$$Q\left[T\left(x - v_x\tau\right)\right] - Q\left[T\left(x + v_x\tau\right)\right] = -\frac{dQ}{dT}\frac{dT}{dx}2v_x\tau$$

combining these two we have

$$j_q = -nv_x^2\tau\frac{dQ}{dT}\frac{dT}{dx}$$

If we now express v_x and dQ/dT in microscopic terms we can find κ. From kinetic theory $v_x^2 = v^2/3$ and $mv^2/2 = (3/2)k_BT$. The heat capacity is defined as

$$c_v \equiv n\frac{dQ}{dT} = \frac{3}{2}nk_B \quad \text{thus}$$

$$\frac{\kappa}{\sigma}\frac{mv^2c_v}{3ne^2} = \frac{3}{2}\left(\frac{k_B}{e}\right)^2$$

which is the famous Wiedemann-Franz law, discovered experimentally in the form

$$\frac{\kappa}{\sigma} = LT$$

where L is called the Lorentz number and was observed to have about the same value for most metals. This was a great triumph of the classical free electron theory of metals.

While this result is approximately correct, the underlying assumptions turned out not to be correct. QM must be taken into account. Above we assumed that each conduction electron contributes $3/2\ k_B$ to the specific heat in accordance with the kinetic theory of gases. The measured contribution is about 100 x less than that. In addition, the specific heat is measured to be dependent on temperature, in contradiction to the classical result.

The answer to this inconsistency is that, within the metal lattice, QM applies and the conduction electron distribution is not MB but rather Fermi-Dirac, FD. See Figure 5.

This circumstance is brought about by the Pauli exclusion principle, i.e. no two electrons can occupy the same quantum state. This forces conduction electrons into the FD distribution with all states filled up to the level needed to accommodate all of the conduction electrons, the energy of the topmost electrons being referred to as the Fermi energy, ε_F. Only electrons within k_BT of ε_F are free to absorb energy and thus do all the work of conducting heat and electric current. From this we can see that it is really ε_F and v_F that should be put into the classical expressions above. From the solution of the Schroedinger equation in the metal lattice environment one finds

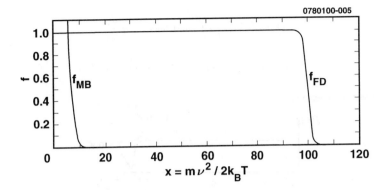

Figure 5. Fermi-Dirac, FD, distribution contrasted with the MB distribution

$$\varepsilon_F = \frac{\hbar^2 \left(3\pi^2 n\right)}{2m}$$

showing that ε_F is governed by the inter electron spacing as one would intuitively expect from its definition. The resulting characteristic electron velocity is

$$v_F = \frac{2\varepsilon_F}{m}, \text{ not the classical } \frac{3k_B T}{m}$$

[Note that $k_B T \sim 0.025$ eV at room temperature whereas $\varepsilon_F \sim 5.3$ eV for Nb] The major consequence of this fact is that electron velocities are much larger than predicted by kinetic theory, e.g. $v_F \sim 10^8$ cm/s or $v_F / v_{classical} = O(10)$ which in turn means that, since $\ell = v_F \tau$, the m.f.p., is much greater than in the classical picture.

To see that the Pauli exclusion principle and consequent FD statistics make quantitative and well as qualitative sense, find c_v. First observe that we can associate a temperature with the Fermi energy, i.e. $T_F = \varepsilon_F / k_B \sim 6 \cdot 10^4$ K for Nb. Because all electrons except those at the top of the Fermi sea are trapped in quantum states, one cannot add energy to them until the temperature had been raised to an amount comparable with T_F. Thus, for a small temperature change, ΔT, only the electrons within and energy $k_B \Delta T$ will be able to contribute to c_v and therefore c_v measured $<< c_v$ as calculated classically. At the top end of the FD distribution, i.e. at the top of the Fermi sea, at $T > 0$, the few electrons there have an MB distribution. The velocity of these electrons is nearly v_F and each one contributes $(3/2)k_B$ to c_v. Only a fraction, $k_B T / \varepsilon_F$ are in this MB tail. See Figure 6.

The actual number of excited electrons, $\#e^* = D(\varepsilon) \cdot k_B T$ where D is the density of states near ε_F and $k_B T$ is the width of the tail. The excitation energy for each is also about $k_B T$ so the energy density is

$$\tilde{Q} = D(\varepsilon) \cdot \left(k_B T\right)^2 \text{ and since } c_v = \frac{d\tilde{Q}}{dT}$$

$$c_v(electronic) = \text{numerical constant} \cdot T = \gamma T$$

as observed in experiment.

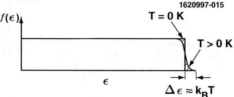

Figure 6. FD distribution at $T = 0$ and $T > 0$ showing electrons that do the work of heat conduction – those in the tail above the Fermi energy.

Now let us revisit the ratio of thermal to electrical conductivity

$$\frac{\kappa}{\sigma} = \frac{mv^2 c_v}{3ne^2}\left(\frac{k_B}{e}\right)^2 T \ ?$$

The QM calculation gives 1/100 of old value of c_v but the new v^2 is 100 x the old value so that these factors essentially cancel. If we do the QM arithmetic properly we get

$$\frac{\kappa}{\sigma T} = \frac{\pi^2}{3}\left(\frac{k_B}{e}\right)^2 = 2.44 \cdot 10^{-8} \ \text{W}\Omega\text{K}^{-2}$$

The previous agreement of the classical theory with experiment was fortuitous, but, it shows us that we can use the kinetic theory if we replace the rms velocity of that theory with the Fermi velocity, a very useful finding.

1.3 Review Thermal Conductivity in the light of QM (phonons)

We have implicitly been dealing with conditions at $T \rightarrow 0$. What about $T > 0$? At $T = 0$ the ions are at rest and do not scatter the electrons and only interstitial impurities contribute to the thermal conductivity. As the temperature rises, thermal excitations of the ion cores (phonons) begin to scatter the conduction electrons. In the metal lattice the thermal vibrations (phonons) are quantized. They can be treated as an assembly of harmonic oscillators subject to the Planck distribution law. See Fig. 7

$$\langle n \rangle = 1 \Big/ \left[\exp\left(\hbar\omega / k_B T\right) - 1\right]$$

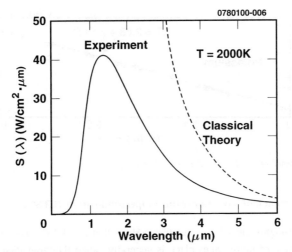

Figure 7. The Planck distribution function for phonons

Before proceeding, note that for the ion core lattice there is a characteristic temperature called the Debye temperature, Θ_D. By definition $k_B \Theta_D = \hbar \omega_D \propto 1/\lambda$ where $\lambda \sim$ lattice spacing and ω_D is the highest possible frequency of lattice vibration. Thus, Θ_D is analogous to T_F for the electrons. (Θ_D (Nb) ~ 0.02 eV

If we look at the low temperature ($T < \Theta_D$) dependence of this, we find that the number density of phonons $\propto T^3 \rightarrow$ m.f.p. $\propto T^{-3}$. Remembering that $\kappa = v^2 \tau c_v / 3$ from the free electron model and that in this model $c_v \propto T$, the phonon part of $\kappa = \kappa_{phonon} \propto T \cdot T^{-3}$ thus

$$\frac{1}{\kappa_{total}} = \frac{1}{\kappa_{impurities}} + \frac{1}{\kappa_{phonon}} = \frac{B}{T} + AT^2$$

1.4 Review Heat Capacity in light of QM (phonons)

$$c_v = c_{v,electronic} + c_{v,phonons}$$

and the heat capacity due to phonons is proportional to their number so that, combining with the previous electronic c_v we have

$$c_v = \gamma T + AT^3 \quad \text{(See Figure 8)}$$

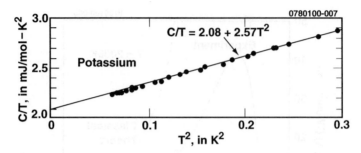

Figure 8. Experimental values of c_v/T vs. T^2 for Potassium

2 The phenomenon of low temperature superconductivity (LTSC)

In 1911, Kamerlingh-Onnes discovered that $\rho(\text{Hg}) \to 0$ at $T = 4.2$ K. The temperature at which this change of phase takes place is generally called the "transition or critical - temperature", T_c. About 20 years later the Meissner effect (magnetic field excluded from the volume of the superconductor for $T < T_c$) was discovered and about 40 years later the BCS theory at last explains LTSC. 62 elements exhibit LTSC as do 100's of compounds. With the recent discover of high temperature superconductivity (HTSC) there are 100's more compounds that exhibit superconductivity, SC. Very few of these elements and compounds possess technically useful parameters.

SC is characterized by a three dimensional "critical surface" as displayed in Figure 9. When the material has current density, J , magnetic field, H , and temperature, T , such that the point corresponding to theses three values lies *inside* the critical surface, the material is in the SC state.

In SC, electrical conduction is effected by a condensate of valence electrons into "Cooper pairs" which can flow through the lattice without scattering. There is a characteristic energy scale associated with the Cooper pairs called the gap energy, $\Delta_{gap} \sim O(\text{meV})$. Very roughly, the mechanism of pairing is this: an electron passing through the lattice attracts the positive ion cores. Their motion is delayed owing to their inertia. The ions moving together in the wake of the first electron "overscreen" the interstitial field so that there is an attraction for a second electron, binding the pair. This attraction is quite long range; $R \sim v_F \cdot (2\pi/\omega_D) \sim$ many 10's nm. Only a small fraction of the e's (i.e. those within $k_B T_c$ of ε_F) participate in this interaction. The quantized pairs (spin up + spin down) obey Bose-Einstein statistics (all can occupy the same energy state) while the remaining e's obey FD statistics. The pair energy $2\varepsilon < 2\varepsilon_F$ due to the attraction which distorts the over all energy level scheme producing a "gap" at the top of the Fermi sea. See Figure 10. To boost a pair out of the sea, i.e. break the pair, one must supply a quantum of energy $\sim O(2\Delta_{gap})$. This bound state pair can move through the lattice with no scattering and hence all electrical resistance is lost.

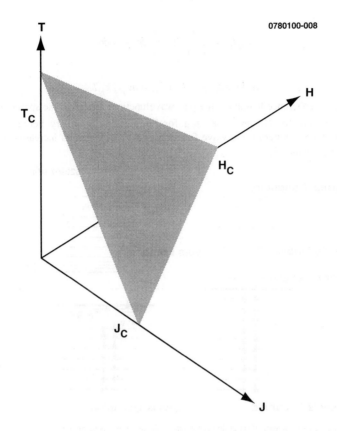

0780100-008

Figure 9 Critical surface for superconductivity. Between the origin and the surface the material is superconducting

At $T = 0$ all of the valence electrons have been paired so they cannot take up heat and thus the specific heat of a superconductor at $T = 0$ is zero. As the temperature rises towards T_c more and more unpaired e's become available to take up heat. See Figure 11

$$c_{e,\text{superconducting}} \propto \exp{-\Delta/k_B T}$$

Note that $\Delta = \Delta(T)$ for $T<T_c$. See Figure 12. For $T>T_c$ superconductivity is lost, the material goes "normal". The BCS theory shows a direct relation between T_c and Δ:

$$\Delta(0) = 1.76 k_B T_c$$

Electrons within $k_B T_c$ of the Fermi level can pair. This plus the Heisenberg uncertainty relation $\delta x \delta p \leq \hbar$ gives us an idea of the spatial extent of the Cooper pair:

262

$$k_B T_c = \delta\left(\frac{p^2}{2m}\right) = \frac{p}{m}\delta p = v_F \delta p$$

$$\delta x \delta p \equiv \xi_o \delta p \approx \hbar \text{ or } \xi_o \approx \hbar v_F / k_B T$$

ξ_o is called the coherence length (of the pair wavefunction) and gives the spatial extent of the pair ~ 39 nm for Nb and ~ 83 nm for Pb. At the boundary between normal conducting and SC material, the wave function of the Cooper pairs increases from 0 to full value in a distance ξ_o .

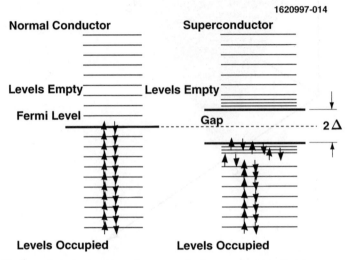

Figure 10. Change in the density of states that accompanies the onset of superconductivity

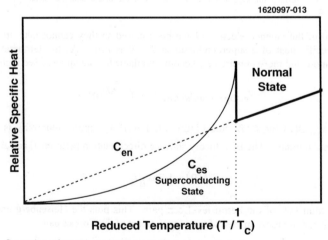

Figure 11. Comparison of normal state and SC state electronic specific heats

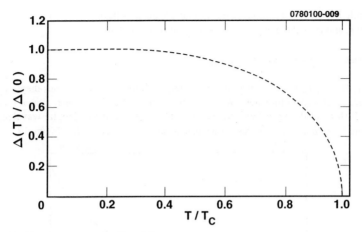

Figure 12. Δ the energy gap as a function of T

In summary we recall some important characteristic energies

√   free particles $\varepsilon \sim k_B T \sim 0.02$ eV at 300K – MB distribution
√   e's in the lattice $\varepsilon_F \sim 10$ eV – FD distribution
√   ions in the lattice $\varepsilon_D \sim 0.02$ eV – Planck distribution
√   SC gap $\Delta \sim 1$ meV – Bose-Einstein distribution

and lengths:

√   m.f.p. in normal metal at 300 K ~ 10 nm
√   lattice spacing in metals ~ 0.4 nm
√   radius of volume containing one conduction e ~ 0.2 nm
√   coherence length, $\xi_o \sim 100$ nm

3   General aspects of superconductivity

3.1 Electrical properties, general

In discussing the electrical properties of SC's it is useful to adopt the "two fluid model" in which the valence e's are divided into tow categories, the SC electron fluid, e_{SC} , and the normal electron fluid, e_{normal} . The e_{SC} correspond to the Cooper pairs and the e_{normal} to the unpaired conduction electrons. Below T_c

$$\text{number of } e_{normal} \propto \exp-\Delta/k_B T$$

Thus, at $T = 0$ all conduction electrons are paired, i.e. number of $e_{normal} = 0$. Note that T_c depends on the density of states and the e – ion coupling strength. Strong coupling goes with high T_c which means that good normal conductors have low T_c, e.g. $T_{c,Al} \sim 1$ K, $T_{c,Nb} \sim 9$ K and $\rho_{Al} \sim 2.6$ and $\rho_{Nb} \sim 14$, both in units of 10^{-8} Ωm.

3.2 Electrical properties, DC Resistance

Below T_c Cooper pairs (CP) carry all DC current. As the pairs are Bosons they are all in the same, macro-quantum state. When $j \neq 0$, all the CP drift together. Since there are many CP they need only a low drift velocity to carry j so the deBroglie wavelength is very long, >> impurity size so there is no scattering and $\rho = 1/\sigma = 0$. See Figure 13

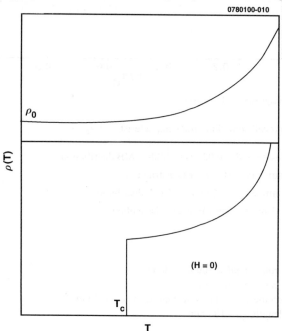

0780100-010

Figure 13. Low T resistivity of normal and superconductors. ρ_o due to impurities

$$j = -n2e\Delta v$$

and the current density can increase until the needed Δv in the applied field leads to an energy increase $> \Delta$ thus beginning to break the CP. The particular j to which this corresponds is called j_c, the "critical current density". Thus

$$j_c = \frac{2en\Delta}{mv_F} \text{ (ideal)}$$

The corresponding H_c will be discussed later.

3.3 RF Resistance

The Cooper pairs can carry current without resistance. However, they do have mass and therefore inertia so that there must be an internal E_{RF} to make $j_{internal}$ change direction for RF currents. The E_{RF} acts on the e_{normal} as well as e_{SC}. The e_{normal} scatter and exhibit ohmic resistance which lead to losses and heat deposition at the surface of the superconductor.

$$E_{RF} \propto \frac{dH}{dt} \propto \omega H \text{ for sinusoidal excitation i.e. } E = E_O \exp(i\omega t)$$

from Ohm's law $\vec{j}_{normal,int} \propto n_{normal}\vec{E}_{RF,int} \propto n_{normal}\omega\vec{H}$ and also

$$P_{dissipation} \propto j_{int} E_{int} \propto n_{normal}\omega^2 H^2$$

This relation is usually parametrized as: $\quad P_{diss} = \frac{1}{2}R_s H^2 \;$ so that

$$R_s = const. \cdot A_s n_{normal}\omega^2 = A_s\omega^2 \exp\left[\frac{-\Delta(0)T_c}{k_B T_c T}\right] \propto \omega^2 \exp\frac{-\Delta}{k_B T}$$

where A_s depends on the material parameters, e.g. $v_F, \lambda_L, \xi_0, mfp$. See Fig. 14.

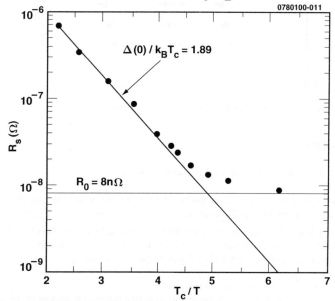

Figure 14 Measured R_s of Nb at 1.5 GHz vs the "reduced temperature"

3.4 London Penetration Depth λ_L

In the SC state it is observed that the external fields penetrate only a small distance into the material, i.e. $H = H_0 \exp\left(-x/\lambda_L\right)$. This is reminiscent of the "skin effect" in normal conductors (see Fig. 15) The derivation of this effect can be found in the references given in the bibliography which show that

$$\lambda_L^2 = \frac{m}{n_s e^2 \mu_0}$$

independent of frequency! This is a consequence of the fact that the mean free path of the superconducting electrons is ∞. For SC elements $\lambda_L \sim O(10' s \text{ nm})$ and for compounds $\lambda_L \sim O(100' s \text{ nm})$. Near T_c

$$\lambda_L(T) = \lambda_L(0)\left[1 - \left(\frac{T}{T_c}\right)^4\right]^{-1/2}$$

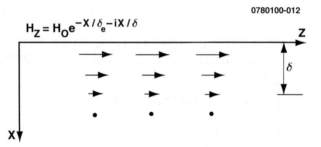

Figure 15 Fall off of internal fields with depth at the surface of a conductor

4 Thermal Properties

4.1 Thermal Conductivity in the Superconducting State

The behavior is quite different from the normal state as can be seen in Fig. 16. The factors that come into play as $T \to 0$ are the exponential decrease in n_{normal}, (Cooper pairs cannot carry heat), the decrease in the number of phonons as T^3, the density of defects, and the grain size. For Nb, the observed thermal conductivities range from 0.1 to 1.0 W/mK at $T \sim 2K$ depending on the material purity and state of cold work.

4.2 Specific Heat in the Superconducting State
Displayed in Figs. 10 and 11, BCS theory predicts that the rearrangement of the levels at the top of the Fermi Sea gives a sharp increase in C_s at T_c. As $T \to 0$, the normal electrons

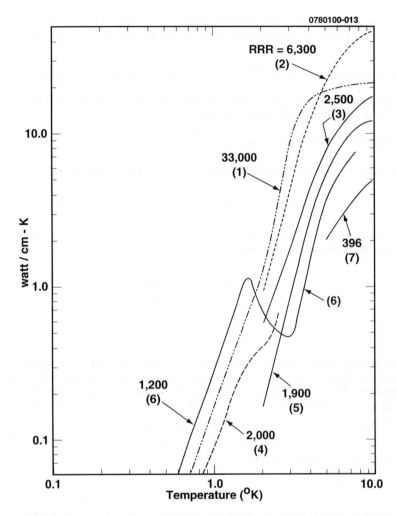

Figure 16 Thermal conductivity vs T for various RRR and grain sizes

freeze out exponentially and the role of phonons can become important even though their density is decreasing as T^3, producing the "phonon peak" sometimes spoken of at low temperature.

5 Magnetic Properties

5.1 The Meissner Effect

As striking and even more elusive than the zero resistance of the SC state is the Meissner effect (1933): If a SC is cooled through T_c in the presence of a (weak) magnetic field, the magnetic flux is suddenly (and reversibly) excluded from the volume of the SC. This is not what a "perfect, i.e. $\rho=0$," normal conductor would do. See Figs. 17a and 17b.

Inside the SC both $B = 0$ and $\dfrac{\partial B}{\partial t} = 0$. Due to impurities there is sometimes flux trapping inside the SC leading to only partial expulsion. This is particularly important for SC cavity operation as will be discussed below.

5.2 Type I and Type II Superconductors

Previously we mentioned two characteristic lengths, λ_L and ξ_o. Their ratio

$$\kappa_{GL} \equiv \lambda_L / \xi_o$$

is called the Ginzburg-Landau parameter and give the boundary between two distinct types of SC behavior: (See Fig. 18)

$$\kappa_{GL} < 1/\sqrt{2} \rightarrow \text{Type I}$$

$$\kappa_{GL} > 1/\sqrt{2} \rightarrow \text{Type II}$$

Elements such as Pb, Sn are *Type I*. Nb, Nb$_3$Sn, NbTi are *Type II*. Below H_c Type I materials expel flux. Above this value of the applied field, Type I materials revert to the normal state. Below H_{c1} Type II materials are in the Meissner state and exhibit flux expulsion as in Type I materials. For $H_{c1} < H < H_{c2}$ Type II materials are in the "mixed" or intermediate state in which flux enters the material. (See Fig. 19) The flux enters the material in an ordered state as shown in Fig. 20. For $H > H_{c2}$ Type II materials revert to the normal state.

5.3 Energy Considerations

The normal to superconducting transition is a phase transition (PT). A material in a certain phase will make a PT to another phase in the free energy of that phase is < that of the initial phase, e.g. water ($T>0°$) $\rightarrow$ ice for $T<0°$. The superconducting state being relatively ordered with respect to the normal state we have

$$F_n > F_s \text{ for } T<T_c \text{ and } H_{ext}=0$$

Consider a SC at $T<T_c$, $H_{ext} \neq 0$

$$F_s(H) = F_s(0) + \mu_o V_s H^2 / 2$$

0780100-014

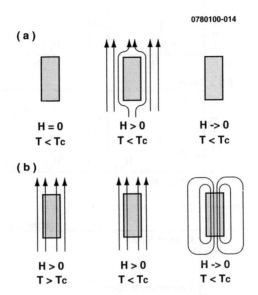

(a)

H = 0	H > 0	H -> 0
T < Tc	T < Tc	T < Tc

(b)

H > 0	H > 0	H -> 0
T > Tc	T < Tc	T < Tc

Figure 17A a, b (a) Screening of external field by a perfect conductor, (b) flux trapping in a perfect conductor

0780100-015

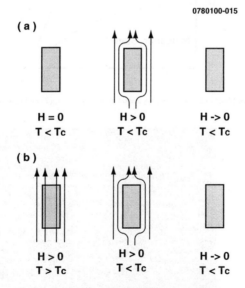

(a)

H = 0	H > 0	H -> 0
T < Tc	T < Tc	T < Tc

(b)

H > 0	H > 0	H -> 0
T > Tc	T < Tc	T < Tc

Figure 17B a, b (a) Screening of an external field by a SC, (b) flux expulsion (Meissner effect) from a SC

270

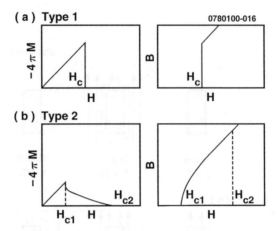

Figure 18 (a) Magnetization of Type I vx H_{ext}, (b) Type II

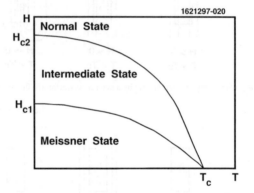

Figure 19 The three phases of Type II SC separated by the curves $H_{c1}(T)$ and $H_{c2}(T)$ which meet at T_c

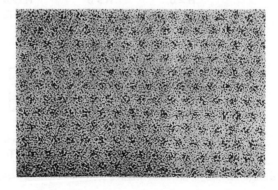

Figure 20 Triangular array of fluxoids in a Type II SC in mixed state.

the second term being the work done on the SC to expel the field from V_s. This implies a value of $H (\equiv H_c)$ for which

$$F_n = F_s(H_c) = F_s(0) + \mu_0 V_s H_c^2 / 2$$

and the phases are in thermodynamic equilibrium so that H_c (or B_c) is termed the "thermodynamic critical field". (See Fig. 21)

0780100-017

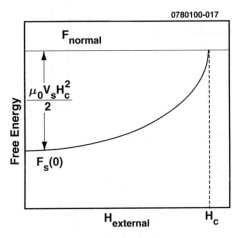

Figure 21 The superconducting state has a lower free energy that the normal state. In the presence of an external field the free energy rises because of the work done to expel the field from the bulk of the SC

BCS theory is able to show that

$$\frac{\mu_0 H_c^2 (T=0)}{2} = \frac{3\gamma T_c^2}{4\pi^2} \left(\frac{\Delta(0)}{k_B T_c} \right) = 0.23 \gamma T_c^2$$

so that high T_c implies high H_c. Since the SC gap, Δ is a function of temperature

$$H_c = H_c(0) \left[1 - \left(\frac{T}{T_c} \right)^2 \right]$$

Consider two slabs of SC material, one with $d > \lambda_L$ and one with $d < \lambda_L$. (See Fig. 22)

In the latter the internal flux does not go to zero so that less work is done to exclude the flux and thus the free energy is lower than for the thicker slab. Question: Is it energetically favorable for flux to enter the SC in the form of thin normal conducting domains? (See Fig. 23) Answer: Sometimes, depending on the relationship of ξ_0 and λ_L. One must examine which situation is more energetically favorable.

272

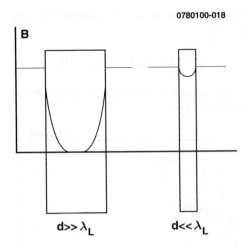

Figure 22 Attenuation of the flux inside both thick and thin SC slabs

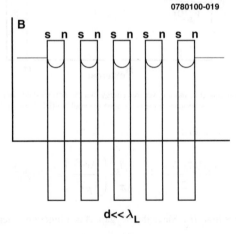

Figure 23 In the presence of an external field it is energetically favorable for the flux to enter in the form of thin normal conducting domains – except for considerations of surface energy associated with the SC / normal boundaries.

It takes some surface energy to form an SC / normal phase boundary:

i) the density of SC electrons is suppressed over the distance ξ_o so the free energy per unit area rises by

$$\mu_o H_c^2 \xi_o / 2$$

ii) in the presence of an external field the field penetrates by λ_L so the free energy per unit area decreases by

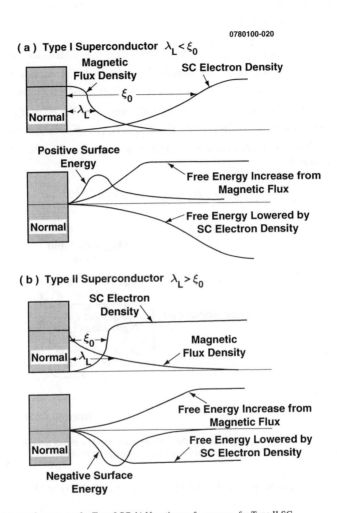

Figure 24 a) Positive surface energy for Type I SC, b) Negative surface energy for Type II SC

$$\mu_o H_{ext}^2 \lambda_L / 2$$

thus the net boundary energy per unit area is

$$\frac{\mu_o}{2} \left(\xi_o H_c^2 - \lambda_L H_{ext}^2 \right)$$

In general SC properties are such that $\lambda_L \neq \xi_O$ so the two contributions do not cancel. (See Figs 24 a and b)

In Type I with $\kappa_{GL} < 1/\sqrt{2}$ the boundary energy is + so Meissner state is favored and flux is excluded up to $H_{ext} \geq H_c$ then the flux enters and SC goes normal. In contrast, Type II with $\kappa_{GL} > 1/\sqrt{2}$ the boundary energy is negative and flux begins to penetrate at $H_{ext} = H_{c1}$. When $H_{ext} \rightarrow H_{c2}$ we have full penetration. For $H_{ext} > H_{c2}$ the material goes normal. (See Figs 25, 26 and 19). For $H_{c1} < H_{ext} < H_{c2}$ the flux is quantized into "fluxoids" and arranged in a regular pattern as seen in Fig. 20. The value of the flux quantum is

$$\Phi_O = h/2e = 2.07 \cdot 10^{-15} \text{ V} \cdot \text{s}$$

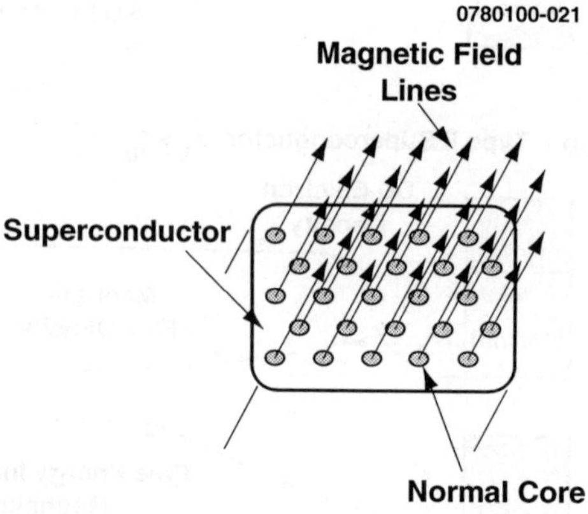

0780100-021

Magnetic Field Lines

Superconductor

Normal Core

Figure 25 For $H_{ext} > H_{c1}$ (type II SC) the fluxoids (vortices) enter in a regular array of single quanta. The cores are normal but space between – several thousand nm – remains superconducting

From the Ginzburg – Landau (GL) phenomenological theory of SC

$$H_{c2} = \frac{\Phi_O}{2\pi\mu_O\xi_O} = \sqrt{2}\kappa_{GL}H_c \text{ for } \kappa_{GL} \gg 1$$

$$H_{c1} \approx \frac{1}{2\kappa_{GL}}\left(\ln\kappa_{GL} + 0.08\right)H_c$$

Below are the properties of some useful materials

Material	Type	$T_c[K]$	$\mu_o H_c[T]$	$\mu_o H_{c1}[T]$	$\mu_o H_{c2}[T]$	$\lambda_L[nm]$	$\xi_o[nm]$
Pb	I	7	0.080	-	-	40	83
Nb	II	9.3	0.37	0.25	0.41	30	40
In	I	3.4	0.028	-	-	-	-
NbTi*	II	~9.3	0.16	0.00	11.0	500	10
Nb$_3$Sn	II	18	0.46	0.034	19-25	200	6

*~ 50% ea

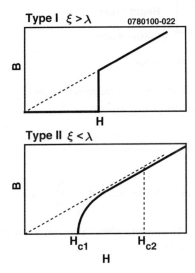

Figure 26 Magnetic behaviors of Type I and Type II SC

6   Superconductivity for RF Cavities

6.1 Cavity Basics

In many accelerators electric power from the mains is converted to kinetic energy of a charged particle beam via an RF cavity. In this work we will consider cavities designed to accelerate relativistic beams only. Figure 27 displays the schematic diagram of such a cavity. The field components are E_s and H_ϕ, the maximum of E_s being on the axis and that of H_ϕ being on the wall at $r \sim 2/3\ a$.

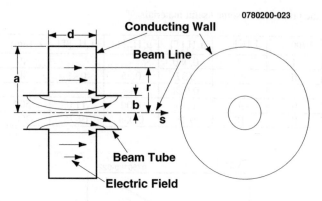

0780200-023

Figure 27 Schematic of cavity accelerating a beam

$E_{axis} = E_0 \exp(i\omega_0 t)$, ω_0 is the resonant frequency of the cavity. In passing through the cavity singly charged particles gain an energy

$$\Delta V = \int_{\text{along beam path}} \vec{E}(s,t) \cdot d\vec{s} \; ; \quad V_c \equiv \Delta V_{\max}$$

Usually the length of the cavity is chosen such that $\quad \dfrac{\lambda_{rf}}{4} < d < \dfrac{\lambda_{rf}}{2}$

For $b/a \ll 1$, $\lambda_{res} \approx 2.61a$. To support a given value of V_c a certain power must be dissipated in the walls of the cavity

$$P_w = \frac{V_c^2}{2R}; \; R \text{ is the "shunt impedance" of the cavity.}$$

R is usually expressed by the combination $\left(\dfrac{R}{Q}\right)Q$ since $\dfrac{R}{Q}$ is a function of cavity geometry only and Q is a function of geometry and R_{surf}, i.e. $Q \equiv \dfrac{\omega U}{P_w} \propto \dfrac{1}{R_S}$ where U is the stored energy in the cavity. $\dfrac{R}{Q} \sim O(10^2)$ Ω for one cavity independent of frequency.

At 1 GHz $Q_{\text{normal copper}} \sim O(10^4)$; $Q_{\text{SC, Niobium}} \sim O(10^{10})$.

6.2 Surface Impedance, the ideal case

In normal conductors at room temperature (RT)

$$Z_{\text{surf}} = \frac{1+i}{\sigma\delta} = R_s\left(1+i\right)$$

the real and imaginary parts having equal magnitudes. At 1 GHz, RT, $R_{s,\text{copper}} \sim 10^{-2}\,\Omega$. This is not the case with a superconductor

$$Z_{surf,\text{SC}} \approx R_{s,\text{SC}} + i\omega\lambda_L$$

$$\text{and } \text{Im}\left(Z_{\text{surf, SC}}\right) \gg \text{Re}\left(Z_{\text{surf, SC}}\right)$$

Figure 28 shown the BCS surface resistance for three materials.

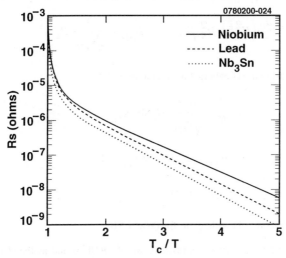

Figure 28 BCS calculated R_s for three materials vs reduced temperature

The BCS results can be approximated by

$$R_{\text{surf, SC}} \approx \frac{2\cdot 10^{-4}}{T}\left(\frac{f[GHz]}{1.5}\right)^2 \exp\left(-\frac{17.67}{T}\right)\Omega$$

with the conditions that $T/T_c < 1/2; f \ll 2\Delta/h \sim 10^{12}$ Hz. At 1.5 GHz, 2K, $R_{\text{surf, Nb}} \sim 10^{-8}$ ohm. Note that, as expected,

$$R_{\text{surf,normal}}/R_{\text{surf,SC}} = Q_{\text{SC}}/Q_{\text{normal}} = O\left(10^6\right)$$

6.3 Limiting Field of SC Cavity (ideal case)

An important figure of merit for SC cavities is the maximum possible value of $V_c \propto E_{max}$. There is no clearly defined limit on E per se but there is a clearly defined limit on H_{surf}. Because $H_{surf} = H_{surf}(E)$ by Maxwell's equations E is limited even in the ideal case. What is $H_{surf, max}$?

Previously we dealt with equilibrium conditions for phase transitions from one SC state to another or to the normal state. It takes time to make such transitions. In particular, in going from the Meissner state to one in which flux penetrates the material, one must "nucleate" fluxoid formation. Thus there may be some "superheated" state which can be maintained briefly – say about one quarter rf period – without nucleation which leads to the concept of a critical field for superheating , $H_{c, SH}$. In Type I SC there is a positive surface energy. When it goes to zero, nucleation can begin, i.e.

$$\frac{\mu_0}{2}\left(H_c^2\xi - H_{SH}^2\lambda_L\right); \;\Rightarrow H_{SH} = \frac{1}{\sqrt{\kappa_{GL}}}H_c$$

or, more generally, using the Ginzburg-Landau theory (1950):

$$H_{SH} \approx \frac{0.89}{\sqrt{\kappa_{GL}}}H_c \text{ for } \kappa_{GL} \ll 1$$

$$H_{SH} \approx 1.2H_c \text{ for } \kappa_{GL} \sim 1$$

$$H_{SH} \approx 0.75H_c \text{ for } \kappa_{GL} \gg 1$$

Figure 29 displays this on a phase diagram for Pb and Nb.

Since the fluxoid nucleation time is known to be of $O(10^{-6})$s and as the rf periods in practical use are of $O(10^{-9})$s we might expect that $H_{c, rf} = H_{SH}$. Figure 30 tells the story. For a well designed cavity $H_{c, rf} / E_{acc} \approx 40 Oe / MV / m$ or $E_{lim} \sim 57$ MV/m, Nb.

Above we have reviewed the ideal properties of rf SC. In reality there are many practical constraints which bring limits of their own.

279

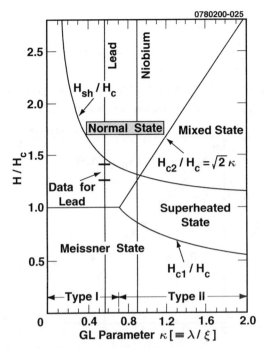

Figure 29 Phase diagram for Meissner, normal, mixed and superheated states

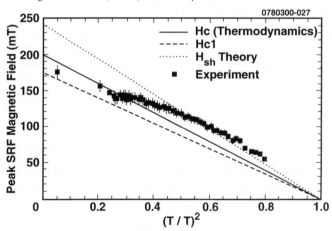

Figure 30 Measured rf critical field for Nb vs reduced temperature

6.4 Additional Constraints on Q and Field limits

Material properties not yet mentioned as well as vacuum-electronic phenomena play a role in limiting the "ideal" cavity performance previously discussed. In the following we emphasize the characteristics of niobium.

280

6.4.1   Residual resistance, R_o

$$R_{surf} = \frac{A}{T} f^2 \exp\left[-\Delta(T)/k_B T\right] + R_O \text{ for } T<T_c/2 \text{ (see Figure 31)}$$

where A is influenced by $\xi, \lambda_L v_F$, mfp, Δ, that is by fundamental properties. A typical low temperature value for Nb is 10 nΩ and the best seen is about 1 nΩ.

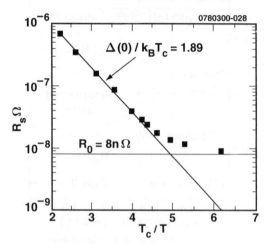

Figure 31 Measured values for R_s for Nb at 1.5 GHz

R_o is influenced by extraneous conditions such as poor joints, chemical residues on the surfaces, foreign inclusions ans trapped magnetic flux as well as material conditions of purity, grain size, surface oxides hydrides near the surface, etc. Great care in preparation of the metal and of the cavity is essential. Very clean conditions are required. Note that R_o can also be H_{surf} dependent.

6.4.1.1   Trapped magnetic flux

Cavities are operated in the Meissner state only but lattice imperfections and/or surface oxides can serve as pinning centers for fluxoids (see section 7). If a cavity is cooled below T_c in a small magnetic field (e.g. the Earth's field) some of the flux is not expelled when the material enters the Meissner state (see figure 32). Trapped H_{ext} over area A breaks up into N fluxoids each carrying flux Φ_o

$$AH_{ext} = N\Phi_O$$

The core of each fluxoid is normal conducting and of radius ξ_o so that the magnetic contribution to the surface resistance of the cavity material is

$$R_{\text{mag}} = N\frac{\pi\xi_o^2}{A}R_n = \frac{H_{\text{ext}}\pi\xi_o^2\mu_o R_n}{\Phi_o}$$

where R_n is the normal state resistance of the material and

$$H_{c2} = \frac{\Phi_o}{2\pi\mu_o\xi_o^2}\text{ and thus}$$

1621297-021

Normal Core   Magnetic Field Lines

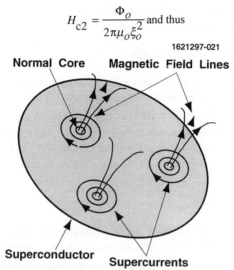

Superconductor   Supercurrents

Figure 32 Field lines and supercurrents for magnetic flux trapped in Type II SC

$$R_{\text{mag}} = \frac{H_{\text{ext}}}{2H_{c2}}R_n$$

When H_{c2} = 2400 Oe, RRR=300, $R_n \sim 1\text{m}\Omega$ at 1 GHz (all for Nb) then

$$R_{\text{mag}}[\text{n}\Omega] = 0.3H_{\text{ext}}[\text{mOe}]\sqrt{f[\text{GHz}]}$$

where the square root comes from the normal state skin effect. This formula leads to the conclusion that the Q of a cavity cooled in the Earth's field ($\sim$ 0.5 Gauss) will be $< 10^9$. Therefore SC cavities must be shielded against external fields to reach their best performance.

6.4.1.2   Residual loss due to hydrides

Surface treatments involving acids may leave significant hydrogen in the niobium lattice near the surface. At concentrations of 2 ppm by weight, hydrides can lower the Q of an otherwise very good cavity to $\sim 10^8$ as the concentration needed to form precipitates decreases with temperature one may avoid the problem by fast cooling. Best, of course, is to avoid hydrogen formation or to remove it by vacuum heating.

6.4.1.3 Residual loss in surface oxide

A natural oxide – mostly Nb_2O_5 – 10 nm thick is universally present. Well formed oxide has been shown to have $R_o < 1$ nΩ so normally it's not a problem. NbO, however, is quite lossy. Fortunately its occurrence is rare.

6.4.1.4 Residual losses at grain boundaries

Grain boundaries are lattice discontinuities and sometimes, effectively, gaps of dimension $O(\xi_o)$. This is particularly so in cavities formed by sputtering Nb onto copper in very fine grains (e.g. 1 μm long and 0.1μm in diameter.) In this case the superconducting currents must be carried by tunneling across the many boundaries. The resistance represented by the tunneling barrier $\propto H_{surf}$ to that R_{surf} rises and Q does down as the cavity field level rises. This same phenomenon occurs in the so called Hi Tc materials which have very fine grains.

6.4.2 Thermal Breakdown at Defects

One of the most important practical limits on achievable field levels is "thermal breakdown". Some "defect" on or near the surface of the cavity is normal conducting and heats up ohmically $\propto R_{surf, normal} H^2$ and beings the neighboring Nb up to $T > T_c$ in a runaway process (see Figure 33)

Many defects of sub mm size have been located and analyzed. Found have been bits of Cu, Ti, Fe, Nb weld spatter spheres S, Ca, K, Cl .... These can be inclusions in the base metal, products of cavity manufacture or residua of cleaning and processing after fabrication or introduction during mounting to the accelerator or test vacuum system. Extreme cleanliness during processing and assembly is required. Clean room techniques such as those used in the semiconductor industry are mandatory.

Using a simple model such as shown in Figure 33 one can estimate the maximum H_{surf} that can be achieved in the presence of such a defect.

$$\dot{Q}_T = \frac{1}{2} R_{surf, normal} H^2 \pi a^2 \quad (a = \text{defect radius} \ll \text{wall thickness})$$

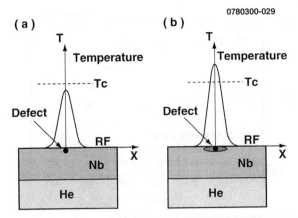

0780300-029

Figure 33 Thermal breakdown of Nb cavity (a) at low field the temperature is higher than ambient but $<T_c$; (b) as field increases T approaches T_c the defect becomes normal and power dissipation increases unstably.

at breakdown this $\dot{Q}_T$ will be transported through the wall of thermal conductivity, κ, across a ΔT of $T_c - T_{bath}$ to give a max allowed H of

$$H_{max} \approx \sqrt{\frac{4\kappa\left(T_c - T_{bath}\right)}{aR_{surf,\,norm}}}$$

Take as an example Nb in a 2 K bath with $R_{suef,\,norm}$ = 10 mΩ, a – 50 μm and RRR = 300 for an average κ of 75 W/m-K corresponds to $H_{max} \sim$ 820 Oe for a limiting E of 20 MV/m. More refined calculations give similar results.

6.4.3   Global heating

In some instances it has been possible to remove all defects of importance. The ultimate limit on H is then due to global heating caused by ohmic heating of the walls $\propto R_o H^2$. Figure 34 shows an example.

6.4.4   Vacuum electronic phenomena

There are two processes, multipacting and field emission, by which electrons originating inside the cavity can gain energy from the EM field and deposit it as heat in the cavity wall and lead to localized thermal breakdown.

6.4.4.1   Multipacting

In this process a free electron (always present at material surfaces) is accelerated by the cavity field and impacts the cavity wall at some point. If the coefficient of secondary

284

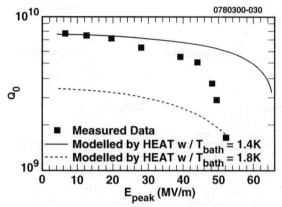

Figure 34 Calculated and measured Q decreases due to global heating.

emission, δ is > 1, more than one electron may be emitted when the first electron strikes a surface. If the electric field is reversing just at that time, the secondaries are accelerated back to the point of origin where they strike and multiply again, leading to resonant buildup of current striking the wall and depositing heat which raises the temperature and leads to thermal breakdown. For many years cavity fields were limited by multipacting.

Figure 35 displays the salient properties of secondary emission. The conditions for δ to exceed 1 are very broad making multipacting a very common phenomenon if special measures are not taken.

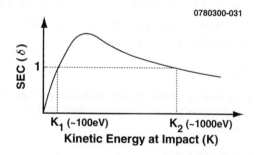

Figure 35 Generic dependence of secondary emission coefficient, δ, vs impact energy. Common material parameters can be found in the CRC Handbook of Chemistry and Physics and in general reference [2]

An elementary example of multipacting is shown in Figure 36. This form is called "two point" multipacting. Application of Newton's 2nd law with the Lorentz force yields

$$V_n = E_O d = \frac{d^2 \omega^2 m_e}{(2n-1)e}$$

where n is the number of rf cycles between electron impacts.

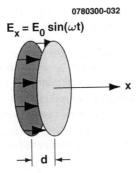

Figure 36 Idealized two point multipacting schematic

One point multipacting in which the electron returns close to its point of origin is also possible as seen in Figure 37.

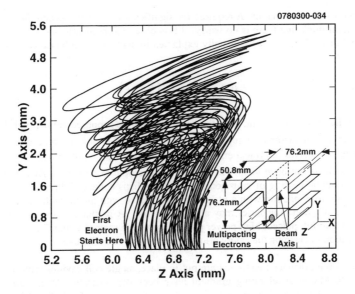

Figure 37 One point multipacting trajectories

In both cases H as well as E play a role in practice. Today we know how to eliminate multipacting in the interior of cavities by curving the cavity wall so that H suppresses the effect. Multipacting in couplers, wave guides and on windows having high values of δ is still a great danger and must be carefully designed against. See the references in the bibliography for details.

6.4.4.2 Field emission (FE)

This process presents a direct limit on surface electric fields independent of the local magnetic field. In this phenomenon, electrons are drawn out of the material by the EM field and accelerated to impact on some cavity surface. Both emission and impact deposit energy in the cavity wall and can lead to thermal breakdown.

FE is a quantum phenomenon. Conduction electrons are normally trapped in a potential well inside the metal. The depth of the well is material dependent and is termed the work function, ϕ which is of the order of 4 eV for common metals. Electrons can escape if thermal energy is added (thermionic emission) or by absorbing a photon (photo-emission). The application of $E \perp$ the surface acts to suppress the ϕ so that even at low temperatures electrons can escape by tunneling through the lowered potential barrier. Fowler and Nordheim derived a formula for the emitted current density

$$j(E)\left[\frac{A}{m^2}\right] = \frac{A_{FN}E^2}{\phi[eV]}\exp\left(-\frac{B_{FN}\phi^{3/2}}{E[MV/m]}\right)$$

$$A_{FN} = 1.54\cdot10^6; \quad B_{FN} = 6.83\cdot10^3$$

According to this formula, the E required for significant I_{FE} is well beyond the $H_{c,rf}$ limit so one might think that there is no worry. FE is observed frequently and follows the FN functional dependence! This is usually explained by adding an "enhancement factor" to the parameterization

$$I(\beta_{FN}E) = A_e j(\beta_{FN}E)$$

where I is he measured current, A_e is the active area of the supposed emitter and β_{FN} is he field enhancement factor of a supposed sharp point on the emitting object. By fitting actual data one arrives at typical empirical parameters of $\beta_F \sim 300$ and $A_e \sim 3\cdot10^{-17}$ m^2. Such points have not been identified through microscopic examination. Evidence seems to point to ohmic heating of emission sites with perhaps enhanced thermionic emission playing some role together with a plasma arc associated with the thermionic gas release. Sharp edges of grain boundaries may also play a role. A partial cure can sometimes be effected by careful, pulsed, high power processing which seems to "burn" away emitters. Great attention to cleaning the cavity prior to operation is also effective.

When all of the practices mentioned are put in place, single cell cavities with maximum fields of 40 MV/m accelerating gradient have been demonstrated. Multi-cell cavities exhibiting 25 MV/m accelerating gradient and Q at the operating field of $> 10^9$ have been measured. Nevertheless some of the limiting mechanisms are still obscure and completely effective quality control practices have not yet been devised.

7   Superconductivity for Magnets

7.1 Introduction

Because superconductors can carry large currents in high magnetic fields it is natural to think of using them for high field, low power consumption magnets. Indeed, many types of superconducting magnets have been successfully made and used: solenoids, hybrid solenoids, toroids, Helmholtz coils, "baseball" magnets, dipoles, quadrupoles, sextupoles, and so on. Here we will mention only dipoles of the type used for bending magnets in particle accelerators. Two types are in common use and are shown in Figure 38. In one case, the so called cosine theta magnet, the conductor distribution sets the shape of the filed. In the other case, the so-called superferric design, there is an iron core in which the shape of the iron pole shapes the field,.

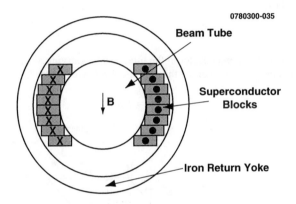

(a) cos θ   Dipole Configuration-iron Return

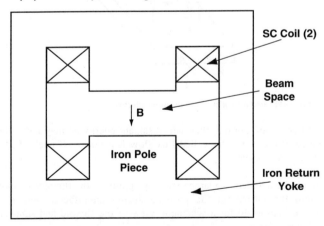

(b) "Superferric" Dipole with Iron Pole and Return Yoke

Figure 38 Two types of superconducting dipole: (a) cosθ, (b) "superferric"

7.2 Magnets from Type I superconductors

Type I superconductors are either in the Meissner state or normal. They are in the normal state if $B>B_c$. The B_c for Type I materials are quite small, e.g. B_c(Pb) ~ 0.08 T – too small to be generally useful. Further, the penetration of the field (current) is λ_L which for Type I is ~ 40 nm so that the wires would need to be impracticably thin to avoid wasting material. Thus, Type I materials are not generally used for winding magnets.

7.3 Magnets from Type II Superconductor

As H_{c2} can be >> 10 T and J_c >> 1000 A/mm^2, it would seem that Type II material would be an easy application to magnets Unfortunately, this is not so. The phenomena of "flux flow" and "flux jumping" are difficulties that must be overcome in making suitable magnet conductors.

7.3.1   Flux flow

Figure 39 shows a superconductor in the mixed state, carrying a current J. There will be an interaction between the B of the fluxoids and the J given by the Lorentz force

$$\vec{F} = \vec{J} \times \vec{B}$$

0780300-036

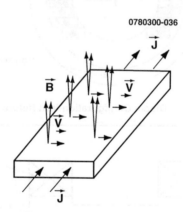

Figure 39 Schematic of fluxoid motion in a Type II superconductor

As the fluxoid cores are normal, they will dissipate energy as they move, leading to a finite resistance of the conductor called flux flow resistance. If unchecked this would render superconductors useless.

This problem can be avoided to a large extent by "pinning" the fluxoids in normal defects introduced into the lattice for the purpose during manufacture and metallurgical processing. See Figure 40 for a schematic view of the fluxoid and supercurrents and normal core that go with it.

Normal Core   Magnetic Field Lines

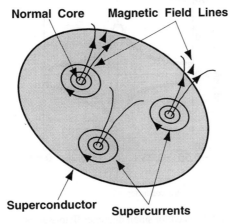

Superconductor   Supercurrents

Figure 40 Field lines and supercurrents for magnetic flux trapped in a superconductor

If normal defects, spaced with the natural fluxoid spacing, are introduced, flux quanta can locate themselves in these naturally normal spots without having the supercurrents and thus with lower energy which leads to a pinning force. When the flux flow force exceeds the pinning force, flux flow begins and the flux flow resistance appears. Thus for a given pinning force there is a critical current density defined by

$$\vec{F}_P = \vec{J}_c \times \vec{B}$$

A superconductor with good pinning is called a "hard" superconductor.

Pinning has its price, i.e. hysteresis and persistent currents. In the Meissner state the magnetization, $M(B)$ is linear and reversible and the shielding current grows in proportion to expel the flux as needed. In the mixed state (hard superconductor) fluxoids are pinned, that is captured at pinning centers, so that if the field is reduced they remain pinned and the material has a "frozen in" magnetization even for $B \rightarrow 0$. See Figure 41. This behavior implies that there will be persistent currents flowing in the material which depend upon the magnetic history and that can have a significant impact on accelerator performance if not properly compensated.

The hysteresis in associated with energy dissipation. If the field is time varying and cyclic then for each cycle the energy loss is

$$Q_{hysteresis} = \oint M(B) dB$$

This energy is supplied by the magnet power supply and leads to heat in the magnet which must be included in the refrigeration calculation.

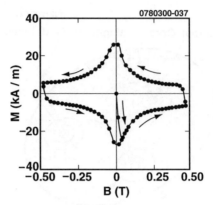

Figure 41 Measured magnetization of a NbTi conductor

7.3.2   Flux jumping and the need for small diameter filaments as conductor elements

Early users of hard superconductors noted that sometimes superconductivity seemed to break down spontaneously. This can be understood via the Critical State Model or CSM which leads to an important result for superconductor engineering.

In CSM one ignores the Meissner state and its surface currents and assumes that there are two current states only: $J=0$ or $J=J_c$. The CSM has been quite successful in predicting superconductor behavior. See Figure 42 for predictions of CSM for three distinct cases.

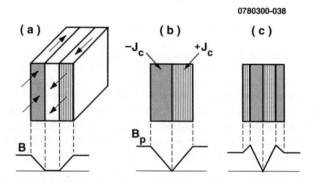

Figure 42 J and B in a slab of hard superconductor in the CSM: (a) small external field; (b) the penetrating field, B_p; (c) external field raised over B_p and lowered again

In the region of current flow the magnetic field exhibits linear behavior in accordance with the Maxwell equation

$$\nabla \times \vec{B} = \mu_o J_c$$

Following Martin Wilson's argument, look at the flux jumping instability (see Figure 43) and make a stability analysis. That is, assume some small change, in this case

$$\Delta Q \rightarrow \Delta T \rightarrow -\Delta J_c$$

and check whether it causes a state switch or runaway; under the forces at work in particular circumstances. From Ampere's law

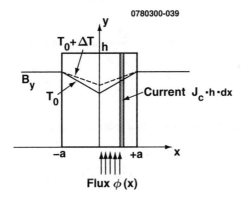

Figure 43 Schematic for stability analysis. A slab of hard superconductor with large B parallel to its surface. If there is a heat input ΔQ there is a corresponding ΔT accompanied by a decrease in J_c.

$$\oint \vec{B} \cdot d\vec{s} = \mu_o \cdot I; \quad \left(I = J \cdot h \cdot dx \right)$$

so

$$B(x) = B_o - \mu_o J_c (a - x) h$$

where J_c is the critical current density at the given field, B_o and the initial temperature T_o. Now suppose that ΔQ per unit volume is put into the slab, causing a certain ΔT and a corresponding drop in J_c, $-\Delta J_c$. What is the effect of this ΔJ_c? B inside the slab will increase. That is smaller J_c leads to less shielding leads to flux change inside the slab leads to a longitudinal voltage leads to heat generation inside the slab because that voltage acts on the J_c flowing there. For the half slab, i.e. $0 < x < a$

$$\phi(x) = \int_0^x B(x') dx' = B_o x - \mu_o J_c \left(ax - x^2/2 \right) h$$

so that the change in flux due to a change in J_c is

$$\Delta\phi(x) = \mu_o \Delta J_c \left(ax - x^2/2 \right) h$$

The joule heat produced in slice dx and height h is just

$$\Delta\phi(x) \cdot J_c \cdot dx \cdot h$$

Integrating and dividing by volume $(a \cdot h)$ we obtain the joule heat per unit volume

$$\Delta g = J_c \int_0^a \Delta\phi(x)dx = \mu_0 J_c \Delta J_c \, a^2 \big/ 2$$

To first order the J_c reduction is $\dfrac{\Delta J_c}{J_c} = \dfrac{\Delta T}{T_c - T_0}$ and thus the total energy; balance is

$$\Delta Q + \Delta g = c_v \Delta T \Rightarrow \Delta Q = \left\{ c_v - \mu_0 J_c^2 a^2 \big/ \left[3(T_c - T_0) \right] \right\}$$

From this we see that Δg is equivalent to a reduction in specific heat, c_v. The "effective" specific heat is

$$\tilde{c}_v \equiv c_v - \mu_0 J_c^2 a^2 \big/ \left[3(T_c - T_0) \right]$$

Evidently if $\tilde{c}_v$ is 0 or negative, the slightest disturbance will cause the superconductor to reduce J_c and expel some of the captured flux. This is called flux jumping. While it will not necessarily cascade to a "quench" it still will cause unstable operation which is untenable for magnet operation. By setting $\tilde{c}_v = 0$ we have a relation for the maximum allowed value of a for prevention of flux jumping, i.e.

$$a < \sqrt{\frac{c_v(T_c - T_0)}{\mu_0 J_c^2}}$$

For a cylinder of radius r instead, we have

$$r < \frac{\pi}{4} \sqrt{\frac{c_v(T_c - T_0)}{\mu_0 J_c^2}}$$

Applying to NbTi at 4.2K and $B_o = 5T$ with $J_c = 3 \cdot 10^9$ A/m^2, $T_c(B_o) = 7.2$ K, $c_v = 5.6 \cdot 10^3$ J/m^3 we find $r_{max} = 30$ μm. In the above analysis, cooling by the liquid helium is neglected so this is referred to as the "adiabatic flux jumping stability criterion". The main lesson to be learned here is that in hard superconductor windings the current carrying elements must be fine filaments of size less than 0.1 mm.

7.3.3 Flux creep and the decay of persistent currents

For a hard superconductor in the mixed state, $R \neq 0$ owing to flux creep. See Figure 44. The basic physics is that of thermal activation. Flux bundles escape pinning potential wells with probability proportional to a Boltzmann factor

$$P_O \propto \exp\left(-U_O / k_B T\right)$$

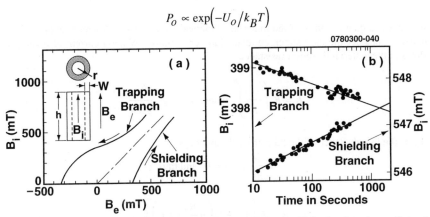

Figure 44 (a) Hysteresis of the internal field in a hard superconductor tube; (b) the time dependence of internal field on trapping and shielding branches. NOTE THE LOGARITHMIC TIME DEPENDENCE OF M

If there is a current flowing then there is a depinning force which lowers the pinning potential and thus raises the rate at which flux moves out of the material leading to a finite resistance and persistent currents with logarithmic decay.

For another view of this, step back and review critical currents of hard superconductors. The state variables are B, T, and J. See Figure 45.

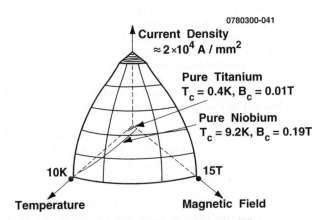

Figure 45 Critical surface of NbTi showing also the behavior of pure Ti and Nb.

At the critical current density (at a given B and T) the transition from superconducting to normal can be fitted by a power law:

$$\rho(J) = \rho_c \left(\frac{J}{J_c} \right)^n$$

where n is a quality index of the material. Practical values of n <50 which leads to slow decay. Consider the hard superconductor tube once more. When a current of density J flows in the material there is a resistance

$$R = R_c \left(\frac{J}{J_c} \right)^n \quad \text{where} \quad R_c = \rho_c \frac{2\pi r}{w \cdot h}$$

If L is the in inductance of the tube, the J decay obeys

$$L \frac{dJ}{dt} + RJ = 0 \quad \text{and thus}$$

$$J \approx J_c \left[1 - \frac{1}{n} \ln \left(n \frac{R_c}{L} t + b \right) \right]$$

where b is an integration constant. This is a logarithmic function as shown in Figure 44.

7.3.4   Quenching

Another important practical consideration is that some disturbance to J, B or T may push the superconductor of a magnet coil beyond the critical surface and lead to a transition to the normal state. Such an event is termed a quench. In a quench the stored energy in the magnet is dissipated in the winding and can lead to damage if the magnet is not properly designed. Good design assures stability against small disturbances and protection of the coil in the event of a large disturbance. This leads to the need for cooling optimization and a bypass to conduct the decaying current in the magnet safely around the superconductor. Practical superconductors have built in normal conductor to help carry the current in case of a quench. See Figure 46. Note that, as in a single filament, there can also be a magnetic instability for the strand as a whole. Similar considerations of flux flow resistance lead to a limiting diameter for strands of about 2 mm.

We saw above that as a hard superconductor becomes normal, the resistance gradually makes a transition from essentially zero to a finite value. As $I \rightarrow I_c$ the superconductor enters the resistive state and transfers part of the current to the matrix conductor (usually high RRR Cu or Al) The "circuit" for this is indicated in Figure 47.

The CSM tells us that the superconducting current will remain at about I_c while the matrix will carry $I_m = I \text{-} I_c$. Both the superconductor and the copper are in parallel and experience the same voltage drop.

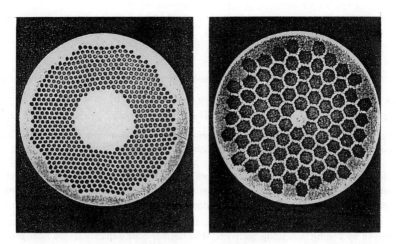

Figure 46 Cross-section of two different NbTi multi-filamentary strands, 0.84 mm in diameter. LH has 636 filaments, RH has 10164 filaments of 5 μm diameter.

0780300-042

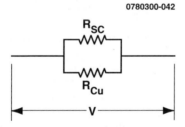

Figure 47 Resistive part of strand impedance

If the dissipated power raises $T > T_c$ the Cu matrix takes the full I. Consider a length of wire, l, cross-sectional area, A, and fractional volume of superconductor, η, initially at temperature, T_o, now raised by some disturbance to $T < T_c$ but high enough to initiate current sharing. Using the relations $V = IR$ and $P = VI$ and $J_c(T) \approx J_c(T_o) \dfrac{T - T_o}{T_c - T_o}$ find

$$G(T) = \rho_m(T) J_c^2 \frac{\eta^2}{1 - \eta} \frac{T - T_o}{T_c - T_o} \cdot A \cdot l$$

is the power generated in l. ρ_m is the resistivity of the matrix material. If the wire is immersed in liquid helium then the heat removed is $Q(T) = h \cdot p \cdot l(T - T_o)$ where h is the heat transfer coefficient, p the wetted perimeter and T_o is the bath temperature. Define the "Steckley" parameter

$$\alpha_{St} = \frac{G(T)}{Q(T)} = \frac{\eta^2 J_c^2 \rho_m A}{(1-\eta)hp(T_c - T_o)}$$

If $\alpha_{St} < 1$ the conductor is completely stable cryogenically and leads to the requirement that Cu/superconductor ratio$>10($ $Cu/Sc = (1-\eta)/\eta)$ for NbTi which is not practical for accelerator magnets owing to the great coil size implied by this criterion. Indeed $\alpha_{St} \sim 22$ for a typical accelerator dipole so tha one must include means for assuring that the conductor temperature never exceeds the damage threshold.

In addition to cooling by the helium bath, there is also conduction of heat along the strands. Imagine a pure superconductor wire without a Cu matrix. The current density is near J_c and a section of length l, heat conductivity, κ has been raised by some disturbance to $T > T_c$. In the now normal section of the material, heat is generated and can only be removed by conduction along the wire. The normal zone expands if the heat generated is greater than the heat removed by conduction along the wire.

$$\rho_m J^2 Al \geq 2\kappa(T_c - T_o)$$

This leads to a lower limit of the length of a normal zone that will propagate and quench

$$l_{mpz} = \sqrt{\frac{2\kappa(T_c - T_o)}{\rho_m J^2}}$$

which for pure NbTi gives $l_{mpz} < 1$ μm so pure superconductor wires are not useful for magnets. If embedded in a copper matrix having $\rho, \kappa \sim 1000$ x superconductor, l_{mpz} ~O(mm). In practical cases the quench threshold is roughly 10 mW/gram of material.

The power density in a normal coil section is $\rho(T)J^2(t)$ which leads to the relation

$$dT = \frac{1}{c_v(T)} \rho(T)J^2(t)dt \text{ so that}$$

$$\int_0^\infty J^2(t)dt = \int_{T_o}^{T_{max}} \frac{c_v(T)}{\rho(T)}dT$$

The RHS can be evaluated from known material properties and the LHS can be measured, giving a basis for engineering the conductor of a magnet and designing its quench protection system. Units of LHS are usually given in 10^6 A·s ("MIITS") Figure 48 shows some typical results. Magnet designers try to keep $T_{max} < 450 - 550$ K .

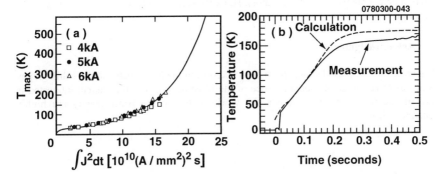

Figure 48 (a) Measured and calculated T_{max} for a 1 m long dipole; (b) Time evolution of T_{max} for a 17 m long dipole.

For added protection, magnet designers invariably add an external quench protection circuit to dissipate some of the energy stored in a quenching magnet. See Figure 49.

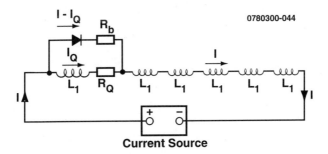

Figure 49 Schematic of a typical quench protection circuit for a series of magnets. Some of the stored energy in the quenching magnet is dissipated externally in the bypass resistor.

8   Bibliography

8.1   General

√ Solid State Physics, N. Ashcroft and D. Mermin, Saunders (1976)

√ Handbook of Accelerator Physics and Engineering, A. Chao, M. Tigner, World (1998)

8.2   Magnets

√ Superconducting Magnets, M.N. Wilson, Oxford (1983)

√ Supercond;ucting Accelerator Magnets, K.Mess, P. Schmueser, S. Wolff, World (1996)

8.3   RF Cavities

√ RF Superconductivity for Accelerators, H. Padamsee, J. Knobloch, T. Hays, Wiley (1998)

9 Acknowledgments

The materials for these lectures have been taken from the books listed in the Bibliography above as have the great preponderance of the illustrations. For the student interested in any of the topics covered above, the works listed in the Bibliography are highly recommended.

SUPERCONDUCTING MAGNETS FOR ACCELERATORS

MARTIN N. WILSON

Oxford Instruments (retired), Brook House, 33 Lower Radley, Abingdon, OX14 3AY, UK.
email: m-wilson@primex.co.uk

Superconducting magnets have been used in all the world's largest accelerators because they allow higher fields to be produced with lower power consumption than conventional magnets. Iron is no longer needed to produce the field, although it is often used to reduce the fringe field. In the absence of iron, a new family of winding shapes has had to developed for generating the field profiles needed in accelerators. At the low temperatures needed to operate superconducting magnets, there are unfortunately many mechanisms which can degrade the performance of a magnet below that which it should be capable of. Strategies for avoiding this degradation are an important part of superconducting magnet technology.

1 Introduction

Superconducting magnets are interesting to accelerator builders for one very simple reason: they abolish Ohm's Law! This means that the magnets:
- consume no power (although they do need refrigeration power)
- can work at high current density
- don't need iron because Ampere turns are cheap and plentiful (although iron is often used for shielding)

As a result of these features, superconducting magnets offer:
- lower power bills
- higher magnetic fields giving reduced bend radii and thus smaller rings with reduced capital cost or new possibilities such as the muon collider
- higher quadrupole gradients, which enable higher luminosities to be achieved.

In this, the first of three linked lectures, we discuss the properties of high field superconductors, coil shapes for accelerators and the problem of degraded performance and 'training' of magnets. For more detail on superconducting magnet technology, the student is referred to [1, 2, 3, 4 and 5]; for information on superconducting materials, see [6, 7, 8, 9 and 10] and for information about cryogenic technology see [11, 12, and 13].

1.1 Materials

The material most commonly used in superconducting magnets for accelerators is niobium titanium, a ductile alloy which may readily be drawn down to the fine filaments needed for optimum performance. Fig 1 shows the critical surface of niobium titanium, plotting current density against magnetic field and temperature; superconductivity prevails everywhere below the critical surface and normal resistivity everywhere above it. Magnets are usually operated at a temperature of 4.2K, the boiling point of liquid helium under atmospheric pressure, but lower temperatures are sometimes used in order to obtain higher performance – notably in the LHC project, presently under construction at CERN.

Niobium tin Nb_3Sn and niobium aluminium Nb_3Al are intermetallic compounds which have much higher performance and are sometimes used for special high field magnets, but their brittle mechanical properties cause many practical difficulties. Many other metallic 'low temperature' superconductors are known, but have not been used in magnets. Of the new high temperature superconductors, BSCCO is the most developed and has been used in a few prototypes, but its brittleness, cost and low current density in field have so far precluded its use in accelerator magnets.

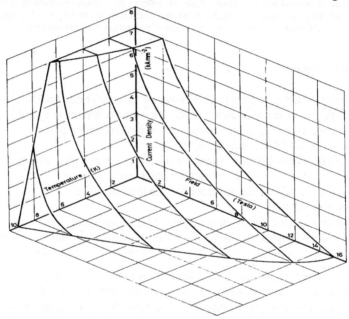

Fig 1: The critical surface of Niobium Titanium; superconductivity prevails everywhere below the critical surface and normal resistivity everywhere above it

The current density shown in Fig 1 is several orders of magnitude higher than that used in the water-cooled copper windings of conventional electromagnets. In practice however, the overall 'engineering' current density of a superconducting magnet winding is ~ $^1/_3$ that of the superconducting material because of the other components, mainly copper and insulation, which must be included in the winding. Fig 2 shows some typical engineering current densities, defined as the critical current per unit cross sectional area of the magnet winding, for different materials at different temperatures.

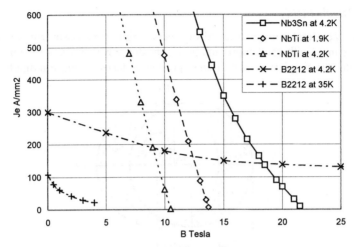

Fig 2: Engineering current densities of some typical magnet windings.

2 Magnet Configurations

The high engineering current densities shown in Fig 2 allow high fields and field gradients to be produced by compact windings without the use of iron (although iron is often used to shield the fringe field). In a conventional electromagnet, the field shape is determined almost entirely by the iron, with the winding shape (usually a simple rectangular cross section) playing a very minor role. Without iron however, the field shape is determined solely by the winding and special new winding shapes are needed to produce the very uniform dipole fields or higher order quadrupole, sextupole etc fields needed in an accelerator. Because the field must be perpendicular to the direction of particle motion, which is always the long dimension of the magnet, the coils must generally be of a 'racetrack' or 'saddle' configuration as shown in Fig 3.

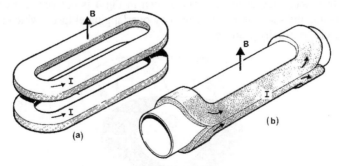

Fig 3: Coils to produce uniform fields perpendicular to their long dimension.

2.1   Dipole Fields

Because accelerator magnets are always very long in comparison with their end turns we may, with good approximation, consider only their cross section.   Fig 4 illustrates the simplest cross section to produce a perfectly uniform dipole field.

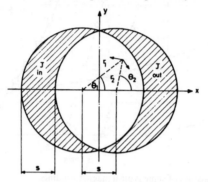

Fig 4: Uniform transverse field produced by overlapping cylinders of current.

To calculate the field, we note that the field inside an infinitely long, cylinder carrying uniform current density J is in the azimuthal direction and of magnitude:

$$B = \frac{\mu_o J r}{2}$$

If we now superpose two such cylinders, carrying currents into and out of the paper, with their centres separated by a distance s, we find the resultant fields:

$$B_y = \frac{\mu_o J}{2}\left(-r_1 \cos\theta_1 + r_2 \cos\theta_2\right) = \frac{-\mu_o J s}{2}$$

$$B_x = \frac{\mu_o J}{2}\left(+r_1 \sin\theta_1 - r_2 \sin\theta_2\right) = 0$$

Of course the central region, where the oppositely directed currents overlap, can be empty space – the magnet aperture – and the separation s is the thickness of the winding.   In practice the aperture shape shown in Fig 4 is not very convenient for accelerators and the shape in Fig 5 formed by overlapping ellipses, which also gives a uniform field, is usually preferred.

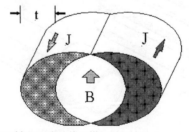

Fig 5: Uniform dipole field formed by overlapping ellipses.

An alternative configuration, which also produces a perfectly uniform field, is given by an annular winding shape in which the current density varies with position as $\cos\theta$, as shown in Fig 6. As with the overlapping cylinders, the central field is given by $B = \frac{1}{2}\mu_o J_o t$, where J_o is the current density at $\theta = 0$. Analysis of the field shape for the ellipses and $\cos\theta$ current distributions is much more difficult than the simple overlapping cylinders; the most convenient approach for these and many other two dimensional field problems is the complex variable method, as described for example by Beth [14, 15].

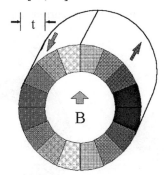

Fig 6: Uniform dipole field produced by a $Cos\theta$ distribution of current density.

Practical winding configurations approximate more or less closely to the ideal shapes shown in Figs 5 and 6; they are usually designed by computational techniques using numerical optimizer algorithms which adjust the coil boundaries to minimize errors coming from the higher order harmonic terms in the field. Spacers are used within the winding to approximate a $\cos\theta$ distribution of current. In general, more spacers allow more degrees of freedom for the optimization and thus a better approximation to uniform field. Fig 7 shows a cross section of the LHC dipole, a typical example of modern accelerator magnet design.

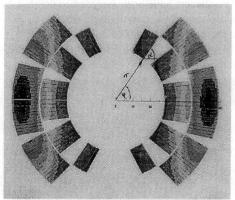

Fig 7: Cross section of coils for an LHC dipole: shading indicates contours of constant field amplitude.

2.2 *Quadrupole and Higher Order Fields*

In addition to dipoles, of course, we need quadrupoles, sextupoles and higher order magnetic fields to focus the beam, compensate chromaticity etc. As shown in Fig 8, a perfect quadrupole may be made by two overlapping ellipses or by a cos2θ current distribution. In the latter case, the gradient is given by:

$$\frac{\partial B_y}{\partial x} = \frac{\partial B_x}{\partial y} = \mu_o \frac{J_o t}{2a}$$

Sextupole field may be made by a *cos3θ* current distribution, octupole field by *cos4θ* and so on.

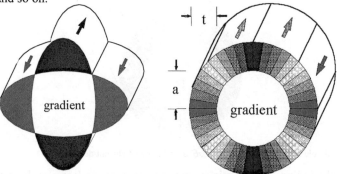

Fig 8: Production of perfect quadrupole fields by overlapping ellipses or a *cos2θ* current distribution.

2.3 *End Turns*

Having designed the two dimensional cross section of the magnet, we must then think about the end turns and design them using a fully three dimensional computer programme. Fig 9 shows the ends of the upper half of the inner coil of an LHC dipole. The spacers have been included to achieve two objectives:

a) to reduce the error terms, particularly sextupole, in the end field.
b) to reduce the peak field at the winding.

Objective b) is special to superconducting magnets, where the current carrying capacity is limited by the highest field seen at any point in the winding.

Fig 9: End turns of an LHC dipole coil

2.4 Electromagnetic Forces

From Ampere's Law, the force per unit volume on a winding in a magnetic field is $F=B.J$; this force is perpendicular to the field and current. Using one of several computer programmes now available for the purpose, one may calculate the force distribution throughout a magnet winding and then proceed to calculate the stress within the winding and the force on the external supporting structure. For quick estimates however, it is often useful to have a simple formula. Fig 10 shows (on the left) the magnetic field lines, with the resultant forces being shown on the right by arrows whose length represents the size of the force per unit volume.

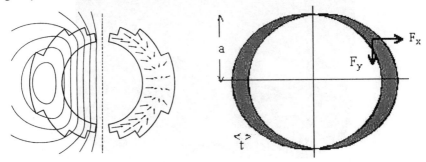

Fig 10: Field lines and forces in a dipole; arrows show direction of force and their length indicates its magnitude; right hand side shows total forces in x and y directions on the top half of the winding.

On the right hand side, Fig 10 shows a simple thin winding with total magnetic forces F_x and F_y. Using his complex variable theory [14, 15] Beth has shown that:

$$F_x = \frac{B^2}{2\mu_o}\frac{4a}{3} \qquad\qquad F_y = -\frac{B^2}{2\mu_o}\frac{4a}{3}$$

where a is the radius of the winding. These formulae are exact for infinitely thin windings and are a good approximation for windings of mean radius a whose thickness is less than ~ half their radius. Note the factor $B^2/2\mu_o$, often known as the 'magnetic pressure'. Note also that the vertical compressive force is just as big as the intuitively more obvious horizontal bursting force. These forces can be substantial, for example in an LHC dipole the total bursting force $(2.F_x)$ is ~350 tonne per metre.

A special structure is needed to contain the bursting forces. It must be strong and stiff, so that the winding does not move appreciably as the field is increased. It must also be manufactured to extremely precise (~10μm) dimensional tolerance so that it will press the winding exactly into place. The technique which has been universally adopted for accelerator magnets, known as coil *collaring*, was developed at Fermilab from the stamping technique used to make the iron laminations of conventional magnets. Fig 10 shows the collars on a pair of LHC dipole coils (note that, as a colliding beam ring, LHC uses its dipoles in oppositely directed pairs). The collars are stamped from sheets of stainless steel or aluminium alloy a few mm thick. They are made in two halves, which are pressed around the coil from top and bottom. The halves are not symmetric and laminations of each type are alternated between top and bottom along the magnet so that they may be locked into place when long steel rods are inserted into the holes shown in Fig 11. The collars are manufactured to be slightly under size and a large hydraulic press must be used to push the two halves together so that the coil is pre-compressed when the rods have been inserted, even when the hydraulic pressure is released. A further advantage of collars is that they do not conduct eddy currents, which would cause field errors, when the field is ramped.

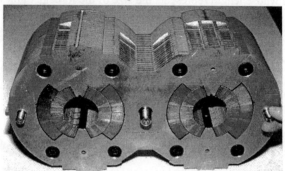

Fig 11: Cross section of an LHC dipole showing the force supporting collars and locking rods.

3   Degraded Performance and Training

3.1   Load Lines

Fig 12 shows a diagram which is commonly used in superconducting magnet

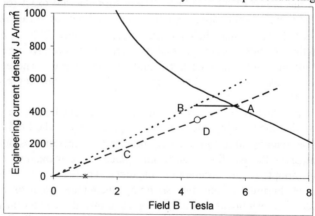

Fig 12: Critical current density of a typical winding and load lines of a magnet (dashed is peak field; dotted is aperture field).

design. The curved line is the 'short sample' critical current of the superconductor (measured on a sample of the superconducting wire or cable by immersing it in a magnetic field and then increasing the current through until a resistive voltage is detected). The two dashed lines are *load lines* for the magnet, relating the field it produces to the current flowing in it. The thick dashed line is for the peak field on the magnet, while the dotted line is for the useable aperture field. We thus expect the maximum current and field that the magnet is capable of to be given by the intersection point A, where the superconductor at the peak field point is carrying its critical current. Note that, at this maximum current, the aperture field point B is well below the critical line – hence the desire of magnet designers to keep peak fields as low as possible.

3.2 Training

Although we would like magnets to go straight to their 'short sample' critical current at point A, in practice they often go into the resistive state at much lower currents, say in the region C of Fig 12. When this happens, the inductive stored energy of the magnet is dissipated as heat in the resistive region and the magnet is said to *quench*. Following a quench, it is necessary to wait many minutes for the magnet to cool down again. When the magnet current is again raised, it will usually quench at a somewhat higher level of current and field; this process is called *training*. Sometimes, after many training quenches, the magnet will eventually reach its short sample critical current, but often it never does.

Over the last 30 years, the problem of training has been ameliorated by many detailed improvements in magnet and conductor technology, but training is still with us and remains an ever present problem for the magnet designer. Fig 13 shows some training curves for a prototype LHC dipole, plotting maximum field achieved as function of number of quenches.

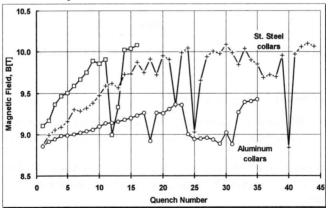

Fig 13: Quench plot for a prototype LHC dipole, comparing the same coils with two different types of force supporting collars. (courtesy of A. Siemko CERN)

308

3.3 Some Reasons for Degraded Magnet Performance

Why do magnets fail to reach the performance they should be capable of? In general it is believed that some kind of sudden release of energy occurs locally at certain points in the winding as the current and field are raised. Specific heats decrease by a factor $\sim 10^4$ as we cool down from room temperature to 4.2K, which means that an energy release which would raise the temperature by only 10^{-3}K at room temperature would cause a 1K rise at 4.2K. Thus, effects which would be quite trivial at room temperature, become catastrophic at low temperature.

Fig 14 examines the effects of local releases of energy in the magnet of Fig 12, assuming the magnet current has been set at the point D, giving a peak field of 4.5T. The solid line plots engineering current density in a field of 4.5T as a function of temperature. At the point D, corresponding to point D in Fig 12, it may be seen that the superconductor is somewhat below its critical temperature and current; this gap is sometimes known as the *temperature margin*; increasing the conductor temperature to 5.3K will cross the critical line. Thus at point D, we have a temperature margin of 1.1K and an energy disturbance causing greater temperature rise will quench the magnet.

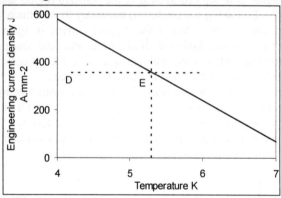

Fig 14: Engineering current density at 4.5T in the magnet of Fig 12, plotted as a function of temperature

3.4 Mechanisms for Energy Release

There are many possible mechanisms for energy release in magnets, some of them observed experimentally and some purely hypothetical; here are the three most common.

a) flux jumping, which has already been covered by Maury Tigner [16].

b) sudden movements of the conductor, which may release energy by frictional heating.

c) cracking of resin or other materials used within the magnet winding to prevent movement of the conductor.

For conductor motion, let us imagine a length of conductor, carrying current density J in a magnetic field B, which suddenly moves a distance δ under the action of the electromagnetic force. It may easily be seen that the field force does work per unit volume $Q = B.J.\delta$. Thus, if a conductor at the point D of Fig 12 moves a distance δ = 5 μm and all the work done by the field force is dissipated as frictional heating the energy released will be:

$$Q = B*J*\delta = 4.5T * 335x10^6 A.m^{-2} * 5x10^{-6}m = 8000 \ J.m^{-3}$$

Because the specific heat $C(\theta)$ varies very strongly with temperature ($\sim \theta^3$) at low temperatures, it is convenient to work in terms of the enthalpy $H(\theta) = \int C(\theta)d\theta$. Fig 15 plots the enthalpy per unit volume of some common materials and the mean enthalpy of a typical magnet winding containing NbTi, copper and epoxy resin.

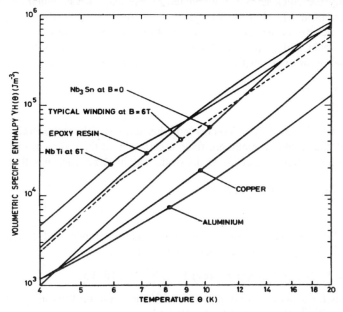

Fig 15: Enthalpy per unit volume of the main materials used in superconducting magnet windings.

For the 'typical winding' in Fig 15, we see that the enthalpy at 4.2K is 3000Jm⁻³, the 5μm movement would thus provoke an increase to 11000Jm⁻³, raising the temperature to 5.8K and thereby quenching the magnet. From this simple calculation, we see how extremely sensitive superconducting magnets are to small movements.

3.5 Impregnation

Engineering to a precision of 5μm is extremely difficult! – and it was not long before the magnet makers hit on the idea of vacuum impregnation to fill the winding with an insulating material like epoxy resin. In this way, every tiny hole remaining is completely filled and the likelyhood of movement, even on a μm scale, should be eliminated. Unfortunately however, the use of insulating materials brings other problems because, as organic materials, they contract much more than the metal of the conductor. This means that, when the mixture of conductor and metal are cooled down, the metal goes into compression and the insulator goes into tension. Furthermore, most organic materials pass through a *glass transition* on cooling, which means they become extremely brittle. A brittle material under tension is likely to crack. When this happens, a large proportion of the strain energy caused by differential thermal contraction, is dissipated as heat. Very often the cracking occurs as the magnet is being cooled down from room to low temperature, but occasionally the crack is only precipitated by the extra stress which comes as the magnet current and field are increased. If the crack releases sufficient heat to take the superconductor above its critical temperature, a quench ensues.

If the resin and metal are glued together in such a way as to create a one dimensional or *uniaxial* tensile strain, the strain energy is given by:

$$Q_1 = \frac{\sigma^2}{2Y} = \frac{Y\varepsilon^2}{2}$$

where σ = tensile stress, Y = Youngs modulus and ε = differential strain. If, on the other hand, the resin and metal are fixed together in all three dimensions, the resin will be put into a state of triaxial tensile strain and the strain energy will be:

$$Q_3 = \frac{3\sigma^2(1-2v)}{2Y} = \frac{3Y\varepsilon^2}{2(1-2v)}$$

where v = Poisson's ratio. Typical values for a mix of copper conductor with epoxy resin cooled down from room temperature to 4.2K are:

$$\varepsilon = (11.5 - 3) \times 10^{-3} \text{ (copper - resin)} \qquad Y - 7 \times 10^9 \text{ Pa} \qquad v = \frac{1}{3}$$

which gives for uniaxial strain $Q_1 = 2.5 \times 10^5$ J.m^{-3} and for triaxial strain $Q_3 = 2.3 \times 10^6$ J.m^{-3}. If all this energy were to released as heat in a typical winding starting at 4.2K, the temperature would rise to $\theta_{final} = 16$K and $\theta_{final} = 28$K for uniaxial and triaxial strain respectively. It is clear that the energy released by resin cracking has the potential to take local areas of the winding far above their critical temperature.

3.6   Ways to Reduce Degradation and Training

Although the main causes of degradation and training have become clear and practical recipes for avoiding those causes have been worked out, there is no universal cure and, as can be seen from Fig 13, training is still with us despite much careful design work. We can however make some general statements about 'good practice' in making magnets to minimize the problem. Firstly, any movement of turns in the winding when the field force comes on must be avoided. In complex systems, a finite element structural analysis should be carried out, with particular attention to the possibility of differential movement at interfaces. If movement is unavoidable, such as at the interface between a winding and a structural member, that interface should be given some kind of low friction treatment. It is not so much a case of getting the absolute minimum friction as of avoiding 'stick slip' sliding behaviour, which can release sudden bursts of energy.

When resin impregnation is used, the volume of resin used should be kept to a minimum. If large volumes of resin are inevitable, they should be filled with a low contraction reinforcing material such as glass fibre. If it is possible to make the winding porous to liquid helium, this should be done because liquid helium has the largest known heat capacity at $4.2K \sim 10^3$ times greater than the materials shown in Fig 15.

In addition to the above precautions, the conductor should be designed so as to avoid flux jumping, which means that it must be made in the form of fine filaments, as described by Tigner [16]. It must also be as stable as possible against moderate energy inputs, a question which we will consider in the next lecture 'Conductors for Accelerator Magnets'.

References

1) Superconducting Accelerator Magnets, KH Mess, P Schmuser, S Wolf. pub World Scientific, (1996) ISBN 981-02-2790-6

2) High Field Superconducting Magnets, F. M. Asner, pub Oxford Science Publications 1999 ISBN 0 19 851764 5

3) Case Studies in Superconducting Magnets, Y Iwasa, pub Plenum Press, New York (1994), ISBN 0-306-44881-5.

4) Superconducting Magnets, MN Wilson, pub Oxford University Press (1983) ISBN 0-019-854805-2

5) CERN Accelerator School: Superconductivity in Particle Accelerators, Hamburg 88, CERN 89-04.

6) Superfluidity and Superconductivity: Tilley DR and Tilley J, pub Adam Hilger 86 ISBN 0-85274-791-8

7) Introduction to Superconductivity, Rose-Innes AC and Rhoderick EH,

8) Superconductivity, the Next Revolution? Vidali G, pub Cambridge University Press 93, ISBN 0-521-37757-9

9) Proc Applied Superconductivity Conference, pub as IEEE Trans Applied Superconductivity, Mar 93 to Mar 99, and IEEE Trans Magnetics Mar 85 to 91,89 87,85,etc

10) Superconductor Science and Technology, published monthly by Institute of Physics (UK).

11) Helium Cryogenics,Van Sciver SW, pub Plenum 1986 ISBN 0-0306-42335-9

12) Cryogenic Engineering, Hands BA, pub Academic Press 86 ISBN 0-012-322991-X

13) Cryogenics: published monthly by Butterworths

14) Beth R. A., Proc 6th Int Conf on High Energy Particle Accelerators, Cambridge Mass (1967) p387.

15) R. A. Beth; J. Appl. Phys, Vol 38, pp 4689 (1967)

16) Tigner M. 'Fundamentals of Superconductivity' Lecture   at this school.

CONDUCTORS FOR ACCELERATOR MAGNETS

MARTIN N. WILSON

Oxford Instruments (retired), Brook House, 33 Lower Radley, Abingdon, OX14 3AY, UK.
email: m-wilson@primex.co.uk

In this second lecture, we consider the special requirements that accelerators impose on magnet design and how they affect the conductor. Firstly, the conductor must be as stable as possible against the sudden releases of energy described in the first lecture. To satisfy this requirement, the superconductor should be intimately mixed with a good normal conductor, such as copper. Secondly the conductor must be stable against flux jumping, should have low losses in ramping fields and should not have too much magnetization, which produces field errors. All these requirements are fulfilled by making the superconductor in the form of fine filaments. Finally, the conductor must be capable of carrying high currents, which calls for the use of cables.

1 Introduction

Accelerator magnets are special in several important respects:

i) although very large in one dimension, the magnets produce field over a small transverse aperture.

ii) they must be ramped from low to high field.

iii) they must maintain an accurate field shape, under both steady state and ramping conditions.

iv) all magnets of a given type must be energized in series to ensure that they exactly carry the same current.

These requirements make specific demands on the conductor:

i) demands a high current density to give a compact winding, preferably such that the winding thickness is less than the aperture; this requires that the conductor should be as stable as possible against flux jumping or mechanical energy disturbances which might prevent the conductor achieving its full current.

ii) & iii) mean that the superconductor magnetization must be small and should not be significantly increased by coupling current effects during the ramp.

iii) means that the windings must be accurately located and not move significantly as the field, and hence the force, is increased.

iv) implies that the operating current must be high and that the magnets must be self protecting (we will return to the latter point in the final lecture).

In the following paragraphs we briefly examine some consequences of these requirements, starting with the requirement for stability against energy inputs.

314

2 Stability

2.1 Quench Initiation

Fig 1 shows voltage traces of the events occurring just before a quench in an LHC dipole magnet.

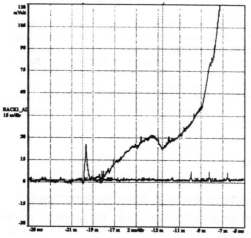

Fig 1: Trace of voltage versus time in the winding of an LHC dipole magnet at the time of quenching. (courtesy of A. Siemko CERN)

First of all we see a spike, which is probably a short length of conductor moving under the field force. After a short time, the resistance starts to grow as a result of the frictional heat input and eventually becomes very large – the quench. An intriguing aspect of Fig 1 is the intermediate region shortly after the start of resistance growth, where the resistance decreases for a time. If only we could somehow persuade the conductor to continue this downward trend in resistance until full superconductivity was recovered, the magnet would be able to survive a movement or other disturbance without quenching. It would be *stable* against disturbances.

2.2 Minimum Quench Energy

A concept often used in discussions about conductor stability is the *minimum propagating zone MPZ*; Fig 2 illustrates. This resistive zone will grow or shrink depending on whether or not the Ohmic heat generation within it is greater than the heat leaking out of it. The balance point between these two states is called the *minimum propagating zone* MPZ; very approximately it satisfies the condition:

$$\frac{2kA(\theta_c - \theta_o)}{l} + hPl(\theta_c - \theta_o) = J_c^2 \rho Al$$

where k = thermal conductivity, θ_c = critical temperature, θ_o = starting temperature, ρ = resistivity, A = cross sectional area of conductor, h = heat transfer coefficient to coolant (if there is any in contact locally) and P = cooled perimeter of conductor.

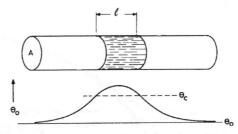

Fig 2: Schematic of the temperature distribution along the conductor in a minimum propagating zone

We may thus write the length l of the MPZ as:

$$l = \left\{ \frac{2k(\theta_c - \theta_o)}{J_c^2 \rho - \frac{hP}{A}(\theta_c - \theta_o)} \right\}^{\frac{1}{2}} \tag{1}$$

The energy to create an MPZ is roughly the energy needed to raise length l of conductor to the critical temperature; it is known as the minimum quench energy MQE. A conductor with large MQE would thus be expected to be more resistant to sudden energy inputs and magnets made from this conductor should be more likely to get to their critical current without training. From eq (1) we see that, to maximize l for good stability, we need to maximize thermal conductivity and minimize electrical resistivity; Figs 3 and 4 plot k and ρ for the materials in question.

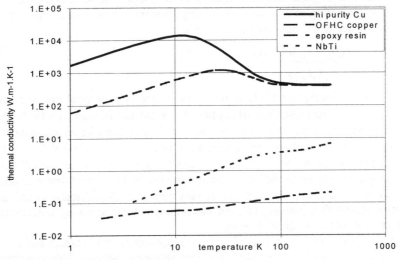

Fig 3: Thermal conductivity at low temperature.

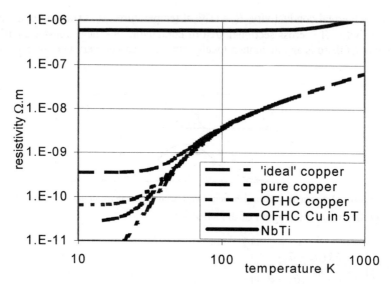

Fig 4: Electrical resistivity at low temperature.

It may be seen that there is a factor ~ 10^3 between copper and NbTi in both thermal conductivity and electrical resistivity, so a conductor with the properties of copper would have an MPZ and MQE about 10^3 greater than pure NbTi. For this reason, high field superconductors are always used in the form of composite wires, with fine filaments of superconductor embedded in a matrix of copper.

An important question is how finely divided the superconductor must be to ensure that it is in good thermal contact with the copper. If we analyze the steady state heat generation with a superconducting filament enclosed in a copper matrix [1], we may derive a characteristic distance d:

$$d^2 = \frac{\lambda k (\theta_c - \theta_o)(1 - \lambda)}{\lambda J_c^2 \rho} \qquad (2)$$

where λ is the fraction of NbTi in the cross section, k = thermal conductivity of copper, ρ = electrical resistivity of copper. For a circular filament, we find thermal instability within the filament when its radius a is $> 2\sqrt{2}d$, but for an adequate safety margin, it is better to choose the filament radius $a < \sqrt{2}d$. Typical figures for a NbTi/copper composite with $\lambda = 0.4$ give $\sqrt{2}d = 30\mu m$. Thus, to get the full benefit of copper in increasing the MQE, we must make the NbTi filament diameter smaller than $60\mu m$. Coincidentally, this size is very similar to the adiabatic criterion for stability against flux jumping [2], but the correspondence is purely coincidental - the factors involved are quite different.

Also from eq (1) we may see that, if the resistance is small enough and the heat transfer is large enough, the MPZ becomes infinitely large. This condition is known as *cryostatic* stability and is used in very large windings, such as bubble chamber or cyclotron magnets. The problem is that the condition can only be fulfilled by using a lot of copper, which makes the fill factor λ very small and means that the overall current density is also small. The winding thickness to produce a given field thus becomes very large. In large magnets, where everything is on a large scale, thick windings are not much of a problem, but for the slender aperture of an accelerator magnet, they would be hopelessly uneconomic. Nevertheless, we may be able to get some advantage from heat transfer, without reducing λ, by making the term *P/A* in equation very large. This has been done in prototype conductors by covering the wire with porous metal, which is then wetted by liquid helium. The fine pore size gives a very large wetted perimeter and thus a large heat transfer, but only for a short time until the liquid has evaporated. Fig 5 presents some experimental measurements of MQE on samples of different types of superconducting cable; the highest MQE being measured on the cable with a porous metal coating. It remains to be seen whether such improvements in heat transfer can improve the performance of magnets, but the need to allow some permeation of the winding by liquid helium is already an established part of accelerator magnet construction.

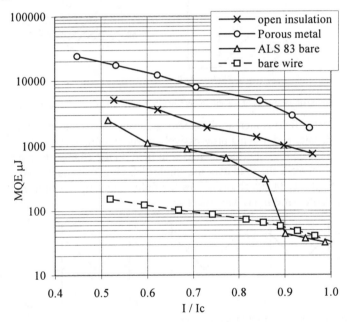

Fig 5: Measurements of MQE on different types of LHC cable (made by pulsing small carbon heaters attached to the conductor)

318

3 Magnetization

3.1 Hysteresis in the Superconductor

When the field is increased or decreased on any technical superconductor, it sets up screening currents which try to oppose the change – just like eddy currents except that they don't decay. We model this process in one dimension (because it's simpler) as shown in Fig 6, which shows a slab conductor infinitely long in the y and z directions, so that the only variation is in the x direction. From one of Maxwell's equations, we thus find that a uniform current density gives a constant field gradient $\mu_o J = dB/dx$.

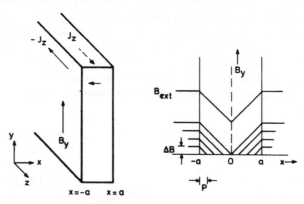

Fig 6: One dimensional model of screening currents in infinitely large superconducting slab.

As the external field is increased, screening currents are induced flowing at constant (critical) density and moving progressively inwards until, at the *penetration field,* they reach the centre. Further increases in field do not change the current pattern, but merely increase the field over the whole slab. Fig 7 shows the pattern when the external field is reduced, showing how the field change moves in from the outside until the whole pattern has been reversed.

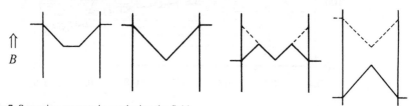

Fig 7: Screening currents in a reducing the field

Screening currents have a magnetic moment, making the superconductor behave like a magnetic material. We may measure this magnetization in a magnetometer and plot it as a function of field as shown in Fig 8. The steeply sloping sections of the curve are those regions where the screening currents are being established or are inverting. The fairly horizontal sections are where the superconductor is fully penetrated; in this situation, the only change in magnetization comes from the variation of critical current density with field. We thus have a hysteresis curve for the superconductor, with a major hysteresis loop as the outer envelope and minor hysteresis loops within it when the field ramp is reversed - just like the hysteresis behaviour of iron, only in this case the magnetization is diamagnetic rather than ferromagnetic.

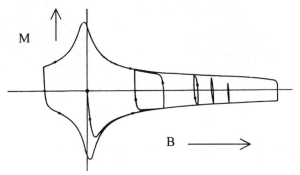

Fig 8: Magnetization of a superconductor when the field is increased and then reduced (arrows show direction of field change).

When the field is fully penetrated, one may easily show that the magnetization, defined as magnetic moment per unit volume, is given (in Tesla) by:

$$M_{slab} = \frac{\mu_o}{2} J_c a \tag{3}$$

In cylindrical geometry a similar, but more complex, calculation gives

$$M_{cyl} = \frac{4}{3\pi} \mu_o J_c a \tag{4}$$

where a is the slab half width or filament radius. Magnetization is a nuisance in accelerator magnets because it produces errors in the field pattern. Equation (4) shows the magnetization may be reduced by making the filament smaller and, for this reason, conductors for use in accelerator magnets are made with the finest possible filaments or superconductor, typically $\sim 7\mu m$ diameter. This size is considerably smaller than that needed for stability against flux jumping [2], or to maintain good thermal contact with the copper (eq 2). Note that, because the hysteresis loop depends on history, superconducting accelerator magnets must always be taken round a full field cycle before being used with particles.

320

3.2   Coupling in Wires

A single 7μm diameter field of NbTi would carry a current of ~ 50mAmp – clearly we need to put many filament in parallel! Superconducting wires for accelerator magnets are therefore made with ~ 10^4 filaments of NbTi embedded in a copper matrix. Unfortunately the copper, which is needed for stability, brings more magnetization problems, as illustrated in Fig 9.

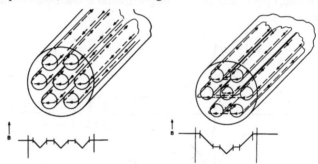

Fig 9: Schematic of a filamentary composite superconductor in an external field, showing how the filaments are coupled together by currents crossing the copper matrix.

The left hand side of Fig 9 shows the situation we would like to have, with each filament having its own screening current, but the right hand shows the situation we actually get, where the filaments are coupled together by currents which cross the copper and thereby produce a much greater magnetization. Of course, these coupling currents will decay with time because part of their path is resistive – but the time constants involved may be many years in a wire more than a few metres long. Fortunately, this problem is readily solved by twisting the wire so that the sense of the screening current must reverse every twist pitch, which greatly reduces their time constant.

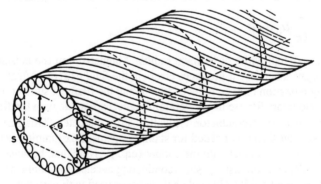

Fig 10: Screening current in a twisted filament composite (dashed line shows a typical current flow line),

By calculating the resistance and flux linkage of a typical current flow like that show in Fig 10, we may show [1] that the time constant of the coupling current is given by:

$$\tau = \frac{\mu_0}{\rho_{et}} P^2 \frac{1}{8\pi^2} \tag{5}$$

where P is the twist pitch and ρ_{et} is the effective transverse resistivity of the composite wire, shown by Carr [3] to be:

$$\rho_{et} = \rho \cdot \frac{1+\lambda}{1-\lambda} \tag{6}$$

where λ is the filling factor (proportion of NbTi) and ρ is the copper resistivity. In tightly twisted composite wires, these time constants are typically a few msec. When the external field is changed field, the coupling currents produce a magnetization, which is in addition to the filament magnetization, of

$$M_c = 2 \frac{\Delta B}{\Delta t} \tau \tag{7}$$

We may detect the coupling current directly by plotting magnetization loops with different sweep of the external field. Fig 11 shows a typical result where the smallest loop, taken at very low sweep rates, corresponds to the filaments alone and the larger loops are the sum of filament + coupling.

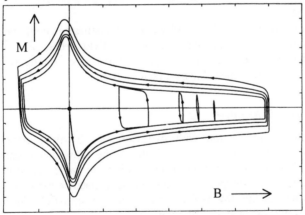

Fig 11: Magnetization loops of a twisted filamentary composite wire, measured with different sweep rates of the external field.

3.3 Coupling in cables

We shall see in the next section how practical accelerator magnets are made from Rutherford cables. Coupling also occurs between the wires of these cables, following diamond shaped patterns like those shown in Fig 12. Looking at the left

322

hand loop, we see current flowing from right to left until it reaches the centre of the cable, where it crosses from top face to bottom face of the cable and flows from right to left until it again crosses from top to bottom face, making a closed loop.

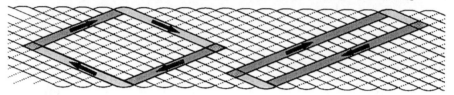

Fig 12: The flow of coupling current in a Rutherford cable.

By considering the resistance and flux linkages of loops like those shown one may calculate a mean time constant:

$$\tau = \frac{\mu_0}{\rho_{ec}} P^2 \frac{\alpha}{60}$$

where ρ_{ec} is the effective transverse resistivity across the cable, α = aspect ratio (= width/thickness) of cable, typically ~ 10 and P = twist pitch, typically ~ 100 mm. So for a solid copper cable, ie good contact between the wires we find $\tau \sim 10$ sec. The magnetization of the cable is increased by this coupling:

$$M_c = \alpha \frac{\Delta B}{\Delta t} \tau$$

Thus, in a storage ring where $\Delta t \sim 1000$sec, the magnetization could be several % of the aperture field, which would cause unacceptable field errors. For this reason, the wires in Rutherford cable are usually given a resistive coating to increase the effective resistivity across the cable by a factor ~ 100.

3.4 Field Errors in Magnets

We conclude this section with two examples of field errors caused by magnetization in the conductor. Fig 13 shows the skew quadrupole term in an experimental high field dipole made from niobium tin cable [4] which, because it was an experimental prototype, had a low contact resistance between strands in the cable. The effect of this low resistance is plain to see: a field error of some %, which increases proportionately with ramp rate.

Fig 14 shows data [5] from a prototype LHC dipole, where the strands of the cable have been treated with an oxidized silver tin alloy to increase the contact resistance. It may be seen that the error is now much smaller and is acceptable for accelerator use – although magnetization still accounts for the largest single error component in superconducting accelerator magnets. Three points are worth noting in Fig 14:

a) The smallest loop width at high field is for the very slow ramp rate, as expected, and is due to the filaments acting independently.

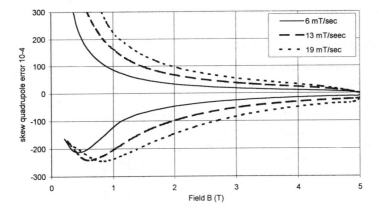

Fig 13: Skew quadrupole component of the field error in the Twente high field dipole magnet.

b) At faster ramp rates, the loop turns 'inside out' at medium fields, indicating that the coupled magnetization is producing an error of opposite sign to the filament magnetization. We can understand this by noting that the high field parts of the magnet will have a small filament magnetization (because J_c is small) but a large coupled magnetization because dB/dt is large.

c) All curves curl upwards at highest fields because the iron shield of the magnet is saturating.

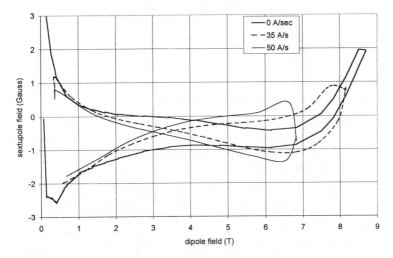

Fig 14: Sextupole component of the field error in a LHC dipole at three different ramp rates.

324

4   Manufacturing Techniques

4.1   Filamentary Composite Wires

Fig 15 shows a schematic of the process used to make filamentary composite wires of NbTi.

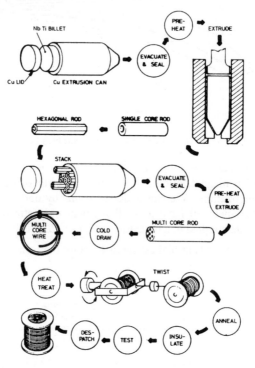

Fig 15: Schematic of a typical manufacturing process for NbTi wires

Production starts with a billet of NbTi alloy which has been prepared by electron beam melting in high vacuum.   This billet is placed in a machined copper container, typically 250 mm diameter, with an end cap which is electron beam welded in place, again under high vacuum. The billet is hot extruded, which bonds the copper to the NbTi, so that the resulting bar may then be drawn down to hexagonal shaped rods. These rods are cut to length, cleaned and stacked in another copper canister, which is then extruded and drawn to wire. Although the drawing process is done cold, a series of heat treatments are applied at suitable intervals to precipitate a second (non superconducting) phase of the alloy, which provides the pinning centres needed for high current density [2]. Twisting and insulation or coating with a resistive surface treatment complete the process.

4.2   Cabling

Small magnets are usually made from single wires, but accelerator magnets must be made from cable to give them a high operating current, so that they may be energized in series without incurring excessive terminal voltages.   Strands within the cable must be twisted to minimize the coupled magnetization and fully transposed to ensure uniform current sharing between them.   By fully transposed, we mean that every wire must exchange places with every other along the length of the cable.   For example, a simple rope made by twisting 6 wires around a central core is OK for magnetization, but is not fully transposed, so that the centre wire does not carry its fair share of the current.   Rutherford cable, shown in Fig 16, is fully transposed and has the advantage that it may be heavily compacted by rolling, without much damage to the strands.

Fig 16: Rutherford cable.

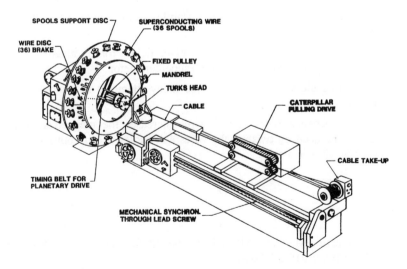

Fig 17: Manufacture of Rutherford cable.

Rutherford cable is manufactured on machines like that shown in Fig 17, in which the spools of wire are mounted on a wheel which rotates as the wire is pulled off the spools and the cable is pulled through the machine by the 'caterpillar' drive. A 'core pin' is situated at the point where the wires come together, so that the wires twist around it to form a hollow spiral tube. A little further downstream, this hollow tube is pulled off the core pin and into the rollers, where it is flattened into the shape shown in Fig 16. The rollers are arranged in a 'Turks head' configuration, which means that they compress all four faces of the cable simultaneously. Careful setting up is needed to optimize this process but, with care, dimensional tolerances of a few μm can be achieved – sufficient to satisfy most of the demands of the magnet makers.

5 Concluding Remarks

In this section we have shown how the demands of accelerator magnets make special demands on the conductor, and these demands have strongly influenced the way conductors are made. To summarize:

- sudden releases of energy in the magnet winding cause quenching
- we can make conductors less vulnerable to these energy releases by increasing the MQE, which means increasing the size of the MPZ
- adding copper helps enormously, but the superconductor must be finely divided to keep it in good thermal contact; fine subdivision is also needed to avoid flux jumping [2]
- measurements of MQE can quantify stability of the conductor against energy inputs
- magnetization of the superconductor gives field errors in magnets and must be reduced by fine subdivision of the superconductor
- practical accelerator conductors need many filaments in a single wire and many wires in a cable
- coupling between filament and wires increases the magnetization, which must be reduced by twisting the wires and cables and by putting resistance between wires in a cable.

References

1) Superconducting Magnets, MN Wilson, pub Oxford University Press (1983), pp108, ISBN 0-019-854805-2
2) Tigner M. 'Fundamentals of Superconductivity' Lecture at this school.

3) Carr W.J. Jnr IEEE Trans. MAG-13 (1) pp192 (1977).
4) Den Ouden A. et al IEEE Trans Appl. Superconductivity 7, No 2 pp733, (1997)
5) Bottura L. CERN: private communication.

QUENCHING, CURRENT LEADS AND A LOOK AT SOME SUPERCONDUCTING ACCELERATORS

MARTIN N. WILSON

Oxford Instruments (retired), Brook House, 33 Lower Radley, Abingdon, OX14 3AY, UK.
email: m-wilson@primex.co.uk

In this third and final lecture, we discuss the problem of quenching, where a part of the magnet suddenly reverts to the resistive state and can get very hot, unless suitable steps are taken to protect it. Current leads, supplying the magnet at low temperature from a power supply at room temperature, must be carefully designed if they are not to introduce too much heat leak. Finally we look at some of the large superconducting accelerators which have been built using the technology described in these lectures .

1   Quenching

1.1   Introduction to the Process

As mentioned in earlier lectures, quenching is what happens when any part of the conductor in a magnet goes from superconducting to resistive state. Because current densities are so very high, the Ohmic heating after a resistive transition is intense. In this regard, it is worth noting that the current density of a typical superconducting wire is higher than the failure current of a domestic fuse wire. When quenching occurs in a short sample of superconducting wire, the rapidly increasing resistance cuts off the current. In a magnet however, the self inductance of the circuit continues to force the current through, despite the increasing resistance. Thus we often run into problems of voltage $V = L \cdot dI/dt$ as well as heating.

The total energy available for dissipation as heat is the inductive stored energy

$$E = \frac{1}{2} L I^2 = \int \frac{B^2}{2\mu_o} dv$$

So for example, a twin aperture LHC dipole, with a self inductance of 0.118 Henry, stores 7.8MJ at its operating current of 11.5kA. This energy is equivalent to the kinetic energy of the 26 tonne magnet if it were travelling at 88km/hour – a very substantial kick! Another way of thinking about it is in terms of field energy, for example a field of 10T stores 40MJ/m^3.

Fig 1 illustrates the quenching process; a small resistive zone R_Q has been initiated, perhaps by a conductor movement. It starts to generate heat which is conducted along the wire, thereby raising the temperature of neighbouring regions and causing the resistive zone to grow.

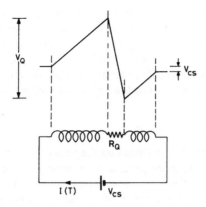

Fig 1: Circuit diagram of a quenching magnet.

The hottest part of the magnet is always the point where the quench starts, because it is here that the heating goes on for longest. Because the rapid variation of resistivity and specific heat make it difficult to calculate this temperature rise directly, we use a simple trick, originally known as the 'fuse blowing calculation'. For the rather short times involved, we may assume adiabaticity, write the heat balance and then move all temperature terms to one side of the equation:

$$J^2(T)\rho(\theta) = \gamma C(\theta)d\theta$$

$$\int_0^\infty J^2(t)dT = J_o^2 T_Q = \int_{\theta_o}^{\theta_m} \frac{\gamma C(\theta)}{\rho(\theta)} d\theta = U(\theta_m) \tag{1}$$

where: J = overall current density, T = time, ρ = overall resistivity, γ = density, θ = temperature, $C(\theta)$ = specific heat. The function $U(\theta)$, which is a function of the material only, is plotted for some common magnet materials in Fig 2.

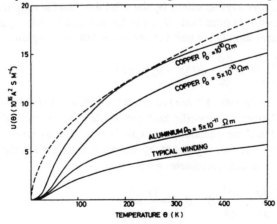

Fig 2: The function U(θ), as defined by equation (1) for some common magnet materials.

Note that, for the 'typical winding' (which is an appropriate mixture of NbTi, copper and epoxy resin) the current density J is defined over the whole cross section of the winding. From eq (1) and Fig 2, we see that the question of predicting the peak temperature rise after a quench simplifies to that of predicting the current decay time T_Q. The remainder of the section will be concerned with how to calculate T_Q and, if it turns out to be too big, how to reduce it.

1.2   Quench Propagation Velocity

The most important factor in determining the decay time after a quench, and hence the peak temperature rise, is the *propagation velocity* of the resistive zone. A fast velocity will ensure that the resistive volume grows quickly with time, so that the higher resistance forces a rapid decay of current; alternatively we may say that the magnetic stored energy is dissipated over a larger volume so that the temperature rise is less.

Fig 3 illustrates the temperature distribution at the boundary between a resistive zone in a conductor and one which is still superconducting.

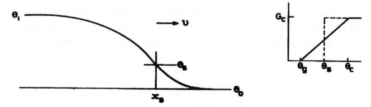

Fig 3: Temperature distribution in a superconductor with resistive zone advancing from left to right.

In a pure superconductor the power generation varies abruptly with temperature, as sketched by the dashed line on the right of Fig 3. In a composite wire with copper, the transition is progressive as shown by the solid line, but for simplicity we approximate by an abrupt transition. Writing the heat conduction equations with a resistive power generation G per unit volume in the left hand region.

$$\frac{\partial}{\partial x}\left(kA\frac{\partial\theta}{\partial x}\right)-\gamma CA\frac{\partial\theta}{\partial t}-hP(\theta-\theta_0)+GA=0 \tag{2}$$

where P = cooled perimeter, A = area of cross section, k = thermal conductivity, h = heat transfer coefficient, C = specific heat and γ = density. In the right hand region, we have exactly the same equation, but with $G = 0$. Let us now assume that the boundary between these two regions at x_s moves to the right at velocity v; we may write (2) in terms of a new coordinate $\varepsilon = x - x_s = x - vt$:

$$\frac{d^2\theta}{d\varepsilon^2}+\frac{v\gamma C}{k}\frac{d\theta}{d\varepsilon}-\frac{hP}{kA}(\theta-\theta_o)+\frac{G}{k}=0 \tag{3}$$

For the simplest case of $h = 0$ the solution of (3) which gives a continuous join between left side $(G \neq 0)$ and right side $(G=0)$ gives the *adiabatic propagation velocity* v_{ad}:

$$v_{ad} = \frac{J}{\gamma C} \left\{ \frac{\rho k}{\theta_s - \theta_0} \right\}^{\frac{1}{2}} = \frac{J}{\gamma C} \left\{ \frac{L_0 \theta_s}{\theta_s - \theta_0} \right\}^{\frac{1}{2}} \tag{4}$$

where, for the second expression we have substituted the Wiedemann Franz Law:

$$\rho(\theta)k(\theta) = L_0 \theta \tag{5}$$

The resistive zone also travels in a direction perpendicular to the conductor; equation (3) applies in all respects except for the thermal conductivity, which is considerably less. Thus we may write α, the ratio of transverse to longitudinal velocities:

$$\alpha = \frac{v_{transverse}}{v_{longitudinal}} = \left\{ \frac{k_{transverse}}{k_{longitudinal}} \right\}^{\frac{1}{2}} \tag{6}$$

Although the transverse velocity is only 1-2% of the longitudinal velocity, it is a very important factor in the quenching process because it produces a three dimensional growth of the resisitive zone, thereby making the resistance grow much more quickly.

1.3   Growth of the Resistive zone.

Here we develop a simple theory of how the resistive zone grows with time. The model makes the following simplifying assumptions:
   a)   the winding is a homogeneous, but anisotropic material.
   b)   temperature rise is given by
$$\int J^2 dt = J_o^2 T_d = U(\theta) \cong U(\theta_o)(\theta / \theta_o)^{1/2}$$
   where we have used a parabola to approximate the shape of Fig 2.
   c)   resistivity increases linearly with temperature.
   d)   the current remains constant at its starting value until all the inductive stored energy of the magnet has been dissipated, then it falls to zero.

As shown in Fig 4, after a time T, the normal zone has grown to an ellipsoid of semi axis $X = vT$. We calculate the total resistance of this ellipsoid by integrating over the set of inner ellipsoids with $x < X$, which are at higher temperatures.

$$R = \int_0^X \frac{4\pi x^2 \alpha^2 \rho(\theta)}{A^2} dx \tag{7}$$

where v = longitudinal velocity, a = ratio transverse/longitudinal velocity and A is the cross sectional area of a single conductor.

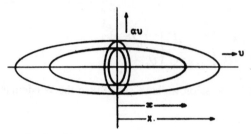

Fig 4: The resistive zone after time T.

The temperature of these inner zones may be related to time by assumption c) above.

$$\rho(\theta) = \rho_o \left(\frac{\theta}{\theta_o} \right) = \rho_o \left(\frac{U}{U_o} \right)^2 = \rho_o \frac{J_o^4 \tau^2}{U_o^2} \tag{8}$$

where τ is the elapsed time since normality; at the centre $\tau = T$, at the edge $\tau = 0$ and in between $\tau = T - x/v$. Thus we find:

$$R = \int_0^{vT} \frac{4\pi \, x^2 \alpha^2 \rho_0 J_0^4 (T - x/v)^2}{A^2 U_0^2} dx = \frac{4\pi \rho_0 \, \alpha^2 J_0^4 v^3 T^5}{30 A^2 U_0^2} \tag{9}$$

where ρ_0 = resistivity at θ_0, $U_0 = U$ function at temperature θ_0 and J_0 = current density at start. Finally, using approximation d), we equate the total Ohmic heating to the initial stored energy.

$$\int_0^{T_Q} I^2 R(T) dT = E \tag{10}$$

from which we may calculate the characteristic quench time:

$$T_Q = \frac{1}{J_0} \left\{ \frac{45 U_0^2 E}{\pi \rho_0 \alpha^2 v^3} \right\}^{\frac{1}{6}} \tag{11}$$

The temperature rise then follows from Fig 2. The current decay and voltage across the resistive zone are easily shown to be:

$$I = I_0 e^{-T^6/2T_Q^6} \tag{12}$$

$$V = L\frac{dI}{dt} = -LI_o \frac{3t^5}{T_Q^6} e^{-T^6/2T_Q^6} \tag{13}$$

which are plotted in Fig 5, and where it may be seen that assumption d), indicated by the dotted line, is not as drastic as it might first have appeared.

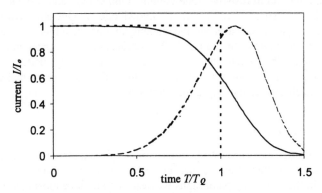

Fig 5: Current decay (solid line) and voltage (dashed line) after a quench, as calculated by eqs 12 & 13.

1.4   Boundaries

Of course the actual quench process in magnets is more complicated than our simple model. One complicating factor is that the resistive zone will hit one or more of the coil boundaries after a certain time. From this time onwards, the rate of growth of resistance is not so great as before and so the decay time increases. Extending the simple theory [1] it may be shown that if, at time T_a, the resistive zone hits a coil boundary which prevents resistive growth in one dimension, the decay time is extended to:

$$t_d \approx \left\{ \frac{1}{3t_a} \right\}^{1/5} \tag{14}$$

where $t_a = T_a / T_Q$, T_d is the extended decay time and the normalized decay time $t_d = T_d / T_Q$.
If the resistive zone hits one boundary at T_a and the next boundary, stopping growth in the second dimension at T_b, the normalized decay time is extended to:

$$t_d \approx \left\{ \frac{2}{15t_a t_b} \right\}^{1/4} \tag{15}$$

Finally, if the resistive zone hits boundaries which truncate its growth in each of the three dimensions at times T_a, T_b and T_c, the normalized decay time is:

$$t_d \approx \left\{ \frac{1}{20t_a t_b t_c} \right\}^{1/3} \tag{16}$$

Although these simple formulae can be very useful in making a first assessment of whether quenching is going to be a problem, the approximations involved can cause substantial errors, particularly when the resistive zone hits coil boundaries with a complicated geometry. In this situation, it is better to make a numerical simulation of the process, using one of the computer programmes such as QUENCH, which step through the build of resistive zone volume, temperature and resistance in a series of finite time steps. To the author's knowledge, no Quench software is available on a commercial basis and the student can only be advised to beg borrow or steal a copy from one of the national laboratories active in this technology. On one point however the student should be quite clear: it is always better to make a prediction of quench behaviour <u>before</u> the magnet's first test than to make a post mortem on the charred remains <u>after</u> the first test!

1.5   Quench protection

If, as is often the case, predictions indicate an excessive temperature rise or voltage after a quench, the magnet designer must use some form of quench protection. In the following paragraphs we describe the three most common.

1.5.1   External dump resistor

The simplest approach, which is often used in large superconducting magnets, is to dump most of the inductive stored energy into a resistor situated outside the cryostat, as sketched in Fig 6.

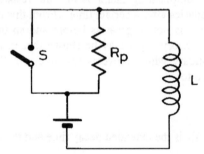

Fig 6: Quench protection by an external dump resistor.

On detecting a quench electronically, the circuit breaker S is opened, forcing the current to decay through the resistor R_p. To a fair approximation, we may neglect the voltage across the power supply and quenched region, which means that the decay time will be:

$$\tau = e^{-\frac{L}{R_p}} \tag{17}$$

Using the $U(\theta)$ plot of Fig 2, the decay time should thus be chosen to keep maximum temperature θ_{max} below the desired value. A very conservative criterion would be $\theta_{max} < 80K$, which would ensure very little thermal expansion, and consequent mechanical stress, in the quenched region. Another popular criterion is

$\theta_{max} < 300$K. Given the unfavourable curvature of $U(\theta)$, choosing $\theta_{max} > 300$K is not recommended, because a small error in τ can produce a large error in θ_{max}. The absolute upper limit on θ_{max} is the softening temperature of the insulation, usually ~ 400K to 500K.

Important design considerations here are that the circuit breaker should be able to open at full current against a voltage $V = IR_p$, which may be many kV in a large magnet, and that R_p is able to absorb the stored energy without excessive temperature rise.

1.5.2   Quenchback heaters.

As may be seen from eq 10, a good way of reducing τ would be to increase v. Effectively, this result may be achieved by means of heaters attached to different parts of the winding, which create a series of new resistive zones, all of which will grow with time like the first zone.   Thus the rate of growth of resistance is increased and the current decay time is decreased.   Fig 7 sketches the arrangement.

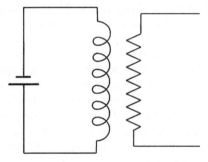

Fig 7: Quench protection by quench back heater.

Important design considerations are that the heater must be in excellent thermal contact with the winding, without compromising the electrical insulation between the two, and that the heater pulsing circuit must be 100% reliable.

1.5.3   Quench detection.

Both the above methods are known as *active* techniques, because they require positive action when the quench is detected. If we are to minimize the heating time of the quench initiation point, it is essential that the action is taken immediately after the quench starts.   Unfortunately, as may be seen from eq 13 and Fig 5, the voltage to be detected at this time is very small. Because superconducting magnets tend to have a high self inductance, the voltage noise produced by quite small current ripples can be large – and of course when the magnet current is ramped up or down, there will be a large dc voltage across the magnet terminals. It follows that an efficient quench detector must be able to reject all the inductive voltage

336

LdI/dt and only accept resistive voltage *IR;* Figs 8 and 9 sketch two simple circuits designed to achieve this objective.

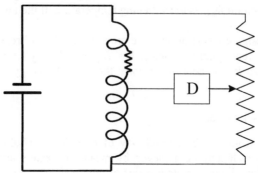

Fig 8: Quench detection by means of a balanced potentiometer circuit, magnet circuit shown heavy, detector light.

Referring to Fig 8, the detector *D* is placed in the sensor arm of the potentiometer circuit so that it sees any out-of-balance current between the top and bottom halves of the magnet. In normal ramping, the two halves are symmetrical and the potentiometer should be adjusted to give zero signal at the detector. A quenched resistive region will then break the symmetry and give a signal. In fact it is not necessary for the middle voltage tap to be in the exact centre of the magnet; it can be anywhere as long as the potentiometer is adjusted to give zero inductive signal. To guard against the remote chance of a resistive region growing symmetrically on either side of the middle voltage tap, one may chose to have two detector circuits, connected to an *and* gate, with their middle taps at different points in the magnet.

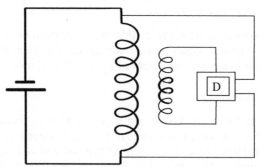

Fig 9: Quench detection using a mutual inductance.

Fig 9 shows how voltage from the magnet self inductance may be opposed by the signal from a mutual inductance, which may for example be a toroid around the current lead:

$$V = L\frac{di}{dt} + IR_Q - M\frac{di}{dt} \tag{18}$$

Within the detector, gains should be adjusted to equalize the self and mutual inductance signals.

1.5.4 Subivision

The big drawback of all active protection systems is the difficulty of guaranteeing 100% reliable operation of the circuits involved throughout the 10 – 20 year working life of a magnet – remember just one failure means death for the magnet! For this reason the method of subdivision is often used – particularly in small magnet systems intended for general purpose use in research. As shown in Fig 10, the subdividing resistors allow current to by-pass the quenched section so that, very roughly speaking, the energy driving current through the resistive zone is the stored energy of a single section rather than the whole magnet. Very often, each of the resistor arms contains a pair of diodes connected 'back to back' so that they only conduct above the breakdown voltage of the diode. In this way, the protection circuit only operates when it is needed and does not interfere with the operational processes of ramping current up and down.

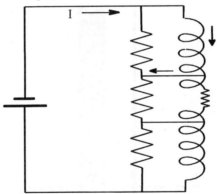

Fig 10: Protection by subdivision.

2   Current Leads

2.1   General problem

Magnets must be kept cold, whereas their current supply is located at room temperature and current leads are needed to connect the two. Because refrigeration at low temperatures is so very expensive, we must design the leads so that they introduce the very minimum heat inleak. This heat comes from two sources: ohmic heating within the lead and conduction down from room temperature. To minimize

338

it we therefore want to minimize the electrical resistivity and thermal conductivity. Unfortunately, nature does not help us here, because eq 5 says that if we choose a material to reduce one, the other must increase. Thus to a first approximation, all (metallic) materials are equivalent and we can only minimize the heat leak by engineering design.

One important point to note here is that helium is a very unusual liquid in that it has a very small latent heat – it does not really want to be a liquid! Thus a small heat leak will boil off a large amount of gas. However, this gas still contains a lot of 'cold'; a heat input of 1 Joule will boil off 48 mg of gas, but a further 74 Joules are needed to warm the gas up to room temperature. For this reason, we look for designs which utilize the boil-off gas to cool the leads.

2.2 Optimizing the Shape of a Current Lead

Fig 11 sketches a current lead, carrying current I from room temperature to liquid helium, where it boils off a mass $\dot{m}$ of liquid per second to produce an upstreaming flow of cold gas, which is then made to exchange heat with the lead.

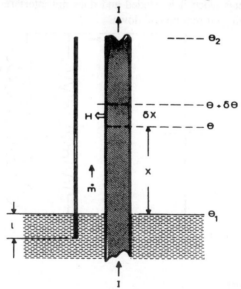

Fig 11: Schematic of a current lead exchanging heat with the boiled off helium gas.

The equation of thermal equilibrium for a section of the lead at distance x above the liquid surface is:

$$\frac{d}{dx}\left(k(\theta)A\frac{d\theta}{dx}\right) - f\dot{m}C_p\frac{d\theta}{dx} + \frac{I^2\rho(\theta)}{A} = 0 \qquad (19)$$

where f is a factor representing the efficiency of heat exchange between the gas and lead. Equation (17) may be solved numerically, taking into account the strong variation with temperature of resistivity and thermal conductivity. However, by transforming the variables and assuming the Weidemann Franz Law, one may show analytically that there is an optimum shape which minimizes $\dot{m}$ [1]. Fig 12 plots the optimum heat leak per unit current versus the efficiency of heat exchange f; the plot applies equally to all leads of optimum shape, all currents and all materials, provided they obey the Weidemann Franz Law.

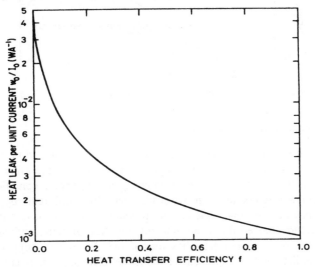

Fig 12: Heat leak per unit current of leads with optimum shape.

The optimum shape does depend on material however; not surprisingly the higher conductivity materials need long thin leads whereas lower conductivity leads need short fat ones. Fig 13 plots the thermal conductivity of two different types of copper.

By numerical integration of the thermal and electrical conductivity, we find the optimum length to area ratio L/A for current leads running from room temperature to liquid helium using the two different copper shown in Fig 13:

- with annealed high purity copper
$$\left\{\frac{L}{A}\right\}_{optimum} = \frac{2.6 \times 10^7}{I}$$

- with phosphorous deoxidised copper
$$\left\{\frac{L}{A}\right\}_{optimum} = \frac{3.5 \times 10^6}{I}$$

Note that it is only the ratio L/A that matters, the actual length used can be adjusted to fit in with practical requirements in the cryostat.

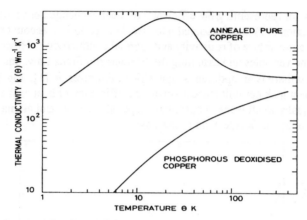

Fig 13: Thermal conductivity of two different current lead materials.

2.3 Stability against Overcurrent

At first sight, it might seem that the best choice of material would be good quality copper, but this is not the case. Although all materials give the same heat leak at optimum shape they respond differently to changes in operating conditions. Fig 14 plots the maximum temperature in two leads, made from different coppers, against current. It may be seen that, whereas at optimum both leads have a maximum of room temperature (indeed this is an inherent condition of being optimum), at 10% overcurrent the pure copper lead has melted but the impure copper has barely changed. Similar relationships can be found for changing other parameters, like heat exchange to the gas, and show that impure materials, because their conductivity does not change so much with temperature, are always much more stable.

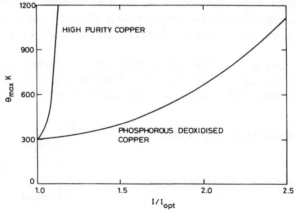

Fig 14: Maximum temperature in the current lead as a function of excess current.

2.4 Health Monitoring

The consequences of overheating in a current lead can be dire. If the lead actually melts and thereby opens the magnet circuit, it will cause an arc, into which will be dumped the whole stored energy of the magnet. Even modest temperature rises, which can degrade the electrical insulation, will cause trouble, for example by creating a short circuit to ground. For a magnet of any size or importance, it therefore makes sense to monitor the maximum temperature and switch off the magnet current if this rises too high. The most convenient way of monitoring temperature is via the relationship, plotted in Fig 15, between voltage drop along the lead and its maximum temperature. This relationship applies to all leads. all currents and all materials, provided they obey the Weidemann Franz Law. It may be used, not only for health monitoring, but also for determining the optimum current of a lead made from unknown material – quite simply, all leads at optimum current have a voltage drop of 75mV.

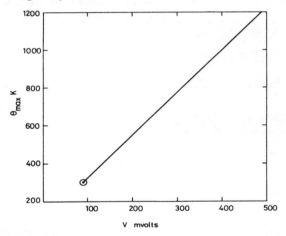

Fig 15: Maximum temperature in a current lead as a function of voltage drop along it.

3 Tour of Superconducting Accelerators

We conclude with a brief tour around some of the world's major superconducting accelerators, all of which have made use of the technology described in these three lectures. Table 1 lists the main parameters of the dipole magnets, which are by far the most numerous type of magnet in any accelerator, and Table 2 summarizes the superconducting cables used. Design of the magnets has evolved steadily with the developing technology; Fig 16 presents a montage of cross sections of the different dipole magnets, all drawn to the same scale. In the following paragraphs, we briefly review the accelerators themselves.

Table 1: parameters of accelerator dipole magnets

	Tevatron	HERA	RHIC	LHC	Helios
max energy GeV	950	820	250 (x 2)	7,000 (x 2)	0.7
max field T	4.4	4.68	3.46	8.36	4.5
max current kA	4.4.	5.03	5.09	11.5	1.04
injection field T	0.66	0.23	0.4	0.58	0.64
aperture mm	76	75	80	56	58
length m	6.1	8.8	9.4	14.2	1.6
operating temperature K	4.6	4.5	4.6	1.9	4.5
number off	774	422	396	1232	2

Table 2: Rutherford Cables for Accelerator Magnets

Accelerator / cable	filament dia μm	cable width mm	twist pitch mm	wire surface
Tevatron	6	7.8	66	zebra
HERA	14-16	10	95	SnAg
RHIC	6	9.7	73	copper
LHC	7	15	115	SnAg pre-ox
Helios	8.5	3.2	40	copper

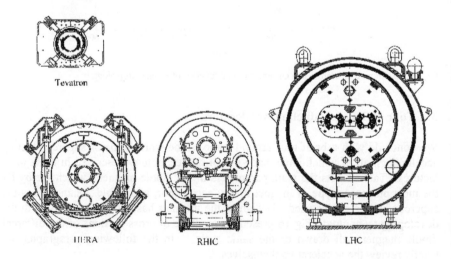

Tevatron

HERA   RHIC   LHC

Fig 16: Cross sections of the dipole magnets used in the world's major superconducting accelerators.

3.1 Tevatron

The first superconducting accelerator to be built, the Fermilab Tevatron [2] was a pioneering enterprise in many respects. It showed that superconducting magnet systems do not have to be complicated and expensive; by careful design and use of mass production techniques costs may be kept down to a very competitive level. Precision stamping techniques, already proven in the production of iron laminations for conventional magnets, were used to make force supporting collars which were both cheap and extremely accurate. The iron yoke was located at room temperature, which made for a very compact design, but brought the possibility of errors coming from misalignment of the cold coils within the warm iron.

3.2 Hera

The Hera collider facility, located in Hamburg Germany, consists of a 30 GeV conventional electron ring and an intersecting 820 GeV proton ring [3]. Force supporting collars are made from aluminium alloy, which has a better contraction match to the coil and therefore maintains a good pre-compression during cooldown. The iron yoke is cold and situated immediately outside the collars. This arrangement gives a very rigid cold mass, which therefore needs fewer supports to room temperature. An additional advantage is that the buss bar for returning the magnet current may be located in the same cryostat, outside the iron yoke where it does not affect the aperture field quality.

3.3 RHIC

The Relativistic Heavy Ion Collider [4] now nearing completion at Brookhaven National Laboratory will accelerate and collide a wide range of particles, from protons to gold ions. The dipoles have just a single layer winding, which is nevertheless able to produce a field quality better than the Tevatron and as good as Hera. There are no collars and the forces are supported directly by the iron yoke, which is separated from the winding by a phenolic spacer. An important first for RHIC was the decision to cold-test only a 20% sample of magnets before installation. This decision was justified by the excellent correspondence between warm and cold field shape and the generous margin (~40%) of quench currents above the operating current. Fig 17 shows the prototype for an interesting series of helical dipole magnets, which will be used in RHIC for particle spin rotation [5].

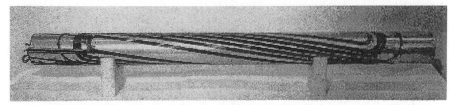

Fig 17: Prototype helical dipole constructed at Brookhaven.

344

3.4 LHC.

The most powerful accelerator so far, the Large Hadron Collider [6] [7] will be built in the tunnel of the existing LEP ring at CERN. In order to achieve the highest energy within these constraints, the dipole field is being pushed up to 8.4T by operation in subcooled superfluid helium at 1.8K. The reduced temperature adds an extra 3T to the capability of NbTi, while the high thermal conductivity and heat capacity of the superfluid, which is subcooled and at atmospheric pressure, improves stability of the magnets. As shown in Fig. 16, the magnets are built to an elegant 'two in one' design in which the fringe field of one aperture slightly assists the field in the other aperture. Advantages this design include better alignment between the two apertures and reduced cost of the cryostat and iron. Fig 18 shows the 'string test' at CERN in which a half cell of the ring is being tested; the complete ring comprises 368 such half cells plus many other magnets, more than 8000 magnets in total.

Fig 18: The string test of an LHC half cell at CERN.

3.5 Helios.

Much smaller than any of the other accelerators in this survey, Helios [8] was nevertheless the only one to achieve a repeat order! Helios 1 was installed at the IBM Advanced Lithography Facility, East Fishkill, NY in 1991 for use as an intense X-ray source in the fabrication of microchips. Helios 2 is being installed at the National University of Singapore, also for work on lithography and microfabrication. The compact ring contains just two dipoles, which must therefore be curved around a 180° arc. Another difficult engineering constraint is that the dipoles must have a completely clear gap along their outer radius mid plane, to allow a clear exit for the X-ray beam produced in the dipole by synchrotron radiation. In this application, the key advantage of superconductivity is compactness, which allows the complete storage ring to be transported intact as shown in Fig 19.

Fig19: Helios 1 arrives by road for installation at IBM East Fishkill.

4   Concluding Remarks

Ideas for a superconducting accelerator were first aired in public at the 'Summer Study on Superconducting Devices and Accelerators', which was held at the Brookhaven National Laboratory in July 1968.   The technology has come a long way in the 30 years since that meeting.   Filamentary superconductors have solved the problems of flux jumping and field distortion due to magnetization of the superconductor.   Techniques have been developed for the economic mass production of coils to high precision with supporting structures which a sufficiently stiff to prevent the coils moving appreciably at the high stress levels generated by electromagnetic forces.   Much progress has been made in reducing the problem of training, but it is still not cured and remains a point of concern in large accelerator systems – when 1000 magnets are connected in series, the strength of the chain is determined by its weakest link. Quench protection is well understood and computer software is available to predict magnet behaviour following a quench; there is now no excuse for burnt out magnets – although it still happens.

Although LHC is still many years away from completion, ideas are already flowing on what the next large accelerator should be.   One possibility is to go for a proton ring at much higher energy than the LHC – the so called VLHC, with 50 TeV per beam. Such a machine, although technically feasible, will be extremely expensive, and new ideas are needed to achieve the necessary economies. The "pipetron" design hopes to reduce costs per GeV by a factor ~10 in low field magnet design where the field shape is dominated by the iron, and the winding is just one turn of a high temperature superconductor HTS cable now being developed for power transmission. An alternative approach is to go for much higher fields of ~12T thereby reducing the perimeter to ~100km and the infrastructure costs accordingly. Very high field conductors such as niobium tin will be needed to achieve these high fields. Given the recent success of several niobium tin prototypes [9] [10], there can be no doubt that such magnets are technically feasible, but it remains to be seen whether the severe materials and stress problems can be solved in a cost effective way.

A quite different possibility is to switch from protons to a different particle, the muon. Unlike protons, muons are fundamental particles and, when they interact, all their energy is available for the production of new states and this energy is well defined. However, they decay very quickly and very high fields are needed to contain them in compact intersecting storage rings so that they have a reasonable chance of colliding before they decay. The combination of high field and high radiation heat input from the decay products of the muon beam make this an ideal application for the new high temperature superconductors, if they can be developed to have adequate current density in high field at temperatures higher than 4.2K.

Whatever the future brings, there can be no doubt that superconducting accelerators will continue to be a lively area of technology which will eventually make some exciting demands on the talents of students from this first Asian Accelerator School.

5 References

1) Superconducting Magnets, MN Wilson, pub Oxford University Press (1983), pp108, ISBN 0-019-854805-2
2) H.T. Edwards, "The Tevatron energy doubler", Ann. Rev. Nucl. Part. Sci. 35 pp605 (1985)
3) K.H. Mess, P. Schmusser, S. Wolf, "Superconducting Accelerator Magnets" published by World Scientific 1996, ISBN 981-02-290-6.
4) P. Wanderer, "Status of RHIC construction", to be published in Proc. MT15 Beijing 1997.
5) E. Willen R. Gupta,A Jain, E Kelly, G. Morgan J. Muratore, R Thomas, "A helical dipole design for RHIC", Proc 1997 Particle Accelerator Conference, Vancouver.
6) LHC study group, "The large hadron collider", CERN report CERN/AC/95-05(LHC) (1995).

7) N.Siegel, "Status of the large hadron collider", IEEE Trans. Appl. Supercon. **7**, 2 pp 252 (1997).

8) M.N. Wilson, A.I.C.Smith, V.C. Kempson, M.C. Townsend, J.C. Schouten, R.J. Anderson, A.R. Jorden, V.P Suller, M.W. Poole, "The Helios compact superconducting storage ring X-ray source, IBM Jnl. Res. Dev. **37** pp351 (1993).

9) Den Ouden A. et al IEEE Trans Appl. Superconductivity 7, No 2 pp733, (1997)

10) AD MacInturff et al "Test results for a high field (13T) Nb_3Sn dipole magnet" Proc. 1997 Particle Accelerator Conf., 12-16 May, Vancouver, B.C., Canada pp3212.

SUPERCONDUCTING CAVITY

TAKAAKI FURUYA

KEK, High Energy Accelerator Research Organization
1-1 Oho, Tsukuba, Ibaraki, 305-0801 Japan
E-mail: takaaki.furuya@kek.jp

1 Introduction

Because of low surface resistance due to superconductivity, the RF wall loss of superconducting (SC) cavities is extremely small. As a result, the quality factor Q_0, which is inversely proportional to cavity loss, is quite high for SC cavities. Furthermore, this small loss allows the SC cavities to operate in continuous-wave (CW) or long-pulse mode at high field gradient. Since the surface loss increases exponentially as the square of the accelerating field, the loss of normal conducting (NC) Cu cavities reaches several MW/m at 10 MV/m, which results in poor efficiency of RF power and problems with cooling. Hence the operation of NC cavities is limited to the short-pulse mode with a low duty cycle (= pulse width × repetition rate) of <0.1% at high field or the CW mode at 1~2 MV/m. On the other hand, the loss of SC cavities at 10 MV/m is of the order of 10 W/m and is reasonably small for cooling by a refrigerator system. Therefore, SC cavities are characterized by two factors: high Q_0 and CW operation at high field gradient. In this lecture, a brief history of SC cavity development and applications will be given first, and then a discussion of fundamentals and the SC cavity system in a real application.

2 Historical aspect

Aiming for CW operation at a high accelerating voltage, SC cavities have been intensively developed, especially in the particle accelerator field, for four decades. The first beam acceleration using an SC cavity was carried out by HEPL at Stanford University in 1965[1]. A 2856-MHz 3-cell cavity accelerated a 1 µA electron beam with a CW field gradient of 2~3 MV/m. This cavity was made of lead-plated copper. After that, the cavity material was switched to pure niobium (Nb) because Nb is the elemental metal with the highest superconducting critical temperature T_c of 9.25 K and it undergoes no degradation of the surface like the oxidization of Pb. A single-cell 8.6-GHz Nb cavity of shown in Fig. 2.1

Figure 2-1: 8.6-GHz Nb cavity at HEPL[2].

achieved a peak magnetic field of 1080 Oe in 1970, which corresponded to an accelerating gradient of 20~30 MV/m[2]. Since then, Nb has been the main material used for SC cavities.

In the 1970s, much fundamental work on SC cavities was done at various laboratories throughout the world. Main work was concentrated on establishing a fabrication procedure to obtain a defect-free Nb surface. However, the maximum field gradient of multicell structures was still 2~3 M/m. In metal research work, Nb$_3$Sn cavities were studied because of their superior SC potential (T_c = 18.2 K, H_c = 5400 Oe), where the Nb$_3$Sn layer was made by vapor diffusion of tin into the Nb surface at 1050°C. But neither surface resistance nor field strength of Nb$_3$Sn did not reach that of pure-Nb cavities[3]. In the research on cavity shape, U. Klein and D. Proch introduced a spherical cavity shape in 1978 to avoid multipacting[4]. The field limitation of 2~3 MV/m had been caused by multipacting discharge in the cavity. So they calculated the electron trajectory and optimized the cavity shape to avoid one-point multipacting. In a spherical cavity the emitted electron is accelerated toward the equator within a few RF periods, where no electric field exists, and stops without any resonant condition, as shown in Fig. 2.2. Since then, the spherical shape has been the standard shape of SC cavities in high gradient applications. This shape has the further advantage that the round and smooth curvature makes it easy to wash the inner surface after chemical treatment. The investigations during this period made it clear that cleanliness was important not only for the cavity surface but also for the whole cavity set-up.

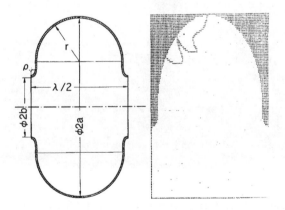

Figure 2.2: A spherical shaped cavity and an electron trajectory[4].

One of the most successful applications in the 1970s was heavy ion linacs, where CW operation is essential to obtain a precise beam, even though the achievable field is 2~3 MV/m. Furthermore, excellent power efficiency is possible, because the cavity loss becomes comparable to the beam power of several watts. To accommodate a very slow particle speed, the accelerating gap is made narrow in a high frequency cavity and it causes discharging. Thus low-β ($= v/c$) structures at a rather low frequency of 50~350 MHz were developed. Figure. 2.3 shows the split-ring structures used in ATLAS at ANL[5], which have drift tubes in the cavity to match the RF period to the β of 0.06~0.16. Further developments on low-β cavities resulted in the quarter-wave resonator (QWR), shown in Fig. 2.4, which became the standard for low-β structures. Because particle velocity changes with beam energy, a variety of structures with a frequency of 99~145 MHz and a β of 0.01~0.3, were combined in ATLAS. Since its commissioning in 1978 ATLAS has operated as the first large-scale application and holds the world record for longest running time. Other SC applications of low-β structures in operation are shown in Table 2.1. In Asia of today, R&D on SC-QWR is successfully progressing in China at Peking University and in India in New Delhi and Bombay.

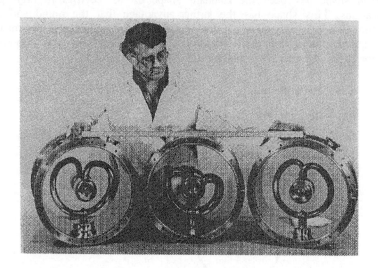

Figure 2.3: Split-ring resonators for ATLAS with various β [5].

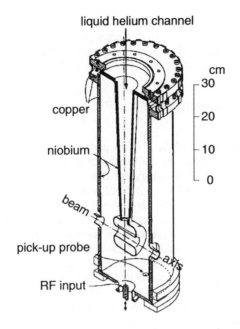

liquid helium channel

cm
30

copper

20

niobium

10

0

beam

pick-up probe

axis

RF input

Figure 2.4: Quarter-wave resonator in JAERI[6].

Table 2.1: Low-β SC systems in operation

Place	MHz	Cavity	
ANL (ATLAS)	49-855	Nb, split ring, QWR	β = 0.009-0.3 since 1978
Stony Brook (SUNY)	150	Pb/Sn, QWR, split loop	β = 0.068-0.1 since 1983
Washington U.	150	Pb/Cu, QWR	β = 0.1-0.2 since 1987
Florida U.	97	Nb, split ring	β = 0.105 since 1987
Kansas U.	97	Nb, split ring	Since 1990
JAERI	130	Nb, QWR × 46	β = 0.08 at 3-5MV/m since 1994
INFN-LNL (ALPI)	80-240	Pb/Cu, Nb, Nb/Cu, QWR × 70	β = 0.056-0.17 since 1994

In the 1980s and 1990s, there was a big change in the application of the SC cavity to high energy accelerators. High energy physics needed large lepton colliders using storage rings, where high CW voltage was required to compensate the energy loss due to synchrotron radiation. Since electrons and positrons lose an energy of

$$\text{Synchrotron loss} = 0.0885 \frac{U^4}{\rho} \ \text{[MV/turn]}$$

at a bending radius of ρ (m), where U is the particle energy in GeV, the required RF voltage increases exponentially as the fourth power of the energy. For example, the total voltage of 218 MV had to be increased to 370 MV for upgrading the energy from 28 GeV to 32 GeV in KEK-TRISTAN, where the use of SC cavities was the only way to save the space for RF cavities and to upgrade the beam energy within the existing RF spaces. Hence, the development of SC multi-cell structures for $\beta = 1$ was strongly persued at many laboratories, such as KEK (TRISTAN/508 MHz), DESY (HERA/500 MHz), CERN (LEP/350 MHz) and Cornell University (CESR/1500 MHz). At CERN a Nb-sputtered Cu cavity was developed, in which a Nb thin film of a few μm was deposited by magnetron sputtering on the inner surface of Cu cavities, in order to increase the thermal stability against local heating due to defects, and to save the cost of Nb materials for a large number of 350-MHz cavities.

One of the most effective developments for improving cavity performance was the improvement of the Nb purity, which was brought about by increasing the melting times in the purifying process. High purity Nb has good thermal conductivity, which increases thermal stability against local heating caused by surface defects, and consequently pushes up the achievable voltage. The residual resistance ratio (RRR), which is defined as the ratio of electric resistivity between at 300 K and 4.2 K

$$RRR = R(300 \ K) / R(nornal \ state \ at \ 4.2 \ K) \ ,$$

is usually used as an indicator of Nb purity and roughly gives the thermal conductivity λ at 4.2 K by $\lambda \sim 0.25 \times RRR$. Multi-melting under precisely controlled vacuum pressure of the furnace improved the RRR of Nb from 40 to 300 in industrial production. Another way to purify Nb is the so-called yttrification or titanification, where Nb sheets or cavities are annealed with Y or Ti in a vacuum furnace at 1200-1400°C to deposit a layer of Y or Ti on the Nb surface. The gas components in Nb move and diffuse into this layer, which is chemically removed after annealing[7,8]. Figure. 2.5 shows the thermal conductivity of Nb samples with various RRRs.

As a result, the maximum field gradient of multi-cell structures was improved to 10 MV/m or higher in all laboratories. Of course the improvement of fabrication technology, such as welding, annealing, polishing, rinsing, assembling, etc., also had a great effect on this result. Another important factor was cleanliness.

Fortunately, because of semi-conductor production, many products for clean work were commercially available, such as a clean room, clean booth, pure water system, and so on. Consequently, we could handle the cavities in a CLASS 100 clean environment.

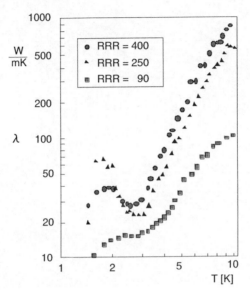

Figure 2.5: Thermal conductivity of Nb with various *RRR*[9].

After the beam tests in the 1980s, construction of SC cavities for large-scale applications started with a design gradient of 5 MV/m in each laboratory. In 1988, the first sixteen 508-MHz 5-cell SC cavities made of Nb (Figs. 2.6 and 2.7) were installed in KEK-TRISTAN and commissioned as the first large-scale SC system for high energy accelerators. The next sixteen cavities were added in the summer of 1989[10]. These 32 cavities supplied a total RF voltage of 200 MV with a gradient of 4-6 MV/m and stored a 14-mA beam. HERA (DESY), LEP (CERN), and CEBAF (J. Lab.) were also completed and commissioned, as listed in Table 2.2. Recently, CERN-LEP recorded a total RF voltage of 3.5 GV using 600-m Nb-Cu SC cavities. CEBAF is a recirculating system for nuclear physics

Figure 2.6: KEK-TRISTAN cavity.

research, which consists of two SC linacs with a total length of 165 m. The SC unit is based on the 1500-MHz 5-cell structure developed at Cornell University. The continuous beam is accelerated by these linacs in five recirculating passes.

Figure 2.7: Cryomodules in KEK-TRISTAN.

Table 2.2: High-β SC systems in operation

	Frequency (MHz)	cavity	
TRISTAN (KEK)	508	Nb, 5-cell × 32 L_{effv} = 48 m	1988 - 1995. V_c = 200 MV, 14 mA, 32 GeV
LEPII (CERN)	350	Nb/Cu , 4 cell × 288 L_{eff} = 600 m	1996 V_c = 3420 MV, 6 mA, 100 GeV
HERA (DESY)	500	Nb, 4-cell × 16 L_{eff} = 20 m	1990 50 mA, 32 GeV
CEBAF (J. Lab.)	1497	Nb, 5-cell × 330, L_{eff} = 165 m	1996 2K, 5-pass recirculating linac, V_c = 800 MV, 100 µA, 1-5.5 GeV

Today, for the future planning of a large SC linear collider, i.e. the TeV Energy Superconducting Linear Accelerator (TESLA), a test facility (TTF) has been established at DESY under a world-wide collaboration, aiming to explore the technology of high gradient 1.3-GHz SC cavities and to demonstrate the feasibility of a SC linac for TESLA. As shown in Table 2.3, TESLA consists of SC linacs of 33 km total length with a gradient of >25 MV/m using 1.3-GHz 9-cell structures (Fig. 2.8). Even for SC cavities with a Q of 1×10^{10}, the 25-MV/m gradient needs a high cooling power of >50 W/m in CW operation, so that the duty cycle has to be reduced to ~1%. Nevertheless, high luminosity operation is expected for TESLA, because, even at 1%, the duty cycle is much higher than that of NC linacs, which is ~0.001%. Furthermore easy transportation and focusing of the beam due to the large aperture size of 1.3-GHz cavities seems to be advantageous for long distance linacs. Recent development of fabrication technology has improved the maximum gradient to 40 MV/m for single-cell cavities and to 27 MV/m for 9-cell structures [11].

Table 2.3: Baseline TESLA Parameters

c.m. Energy (GeV)	500	800	1600
Total length (km)	33	33	62
Gradient (MV/m)	25	40	40
N_b per rf pulse	1130	2260	2260
Bunch spacing (ns)	708	283	283
Repetition rate (Hz)	5	3	3
N_e/bunch (10^{10})	3.63	1.82	1.82
$\varepsilon_x^* / \varepsilon_y^*$ (10^{-6}m rad)	14/0.25	12/0.025	12/0.025
β_x^* / β_y^* (mm)	25/0.7	25/0.5	25/0.5
σ_x^* / σ_y^* (mm)	845/19	618/4.0	436/2.8
σ_z (mm)	0.7	0.5	0.5
δ_E (%)	2.5	2.2	6.7
Number of klystrons	616	1232	2464
P_b (2 beams) (MW)	16.3	15.6	31.2
P_{AC} (2 linacs)(MW)	99	115	230
Luminosity (10^{33}cm^{-2}s^{-1})	6	11	20

356

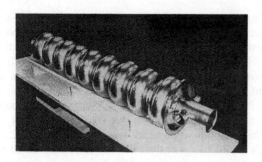

Figure 2.8: TESLA 1.3-GHz 9-cell Nb cavity[11].

A new application of SC cavities for high energy physics is based on the superior characteristics for a high intensity beam due to the high CW field, high stored energy, and low R/Q of SC cavities, which will be discussed in the last section. In the 1990s, not only an energy frontier but precise experiments using high intensity storage rings were considered, i.e., a B-meson factory, ϕ-meson factory and τ-charm factory, in which a beam of several amperes was designed to achieve a goal luminosity of 1-10 × 10^{33} cm^{-2} sec^{-1}. In such rings all higher-order modes (HOM) should be damped sufficiently to avoid beam instabilities that are excited mainly in RF cavities. From this point of view, a smaller number of cavities using SC is advantageous to reduce the total HOM impedance of the RF system. For B-factory storage rings at KEK and Cornell University, single-cell HOM-damped SC cavities have been developed and have been commissioned since 1997 at CESR (Cornell) and 1998 at KEKB (KEK). In CESR, the maximum total current of 0.35 A which had been limited by the instability caused by HOMs in NC cavities was improved to 0.65 A by replacing the NC cavities with four SC damped cavities, shown in Fig. 2.9[12]. All HOMs can propagate out of the cavity through large apertures along the beam axis and be damped by the absorbers bonded on the beam ducts on the room-temperature side. In KEKB, four SC damped cavities supply the 4-8-MV RF voltage, delivering the 1-MW RF power to the 0.52-A beam in the usual operation mode. The maximum beam power of 380 kW has been achieved by each cavity so far. In the summer of 2000, the next four cavities are to be added for the goal intensity of 1.1 A and luminosity of 1 × 10^{34} cm^{-2}sec^{-1}[13]. LHC is a proton-proton collider with an energy of 7 TeV planned as the post-LEP at CERN. For the 0.53-A beam, eight single-cell 400-MHz SC cavities are to be used in each ring to supply the voltage of 16 MV[14].

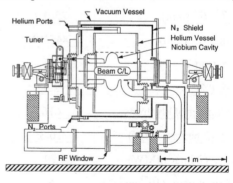

Figure 2.9: HOM-damped SC cavity for CESR at Cornell University[12].

The technology of HOM-damped SC cavities has become attractive for upgrading middle-sized storage rings that have limited RF space. In China, IHEP in Beijing has started the BEPCII project for upgrading the luminosity using SC cavities. In Taiwan, the SC project of the SRRC photon factory has started to increase the beam current to 0.5 A by replacing the existing NC cavities with one SC cavity. Daresbury Laboratory in the U.K. is also considering a SC system for their third generation photon factory, DIAMOND.

Another recent application of SC cavities is in a proton linac for neutron physics and a nuclear waste transmutation system. In this application, the use of SC cavities is attractive to reduce the total length of linacs and the cost, because a high duty or CW beam of 1~2 GeV is essential for obtaining the highest overall efficiency of the plant. Since the β at this energy reaches >0.8, the technology developed for high-β structures can be adopted for this linac. For instance, the high energy section of the nuclear waste transmutation system in Italy is designed as a combination of SC linacs of three sections of β = 0.5, 0.65 and 0.85, which has a total length of 700 m for accelerating a proton beam from 100 MeV to 1600 MeV[15]. In Japan, the design and development of a SC proton linac has been going on since 1995 as part of JHF project between KEK and JAERI. In the most recent design, a 972-MHz SC proton linac accelerates a proton beam from 400 MeV to 600 MeV, which will be extended to 1 GeV in the future, for the fundamental study of nuclear waste transmutation.

Besides beam acceleration, SC cavities can be used for beam deflection, for which a continuous high deflecting voltage is required to kick the high energy particles. CERN developed a K-meson separator in the 1970s using 5.5-m 2856-MHz Nb cavities[16]. At KEK the development of 508-MHz crab cavities has continued for KEKB. In KEKB the beams of electrons and positrons collide with a finite angle of ±11 mrad that may cause an additional beam-beam interaction. In a crab-crossing scheme, the head and tail of each bunch are horizontally kicked in opposite directions by the RF cavity located just before the collision point so as to cancel the crossing angle and to cause a head-on collision, and are kicked back after passing through the collision point by another cavity[17]. For this scheme, a deflecting voltage of 1.4 MV has to be located on both sides of the collision point in each beam line, which can be provided by just one SC crab cavity. This minimum number of cavities has the advantage not only of less HOM impedance but also of a simple arrangement of components around the collision point.

The many efforts on SC cavities in these four decades have achieved great progress in SC cavity technology and its applications. Many cavities are operating in accelerators in various fields. These efforts have been described at the international workshops on SC-RF held about every two years at various laboratories in the world. The proceedings of these workshops[18] are very useful for learning about SC cavities, as well as lecture notes of accelerator school [19,20,21] and a textbooks[22].

3 Shunt impedance of accelerating cavity

3.1 Accelerating voltage

Consider the accelerating RF voltage in a cylindrical pill-box resonator, shown in Fig.3.1. Field components of TM_{mnp} mode are given as

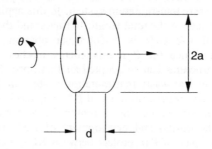

Fig. 3.1: A pill-box cavity.

$$E_r = -\left(\frac{p\pi}{k_c l}\right) E_{mnp} J_m'(k_c r) \cos(m\theta) \sin(k_z z),$$

$$E_\theta = \left(\frac{mp\pi}{k_c^2 rl}\right) E_{mnp} J_m(k_c r) \sin(m\theta) \sin(k_z z),$$

$$E_z = E_{mnp} J_m(k_c r) \cos(m\theta) \cos(k_z z),$$

$$H_r = -\left(\frac{i\omega\varepsilon m}{k_c^2 r}\right) E_{mnp} J_m(k_c r) \sin(m\theta) \cos(k_z z),$$

$$H_\theta = -\left(\frac{i\omega\varepsilon}{k_c}\right) E_{mnp} J_m'(k_c r) \cos(m\theta) \cos(k_z z),$$

$$H_z = 0,$$

$$k_c = \frac{\rho_{mn}}{a}, \qquad k_z = \frac{p\pi}{l}, \tag{3-1}$$

where $J_m(x)$ is a Bessel function and ρ_{mn} is the root of J_m. In most applications the lowest frequency mode, TM_{010}, is used. Each component of TM_{010} becomes simple, as

$$E_r = E_\theta = 0,$$

$$H_r = H_z = 0,$$

$$E_z = E_0 J_0(k_c r),$$

$$H_\theta = -\left(\frac{i\omega\varepsilon}{k_c}\right) E_0 J_0'(k_c r),$$

$$k_z = 0, \quad J_0(k_c a) = 0, \quad k = k_c, \tag{3-2}$$

and the resonance frequency f_a is

$$f_a = \frac{\omega}{2\pi} = \frac{c}{2\pi} k = \frac{c}{2\pi} \frac{\rho_{01}}{a}, \tag{3-3}$$

where c is the velocity of light and $\rho_{01} = 2.405$. The electric field of TM_{010} has only E_z, which depends on r but is uniform along z. A transit particle on the axis $(r = 0)$ sees an oscillating field gradient of $E_o e^{i\omega t}$, therefore the real gain by the particle is described as

$$\text{accelerating voltage} = \int_{gap} E_0 e^{i(\omega t + \phi)} dt.$$

By choosing the injection phase ϕ, we can write a peak accelerating voltage V_c for a particle with $v = c$ as

$$V_c = \left| \int_0^d E_0 e^{ikz} dz \right|.$$

The ratio T

$$T = \frac{\left| \int_0^d E_0 e^{ikz} dz \right|}{\int_0^d E_0 dz}, \tag{3-4}$$

is known as the transit time factor. Thus the accelerating voltage V_c includes this transit time factor T and gives the real peak voltage gain passing through the cavity. From this V_c, we can obtain the effective accelerating field E_{acc}, which is defined as

$$E_{acc} = \frac{V_c}{L_{eff}}, \tag{3-5}$$

where L_{eff} is the net gap length of the structures.

3.2 Shunt impedance

The shunt impedance R_0, defined below, is one of the most important figures to consider in the cavity performance in accelerators, which relates the accelerating voltage V_c to the power dissipation P_c in the cavity wall as a Joule loss. To achieve a high accelerating field and good power efficiency, cavities with high shunt impedance are essential. The shunt impedance is defined as

$$R_0 = \frac{V_c^2}{P_c} = \frac{V_c^2}{\frac{1}{2}\int_s R_s H^2 ds} \quad [\Omega].$$
(3-6)

Note that the other expression, $R_0 = V_c^2/2P_c$, is used in lumped-circuit theory.

On the other hand the quality factor Q_0, which is the ratio of stored energy and power loss per RF cycle, is written as

$$Q_0 = \omega \frac{U}{P_c} = \omega \frac{\frac{1}{2}\mu \int_v H^2 dv}{\frac{1}{2}\int_s R_s H^2 ds} .$$
(3-7)

This can be altered for a cavity assuming uniform R_s as

$$Q_0 = \frac{\omega\mu}{R_s} \frac{\int_v H^2 dv}{\int_s H^2 ds}$$

$$= \frac{\Gamma}{R_s} \left(\Gamma = \omega\mu \frac{\int_v H^2 dv}{\int_s H^2 ds} \right) \quad [\Omega],$$
(3-8)

where the factor Γ, which depends only on the cavity shape and not on the material properties, is called the geometrical factor and is useful for obtaining the surface resistance R_s from the measured Q_0.

From these two parameters, the ratio of R_0 and Q_0 is given as

$$\frac{R}{Q} = \frac{V_c^2}{\omega U} \quad [\Omega] .$$
(3-9)

This R/Q is a figure of merit that relates the accelerating voltage to the stored energy of the cavity and depends on neither the cavity material nor frequency. From these, the shunt impedance R_0 is written as

$$R_0 = \frac{R}{Q}Q_0 = \frac{R}{Q} \cdot \frac{\Gamma}{R_s}. \tag{3-10}$$

For a pill-box with a gap of a half wavelength ($d = \lambda/2$),

$$Q_0 = 257\frac{1}{R_s}, \qquad \Gamma = 257 \ [\Omega] \ ,$$

$$R_0 = 5.14 \times 10^4 \frac{1}{R_s} \ [\Omega] \ , \qquad T = \frac{2}{\pi} = 0.637, \tag{3-11}$$

and

$$\frac{R}{Q} = 200 \ [\Omega] \ .$$

The difference between SC and NC cavities is due to the surface resistance R_s. For NC cavities, R_s is given using the DC conductivity σ as

$$R_s = R_n = \sqrt{\frac{\omega\mu}{2s}} \qquad \text{(for normal conducting)} \ [\Omega]. \tag{3-12}$$

On the other hand, the theoretical surface resistance of SC cavities is given by BCS theory as

$$R_{BCS} = A\frac{\omega^2}{T}exp\left(-\frac{\Delta(0)}{k_BT_c} \cdot \frac{T_c}{T}\right), \tag{3-13}$$

where k_B is a Boltzmann constant, and Δ and T_c are the superconducting gap energy and the critical temperature of the cavity material, respectively[23]. A is a material parameter that includes the penetration depth (λ_{L0}), the coherence length (ξ_0), the Fermi velocity (v_F) and the mean free path. Details of these will be given in another lecture on superconductivity. Fundamental properties of typical superconducting materials are shown in Table 3.1.

Table 3.1: Fundamental properties of typical superconducting materials.

Material	T_c (K)	Δ/k_BT_c	H_c (Oe)
Pb	7.2	2.2	800
Nb	9.2	1.9	2000
Nb$_3$Sn	18	2.2	5400
YBaCuO	93	~2	>10000

As a convenient expression, the Eq. (3-13) is approximately written for Nb and $T < T_c/2$ as

$$R_{BCS} = 10^{-4} \frac{f^2}{T} \exp\left(-\frac{18}{T}\right) \quad [\Omega], \qquad (3\text{-}14)$$

where f is the RF frequency in GHz and T is in K.

For instance, a 500-MHz copper cavity ($\sigma = 0.58 \times 10^8 \; \Omega^{-1} \cdot m^{-1}$) has a surface resistance of 5.8 mΩ, whereas the surface resistance of SC cavities is extremely small and is typically 10~100 nΩ for Nb. From Eq. (3-11), Q_0 and R_0 of a pill-box at 500 MHz are 4.4 $\times$ 10^4 and 8.9 MΩ for a copper cavity, and 3.1 $\times$ 10^9 and 627 GΩ for a Nb cavity at 4.2 K, respectively. Consequently, not only the Q but also the shunt impedance is larger by a factor of ~10^5 for SC cavities.

4   Superconducting cavity

Because of the small surface resistance, the power loss of SC cavities is negligibly small. For example, the wall loss of the 508-MHz KEK-TRISTAN SC cavities is only 50 W for an accelerating voltage of 7.5 MV at 4.2 K. To obtain the same voltage, conventional cavities dissipate power of more than 5 MW on the cavity walls. Thus the RF power efficiency of SC cavities is high, i.e. all RF power from the power source can be delivered to the beam. On the other hand, at such a low power level, the additional loss caused by surface defects, impurities, loading by emitted electrons, etc., is a serious problem for SC cavities, even though it is the order of ~10 W and is negligibly small for NC cavities. From the London penetration depth, the loss mechanism on the cavity surface occurs within an ~100 nm layer. Hence, SC cavities require special care for the cavity surface in the entire fabrication process from the choice of materials to the final assembling.

4.1   Surface resistance

From Fig. 4.1, which shows the surface resistance of a 508-MHz Nb cavity at low field gradient, it is evident that the real surface resistance can be written as the sum of the theoretical resistance R_{BCS} and an additional term R_{res} which gives the lower limit of the surface resistance. That is,

$$R_s = R_{BCS} + R_{res} \, , \qquad\qquad (4\text{-}1)$$

where R_{BCS} is given by Eq. (3-13). The R_{res} is called the residual resistance and becomes dominant in the low-temperature region.

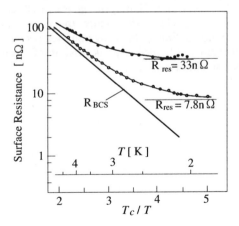

Figure 4.1: The surface resistance of a 508-MHz Nb cavity. T_c is 9.25 K.

Many mechanisms contribute to R_{res}, but details are still not clear so far. One of the loss mechanisms is the contribution of flux trapping of the residual magnetic field due to geomagnetic field or magnetic impurities. A flux trapped on the surface during cooling forms a normal conducting zone of $\pi\xi^2$, where ξ is the coherence length and has the normal conducting resistance R_n given in Eq. (3-12). The number of fluxoids N is proportional to the residual field B as

$$N = \frac{B}{\pi B_{c2} \lambda^2} \, ,$$

where λ is the penetration depth. Thus the additional loss is

$$R_{mag} = N\pi\xi^2 R_n = \kappa^{-2}\left(B / B_{c2}\right)R_n \, ,$$

where $\kappa = \lambda/\xi$ is the Ginzburg-Landau parameter and is ~1 for Nb. The ratio R_n/R_s is roughly ~10^5. Thus the additional resistance R_{mag} reaches R_s at $B = 10^{-5}B_{c2}$. Namely for a material with $B_c = 200$ mT, the B should be shielded below 20 mGauss. Because R_n has a frequency dependence of $f^{1/2}$, this trapping effect is more serious for higher frequency cavities. Furthermore an additional loss due to flux oscillation should be considered at higher RF frequency.

Another mechanism is the loss due to hydride, which is known as "Q-disease". In 1991, Q-degradation related to the cavity cooling procedure was found at Saclay[24]. Figure 4.2 shows their cooling procedure and the Q-E plot at each cooling cycle. The results show that the Q degrades more seriously with longer time of keeping the cavity at 120–170 K. Now this phenomenon is understood as the nucleation of Nb-hydride in the surface, which results in a weak superconductor and increases the surface resistance. Hydrogen moves in Nb at temperatures above 60 K and accumulates at nuclear sites under the appropriate concentration of hydrogen. Therefore, rapid cooling in the dangerous temperature region 120–170 K can avoid this degradation. Alternatively, degassing of hydrogen in a vacuum furnace at >700°C is also effective.

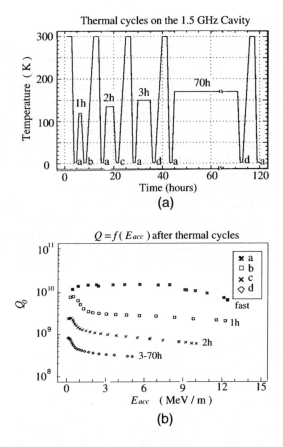

Figure 4.2: The Q-desease observed at Sacley. A variety of cooling
patterns (a) gave different results for the Q-E plot (b)[20].

Other surface conditions, such as dislocations, impurities, roughness, stress,
chemical residue, and others, may contribute to R_{res} but their effects are not clear.
This is an important area of SC cavity research. Although a well-treated surface has
a R_{res} of 1~10 nΩ, the resistance increases, typically to 10~100 nΩ, at high field
level, because of the additional loss due to heating or the effects of emitted electrons.
From the frequency dependence in Eq. (3-13), the R_{BCS} becomes dominant at 4.2 K
for >1 GHz and is decreased to a level comparable with R_{res} by cooling the cavity.
This is the reason that Nb cavities of high RF frequency are cooled below 2 K,
whereas cavities with a frequency below 1 GHz are usually operated at 4.2 K.

4.2 Field limitation

The theoretical limitation of the RF field is determined by the superheating critical magnetic field, H_{sh}, which is a little higher than the DC magnetic field, H_c, namely $H_{sh} = 1.2H_c$ for Nb, where $H_c(T) = H_c(0)[1-(T/T_c)^2)]$. Since the typical shape of SC cavities has an H_{sp}/E_{acc} ratio of ~40 Oe/(MV/m), a gradient >50 MV/m is expected for the H_{sh} of Nb cavities, and greater for high T_c materials. However, heating and quenching due to various phenomena on the cavity surface prevent the gradient from reaching this theoretical limit. Up to now, the maximum gradient has been achieved by pure Nb cavities and not by alloys or high T_c materials. Even in Nb cavities, the maximum gradient is still limited to 40 MV/m for single-cell test cavities and <30 MV/m for multi-cell structures.

To understand the phenomena in SC cavities, RF loss is measured as a function of the field gradient. As described in Section 3, the shunt impedance R_0 connects the RF loss with the cavity field, and is a product of R/Q and Q_0 which indicates the surface resistance of the cavity. Therefore a Q-E diagram is always used to evaluate the cavity performance, and gives a lot of information. For no degradation of the surface resistance, the Q_0 always is kept constant at various field gradients.

4.2.1 Multipacting

An electron emitted from the cavity surface is accelerated by the RF field and impacts the surface again, producing the next generation of electrons. The secondary electrons are also accelerated and produce the third generation at the next impact. If this cycle is continued resonantly and the secondary emission coefficient is larger than unity, the number of electrons increases exponentially and quickly. This electron avalanche dissipates the RF power, i.e. degrades the Q-value suddenly at the resonance level, heats up the impact spot, and eventually leads to SC breakdown. Since this process contains the many parameters of impact energy, the secondary emission coefficient, RF phase and amplitude, the direction of trajectories, and others, prediction of the resonant field levels by a trajectory calculation is difficult. However, a simulation on a simple model shows that resonance happens at narrow discrete field levels, and the onset field level is proportional to the RF frequency for one-point multipacting and f^2 for two-point multipacting. From this point of view, high frequency is advantageous to achieve a high accelerating field.

As described in Section 2, because of the spherical shape, one-point multipacting is no longer a serious problem for cavities, but it is still serious for other parts such as input couplers and HOM couplers.

4.2.2 Thermal breakdown

Another field limitation is caused by local heating and quenching due to surface defects. Now estimate the size of a defect that causes breakdown. Consider a half sphere of NC metal ($2a$ in diameter) embedded in a SC surface of thickness d (Fig. 4.3a). Power dissipation at the defect is

$$\dot{Q} = \frac{1}{2} R_n H^2 \pi a^2, \tag{4-2}$$

where R_n is the surface resistance of the NC metal given in Eq. (3-12). Under the assumption of $a \ll d$, the model is equivalent to a spherical defect surrounded by a sphere of SC metal of diameter $2b$, as shown in Fig.4.3b. Here, we can assume that $a \ll b$. Then the spherical defect dissipates power of $2dQ/dt$, and heat flow to the SC metal of thermal conductivity κ is

$$-4\pi r^2 \kappa \frac{dT}{dr} = 2\dot{Q}.$$

By integrating from a to b,

$$\int_a^b \frac{dr}{r^2} = -\frac{2\pi\kappa}{\dot{Q}} \int_{T_a}^{T_b} dT,$$

where T_a and T_b are the temperature at the defect surface and at the outer surface of the SC metal cooled by liquid He, respectively. Since $a \ll b$,

$$\frac{1}{a} = \frac{2\pi\kappa (T_b - T_a)}{\dot{Q}}.$$

Using Eq. (4-2),

$$H = \sqrt{\frac{4\kappa (T_a - T_b)}{aR_n}}.$$

As the temperature of the defect reaches T_c, i.e. quenching, H reaches its maximum, H_{max},

$$H_{max} = \sqrt{\frac{4\kappa (T_c - T_b)}{aR_n}}. \tag{4-3}$$

As an example, an impurity of $a = 100$ μm and $R_n = 10$ mΩ on the Nb surface with $\kappa = 50$ W/m·K (this K corresponds to Nb of $RRR = 200$) and $T_c = 9.2$ K gives $H_{max} = 3.2 \times 10^4$ A/m at $T_b = 4.2$ K. In other words, a copper particle on the Nb surface of $RRR = 200$ limits the maximum gradient to 10 MV/m at 4.2 K if the particle is located at the peak magnetic field $H_{sp} = H_{max}$, (E_{acc} is calculated from the ratio H_{sp}/E_{acc} of typically 3200 (A/m)/(MV/m) in a spherical cavity. The power dissipation of only 0.16 W is enough to limit the cavity field. From this estimation, one can understand that cleanliness of the surface and the thermal conductivity of the cavity material are important, and efforts must be concentrated on reducing surface defects.

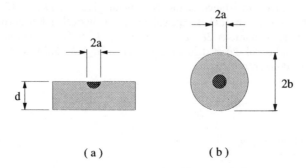

(a)   (b)

Figure 4.3: A half-sphere defect embedded in the SC surface (a), and a spherical defect surrounded by a sphere of SC metal (b).

Figure. 4.4 shows the *Q-E* of a 508-MHz Nb cavity at 4.2 K. At the first cooling, the cavity quenched at 8 MV/m, showing local heating at the equator. After grinding off a 2-mm-long, 0.2-mm-deep defect and slight electropolishing, the maximum field was improved to 13 MV/m.

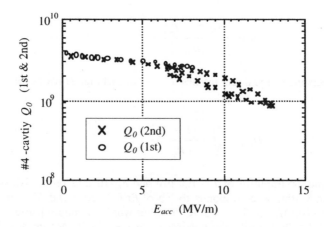

Figure 4.4: The *Q-E* plot of the KEKB 508-MHz single-cell Nb cavity.

4.2.3   Field emission

Field emission is also a limitation due to emitted electrons in the cavity. Electron multipacting increases the number of electrons in a resonance process, but the increase follows the Fowler-Nordheim law[25] in this limitation. The field emission current from metal surfaces is described as

$$I \propto (\beta E)^{2.5} \exp\left(\frac{-\beta \phi^{1.5}}{\beta E}\right),$$

in the RF field, where E is the surface electric field, ϕ is the work function, and β is an enhancement factor that may be related to the emitter shape, although this is not clear. The electron current increases exponentially as the field strength and degrades the Q_0 by dissipating the RF power (electron loading) and consequently heating the cavity wall. This phenomenon becomes serious at a higher field level, and is dominant in the limitation of the recent high field cavities. Figure. 4.5 is a modified Q-E plot of $1/Q$ vs. E^2, where the vertical axis is the power loss. The loss proportional to E^2 is understood to be the loss related to the resistance, and the exponential increase is due to the electron loading.

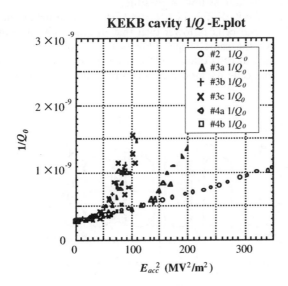

Figure 4.5: $1/Q$ vs. E_{acc}^2 of KEKB 508-MHz single-cell Nb cavity.

4.3 Fabrication and Measurements

In the previous sections, we showed the important factors for realizing high performance SC cavities, namely magnetic shielding, high thermal conductivity of the cavity material, and a defect-free surface. A 100-μm impurity causes a heating spot and becomes a source of emitted electrons, which degrade the achievable field gradient seriously. However in most cases the defects cannot be detected by the naked eye. Thus we have to concentrate our efforts on keeping the surface clean.

4.3.1 Cavity shape

In the design of NC cavities, the cell shape is optimized to obtain the highest R/Q or shunt impedance of R_0 to minimize the wall loss, so that an aperture with a small diameter is chosen together with a nose cone. On the other hand, in a SC cavity for a high-β structure, a simple spherical shape with a large aperture size is optimal at the operating frequency, because it avoids multipacting and makes it easy to polish and wash during the surface treatments. This shape decreases the R/Q by 1/2~1/3 compared to that of NC cavities, but the advantage of easy fabrication guarantees a high R_0 by realizing a high Q-value. In the optimization process, the field ratios E_{sp}/E_{acc} and H_{sp}/E_{acc} must be kept as low as possible.

Figure. 4.6 shows the KEK-TRISTAN 508-MHz 5-cell structure. In general, the field strength of both end cells is lower than that of other cells in a multi-cell structure, if all cells have the same shape. This is due to the asymmetry of the boundary condition of the end cells caused by the beam pipes. Thus the length and the iris diameter of the end cells are modified to compensate this asymmetry and to obtain a flat π-mode.

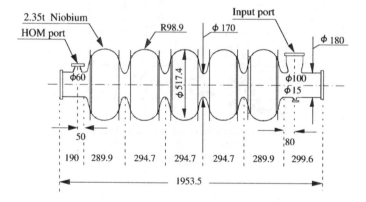

Figure 4.6: The shape of the KEK-TRISTAN 508-MHz 5-cell Nb cavity[26].

4.3.2 Fabrication

It is difficult to define a "standard" procedure for fabrication processes; each laboratory has established its own. The procedure established for KEK-TRISTAN cavities, which is based on an electropolishing process[27] can be outlined as follows.

- Cutting Nb sheet metal of *RRR* 150 – 200.
- Hydroforming of half cells.
- Mechanical polishing and dipping in a HCl bath to eliminate metal impurities. Rust checking by dipping in a water bath.
- Electron beam welding at the equator and the irises.
  Grinding off the inner seams, on which impurities or micro cracks might be left.
- Electropolishing and removing 80 μm of the surface.
- Heat treatment at 700°C for 1.5 hr to reduce the surface stress and hydrogen contamination during electropolishing.
- Frequency tuning by giving an inelastic deformation to the cavity length.
- Final slight electropolishing removing 10 μm.
- Thorough rinsing with ~5000 liters of pure and ultra pure water for > 4 hr.
- Assembling in a CLASS 100 clean room.

Improvement and development of these processes are being continued to achieve a higher field gradient, to improve the reliability, and to lower fabrication costs as shown in references[18].

Material

High purity of Nb is desired, but it lowers the mechanical strength. Typically, the yield strength of post-purified Nb sheets is decreased to ~40 MPa. Thus careful structural analysis of the cavity shape and choice of material thickness is needed. Rather deep scratches on the material surface are mechanically ground off one by one, because impurities might be embedded during the sheet rolling process and hiding behind the scratches. An eddy current scanning system was developed to check the materials for TESLA cavities, by which iron inclusions in the surface were successfully detected[28].

Welding

The most dangerous process in cavity fabrication is welding, because many impurities may melt into the welding seams. Therefore only electron beam (EB) welding is used to weld the Nb in order to get defect-free welding seams. The optimum condition for making a smooth and non-sputtered seam is sensitive to the welding parameters of voltage, current, speed, and the tolerance of the welding materials. In general the tolerance of mismatching between the materials should be within 10% of thickness. In this sense, the forming of half cells is important. Usually the welding surface is smoother on the beam side than on the opposite side.

Thus welding from the inside using a small electron gun is desired. R&D on seamless cavity formation, where by a multi-cell structure is formed from a seamless Nb tube by spinning, is being rigorously carried out at various laboratories.

Surface treatment

In chemical polishing, a 1:1 mixture of concentrated hydrofluoric acid and nitric acid is used. For a large cavity, phosphoric acid is added to slow down the reaction speed. In electropolishing, the Nb surface is polished in a mixture of concentrated sulfuric acid and hydrofluoric acid in a ratio of 85:10 at a DC voltage of 25-30 V between the cavity and cathode inserted in the cavity to obtain a current density of ~50 mA/cm^2. In contrast to the orange peel like surface obtained by chemical polishing, a mirror like surface is obtained by well-controlled electropolishing, which seems favorable for a high field gradient. In the last decade, it was thought that there is no difference between these polishing methods, but recently this difference has again become the object of study for achieving a higher field gradient.

5 Application to high current accelerators

Nowadays a high intensity beam is required for various fields of study. One of the recent problems in high energy physics is the study of CP violation and topics relevant to B-meson decays, where a very small asymmetry between the properties of matter and antimatter has to be detected. To support this research program, the accelerator has to store high intensity electron and positron beams of ampere class current and make them collide to obtain sufficiently high luminosity. KEK and Cornell University have developed newly designed SC damped cavities and have already commissioned them in their B-factory storage rings. Furthermore recent plans for a proton linac for neutron physics and third-generation photon factory accelerators also need high intensity beams. In these projects, SC cavities are considered to be an attractive solution to obtain a high CW field with fewer cavities. Now the high current accelerator is the new field for SC cavity application.

5.1 SC cavities for high intensity storage rings

When a beam passes through a structure such as an RF cavity, it excites the RF field of higher-order modes (HOMs), which accumulates in multi passes and finally causes instability and loss of the circulating beams. Since the beam-induced field is proportional to the beam intensity, the problem of HOMs becomes more serious in high intensity accelerators. In addition, the beam in high intensity storage rings is distributed in a large number of bunches, so that all HOMs have to be damped sufficiently to avoid multi-bunch instabilities.

In general, the main source of the ring impedance is RF cavities, therefore, an RF system using a minimum number of cavities with sufficiently damped HOMs is essential in high intensity rings. From this point of view, a system of SC cavities is advantageous to reduce the number of cavities, because fewer cavities are enough to supply the required RF voltage. As an example, the SC and NC systems are compared in Table 5.1 for KEKB-HER. In the SC version, the number of cavities is <1/3 of that in the NC version and the total RF power supplied from klystrons can be reduced to half of the NC case. Of course it must be considered that the SC cavities need another electric power source for a liquid He refrigerator.

Table 5.1: Comparison of the RF systems for the SC and NC cases in KEKB-HER.

	NC (ARES)	SC
Beam energy	8 GeV	
Beam current	1.1 A	
Beam power	4 MW	
RF voltage	16 MV	
No. of cavities	36	10
R/Q	15	93
Coupling β	2.4	>5000
Cavity voltage V_c	0.45 MV/cavity	1.6 MV/cavity
Wall loss	119 kW	-
Frequency detuning	8 kHz	5 kHz
No. of klystrons	18	10
Beam power P_b	111 kW/cavity	400 kW/cavity
Input power P_g	230 kW/cavity	400 kW/cavity
Total RF power	8280 kW	4000 kW

Another advantage of SC cavities in high current applications is the low R/Q of the accelerating mode. Consideration of beam–cavity interactions indicates that the resonance frequency of the cavity should be lowered as the beam intensity increases so as to minimize the generator power. The amount of this frequency detuning is proportional to the beam intensity as

$$\Delta f_0 = -\frac{I_b f_0}{2V_c}\left(\frac{R}{Q}\right)\sin\phi_s = -\frac{P_b \tan\phi_s}{4\pi U} \, , \qquad (5\text{-}1)$$

where f_0, I_b, V_c, and ϕ_s are the resonant frequency, the beam current, the accelerating voltage and the synchronous phase. P_b is RF power delivered to the beam, i.e $P_b = I_b V_c \cos\phi_s$, and U is stored energy in the cavity. This frequency detuning has to be kept within the revolution frequency f_{rev}, otherwise multi-bunch instability is excited by the accelerating mode when the detuning reaches f_{rev}. In a large storage ring, the revolution frequency, $f_{rev} = c/C$ (C is the ring circumference and c is the speed of light), becomes small and gives the limitation to the frequency detuning.
A long bunch gap, that is, the space between the bunch trains during beam injection, or the empty buckets needed to suppress an ion-trapping phenomenon, leads the phase oscillation of the accelerating mode and consequently modulates the longitudinal bunch position. This phase oscillation $\Delta\phi$ is

$$\Delta\phi \propto \frac{\pi f_0}{V_c}\left(\frac{R}{Q}\right)I_b T_{gap} = \frac{P_b T_{gap}}{2U\cos\phi_s}, \qquad (5\text{-}2)$$

where T_{gap} is the length of the bunch gap.

In both equations, it is evident that RF cavities with higher voltage and lower R/Q, i.e. large stored energy, are advantageous in high current applications. In SC cavities, not only a high RF voltage but also a cavity shape with rather low R/Q is possible. In KEKB, the frequency detuning of conventional Cu cavities is 150 KHz and 350 kHz for peak currents of 1.1 A (HER) and 2.6 A (LER), which are larger than the f_{rev} of 100 kHz. Thus a new NC cavity, i.e. ARES (Accelerator Resonantly Coupled with Energy Storage), which has a reduced R/Q of 15 Ω, has been developed by skillfully connecting a large energy storage cavity to the accelerating cavity so as to increase the net stored energy U. On the other hand, the detuning of the SC cavity in HER is only 17 kHz, because of the high V_c and the low R/Q of its shape.

5.2   Cavity shape and HOM damping

In general, the beam aperture of NC cavities is small in order to keep the high R/Q of the accelerating mode; as a result the trapped HOMs with high impedance remain in the cavity. Hence, a set of HOM couplers surrounds the accelerating cell to extract them. Figures. 5.1 and 5.2 are typical normal conducting cavities developed for high current applications.

Such HOM couplers, which are directly connected on the cavity wall, are not suitable for SC cavities, because the coupling holes on the cavity wall may cause multipacting discharge. In SC cavities, all couplers including the input coupler are usually located on the beam pipe, which makes an external Q of ~10,000 possible. In this case, an RF filter structure is needed to prevent leakage of the accelerating mode through the HOM coupler. To achieve a Q of HOMs of 100 or less, a new scheme of HOM damping, namely an HOM-damped cavity, has been developed for B-factory machines.

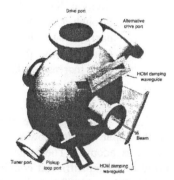

Figure 5.1: Normal conducting RF cavity for PEP- II at SLAC[29].

Figure 5.2: Normal conducting RF cavity for KEKB, ARES. This system consists of three cavities, i.e., the energy storage cavity of TE mode is coupled with an accelerating cavity through a coupling cavity in order to increase the stored energy, that is, to reduce the R/Q of the accelerating mode. Consequently, the R/Q is reduced to 15 Ω without any large increase of the loss of the accelerating cavity.

Figure. 5.3 is a drawing of the 508-MHz SC cavity for KEKB[30]. In order to reduce the input coupler power and the number of HOM modes, a single-cell structure was adopted. The large diameter of the beam aperture (220 mm) was optimized to obtain sufficient coupling of the beam pipe for the monopole modes, TM_{011} and TM_{020}. A 300-mm-diameter cylindrical beam pipe (LBP) is attached on one side to extract the lowest dipole modes, TE_{111} and TM_{110}. In this scheme, all HOMs propagate out of the cavity along the beam pipes and are damped by HOM absorbers located outside the cryostat. This simple damping system is possible only for SC cavities, because in NC cavities rather low R/Q of this shape causes serious power dissipation on the wall. On the other hand, the shunt impedance is kept sufficiently high by the high Q value in SC cavities. At Cornell University, a fluted beam pipe is used instead of a large cylindrical beam pipe[31,32].

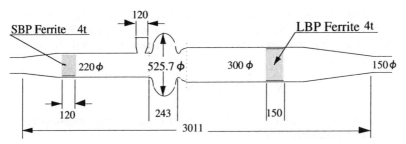

Figure 5.3: Optimized cell shape and ferrite absorbers of the KEKB SC cavity[33].

5.2.1 HOM impedance

Consider the maximum permissible impedance of HOMs in a storage ring. Deriving the formula of beam-cavity interaction is out of the scope of this lecture and only the results are shown here. These will be given in the lecture by K. Akai of this school. To avoid multi-bunch instabilities, the growth time (= 1/growth rate) of mode due to each HOM must be longer than the energy damping time τ of the ring. For longitudinal modes, the threshold of the impedance is

$$N_{cav} \cdot R_{long} < \frac{4 \cdot \tau_{long}^{-1} \cdot v_s \cdot E/e}{I_b \cdot f_{HOM} \cdot \alpha} ,$$

and for transverse modes it is

$$N_{cav} \cdot R_{trans} < \frac{8 \cdot \pi \cdot \tau_{trans}^{-1} \cdot E/e \cdot f_\beta/f_{rev}}{I_b \cdot c} .$$

In these equations, N_{cav} is the number of cavities, E/e is the beam energy in V, v_s is the syncrotron tune, α is the momentum compaction factor, f_β is the betatron frequency, and I_b is the beam current in A. For KEKB-HER, the ring parameters $v_s = 0.02$, $E/e = 8$ GV, $\alpha = 2 \times 10^{-4}$ and $I_b = 1.1$ A, give an impedance limit of

$$f_{HOM} \cdot R_{long} \cdot N_{cav} < 145 \ [k\Omega \cdot GHz],$$

for a longitudinal energy damping time of 20 ms. Assuming $N_{cav} = 10$, $f_{HOM} = 1$ GHz and R/Q of 10 Ω for the most dangerous mode of TM_{011}, the maximum permissible Q becomes 1.45×10^3. However, in the pessimistic case, the HOM power of

$$P_{HOM} = I_b^2 \cdot (R/Q) \cdot Q_{HOM} = 17.5 \ [kW],$$

378

is dissipated by the dampers of each cavity module. Since each damper can absorb at a few kW of HOM power, a Q of ~100 is needed to make the HOM power sufficiently small. For transverse modes, the maximum permissible impedance for $f_\beta/f_{rev} = 45$ and $\tau_{trans} = 40$ ms is

$$R_{trans} \cdot N_{cav} < 690 \ [\text{k}\Omega/\text{m}].$$

Again assuming $N_{cav}=10$ and $(R/Q)'=250\Omega/\text{m}$ for the most dangerous mode, TM_{110}, the maximum permissible Q is 270.

5.2.2 Ferrite damper

In the application to SC cavities, besides superior characteristics of RF absorption, less out-gassing is important for the damper materials, because the gas condenses on the cavity surface and causes discharging and RF trips.

In KEKB, HOM dampers have been optimized with IB-004 ferrite because of its superior RF properties around 1 GHz. The damping characteristics of absorbers depend strongly on the geometrical parameters, such as the distance from the cavity, the length and the thickness of ferrite, and the tapering between dampers and beam duct. Figure 5.4 shows that the Q of TM_{011} depends strongly on ferrite thickness and distance from the cavity. These parameters are optimized by using the SEAFISH code which can calculate the Q of monopole modes for the cavity including resistive materials, and by experiments using a model cavity.

For fabricating 4-mm-thick cylindrical ferrites and bonding them perfectly on to the Cu beam pipes, KEK has developed a fabrication procedure based on the HIP (Hot Isostatic Pressing) process. As shown in Fig. 5.5, the IB-004 ferrite powder is packed into an iron vessel together with a copper cylinder and heated to 900°C in a furnace at 1500 bar after the vessel is evacuated. In the vessel the ferrite is sintered and bonded on to the copper cylindrical wall simultaneously under vacuum[34]. The HOM spectra with and without ferrite dampers are compared in Fig. 5.6, where the Qs of both monopole and dipole modes are reduced to ~100. The measured frequency and the Q value of HOMs are listed in Table 5.2.

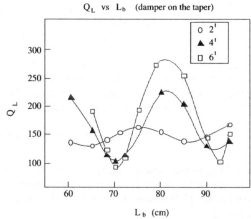

Figure 5.4: The Q of TM_{011}. The horizontal axis is the distance between the ferrite damper and the cavity.

379

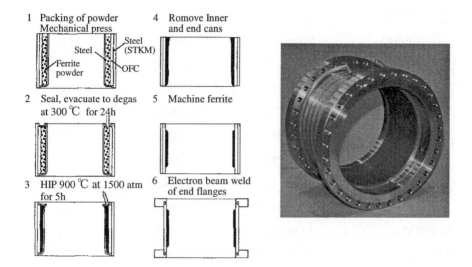

Figure 5.5: Fabrication processes based on HIP, and completed ferrite damper.

a) Al model cavity without Ferrite

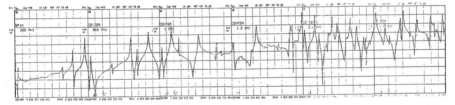

b) Nb cavity with Ferrite

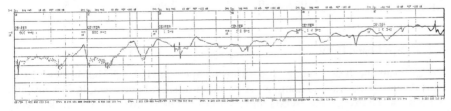

Figure 5.6: HOM spectra of KEKB cavity with (b) and without (a) ferrite damper.

Table 5.2: HOMs of KEKB cavity

(a) Monople

mode	Frequency	R/Q	Q
LBP/TM$_{01}$	782.7460	0.29000	196
LBP/TM$_{01}$	834.3830	0.32200	100
LBP/TM$_{01}$	918.9360	1.20460	60
LBP/TM$_{01}$	1002.5400	5.79200	42
TM$_{011}$	1018.3800	11.58400	170
TM$_{020}$	1032.7500	1.48260	15
SBP/TM$_{01}$	1065.8199	1.23480	106
SBP/TM$_{01}$	1130.9900	2.20000	74
TM$_{030}$	1607.5200	6.48800	300

(b) Dipole

Mode	MHz	$(R/Q)'$	Q
LBP-TE$_{11}$	606.2090	1.84	97
LBP-TE$_{11}$	628.6900	33.78	90
LBP-TE$_{11}$	654.0472	39.59	129
LBP-TE$_{11}$	684.9373	152.47	86
TM$_{110}$	701.3156	245.03	150
SBP-TE$_{11}$	813.2880	6.34	74
TE$_{11}$	1023.6659	2.96	50

5.2.3 Broad-band HOM

To estimate the heat loading of a ferrite damper, broad-band HOMs have to be considered, which become dominant for a beam of short bunch length. For instance a 4-mm bunch excites the HOMs up to 20 GHz. The HOMs in the high frequency region do not stay in the cavity but go into the dampers.

A charge q leaves a power of $k_m q^2$ for the m-th HOM, where k_m is the loss factor and has a unit of Volt/Coulomb. Thus we have to sum up the k_m up to 20

GHz to obtain the total loss factor k. The total power loss is obtained by using the k and the number of injecting bunches per second, i.e. I_b/q, as

$$P_{total} = k \cdot q \cdot I_b \cdot 10^{12} \quad [W],$$

where k is in V/pC, q is the bunch charge in C and I_b is the beam current in A. A study of the KEKB cavity using the ABCI code showed the total loss factor of a full cavity module to be 2.3 V/pC, in spite of 0.66 V/pC for a cell without tapers. The increase in loss factor is caused by a pair of tapers on both sides and is also related to the diameter of the beam ducts. For a 145-mm duct the loss factor is 1.5 V/pC and can be reduced to 1.2 V/pC by using long 60-cm tapers (Fig. 5.7). In addition, the experimental results on the loss factor of the ferrite pipe show that the damper itself has an additional loss of 0.3 V/pC. From these results, the SC module for KEKB has a loss factor of 1.8 V/pC for a 4-mm bunch, which results in absorbed power of 4.2 kW for a pair of ferrite dampers at the design current of 1.1 A in 5000 bunches. An additional loss due to the trapped HOM of ~1 kW should also be of concern. At Cornell, the cavities are connected to each other by 240-mm beam ducts to reduce the number of taper sections. It should be mentioned that k increases rapidly as the bunch length is shortened below 1 cm (Fig. 5.8).

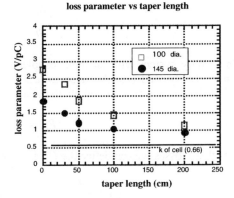

Figure 5.7: Loss factor k of KEKB cavity module as a function of taper length.

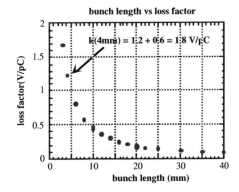

Figure 5.8: Loss factor k vs. bunch length of KEKB cavity.

5.3 Input coupler

5.3.1 Coupling constant

Consider the coupling strength of an input coupler. From the definition of Q, the stored energy $U(t)$ is written as

$$\frac{dU(t)}{dt} = -\frac{\omega}{Q}U(t).$$

In a cavity system with an input coupler, the stored energy is dissipated not only on the cavity wall (P_c), but also as leakage through an input coupler (P_{rad}). Thus, when the RF power is turned off, the total power loss is described as

$$P_{tot} = P_c + P_{rad}.$$

For these losses we can define the Q values as

$$Q_L = \frac{\omega U}{P_{tot}}, \quad Q_0 = \frac{\omega U}{P_c}, \quad Q_{ext} = \frac{\omega U}{P_{rad}},$$

and

$$\frac{dU(t)}{dt} = -P_{tot} = -\left(P_c + P_{rad}\right)$$

$$= -\omega \frac{1}{Q_L}U(t) = -\omega\left(\frac{1}{Q_0} + \frac{1}{Q_{ext}}\right)U(t),$$

where Q_L Q_0 and Q_{ext} are loaded Q, unloaded Q and external Q. Thus, by defining the coupling constant β as

$$\beta = \frac{P_{rad}}{P_c} = \frac{Q_0}{Q_{ext}}, \tag{5-3}$$

the β indicates the coupling strength, and we can obtain the relation between Q_0 and Q_L as

$$Q_0 = (1 + \beta)Q_L. \tag{5-4}$$

As shown in the chapter of T. Akai and also in the note of "Hands on Training of SC Cavity", only at $\beta = 1$, no reflected power returns from the cavity; this is called the optimum coupling. When a beam takes power P_b away from the cavity, the input coupler sees a total loss of $P_c + P_b$ instead of P_c. For most applications the coupling constant β is chosen as unity at maximum beam loading to minimize the generator power, namely no reflect power. Hence, β should be set larger than unity. In SC cavities, P_c is much smaller than P_b, thus

$$\beta = \frac{P_c + P_b}{P_c} \approx \frac{P_b}{P_c} \gg 1, \qquad (5\text{-}5)$$

and

$$Q_L = \frac{Q_0}{1+\beta} \approx Q_{ext}. \qquad (5\text{-}6)$$

In contrast to the β of 1~2 for NC cavities, that of SC cavities is >1000 and Q_L is equal to the external Q of the input coupler. From Eq. (5-5) we can find the optimum coupling Q_{opt}, recalling $R_0 = (R/Q)Q_0 = V_c^2/P_c$,

$$Q_{opt} = Q_L = \frac{V_c^2}{P_b(R/Q)}, \qquad (5\text{-}7)$$

where P_b is the power delivered to the beam. For $V_c = 1.5$ MV, $R/Q = 93\ \Omega$ and $P_b = 250$ kW, the Q_{opt} is 1×10^5 and the β is 10,000 for a Q_0 of 1×10^9.

5.3.2 Coupler power

Acceleration of a high intensity beam with fewer cavities imposes a heavy loading on the input couplers. Actually, several hundred kW of power has to be supplied to the cavity in B-factory machines. Therefore it is necessary to minimize the generator power. From the phasor diagram for a cavity with shunt impedance $R_0 = (R/Q)Q_0$ and coupling constant β, the generator power P_g to keep the voltage V_c under the beam of I_b is

$$P_g = \frac{V_c^2}{R_0}\frac{(1+\beta)^2}{4\beta}\frac{1}{\cos^2\psi}\left[\left(\cos\phi_s + \frac{I_b R_0}{V_c(1+\beta)}\cos^2\psi\right)^2 \right.$$
$$\left. + \left(\sin\phi_s + \frac{I_b R_0}{V_c(1+\beta)}\cos\psi\sin\psi\right)^2\right], \qquad (5\text{-}8)$$

where ψ is the tuning angle and is related to the frequency detune Δf from the resonance frequency f as

$$\tan\psi = -2Q_L\frac{\Delta f}{f}. \qquad (5\text{-}9)$$

The cavity wall loss P_c and the power delivered to the beam P_b are

$$P_c = \frac{V_c^2}{R_0} \quad and \quad P_b = I_b V_c \cos\phi_s,$$ (5-10)

then the power

$$P_r = P_g - P_c - P_b \ ,$$ (5-11)

returns from the cavity. As shown in Fig. 5.9, P_g reaches a minimum at some tuning angle ψ, that is the power at Δf_0 described in Eq. (5-1). In the figure, with the Q_L of 5×10^4 the input power P_g for both 1.1 A (solid-line) and 0.5 A (dashed-line) is plotted as a function of tuning angle ψ at the cavity voltage of 1.6MV. The power of >1MW is reduced to 400 kW by giving appropriate shift of resonant frequency toward lower frequency side.

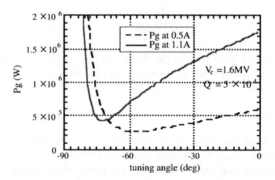

Figure 5.9: Generator power P_g vs tuning angle ψ.

On the other hand, from wakefield theory the detuning has a limit due to the multi-bunch instability caused by the accelerating mode $\mu = 1$, whose growth time τ is a function of Δf:

$$\tau^{-1} = \frac{\alpha \cdot f \cdot I_b}{2 \cdot E/e \cdot v_s} h\left(R^+ - R^-\right),$$

where

$$R^+ = \frac{1}{2} \frac{R/Q \cdot Q_L \cdot N_{cav}}{1 + 4Q_L^2 \delta_+^2},$$

$$R^- = \frac{1}{2} \frac{R/Q \cdot Q_L \cdot N_{cav}}{1 + 4Q_L^2 \delta_-^2},$$

$$\delta_\pm = \frac{1}{h}\left(1 - v_s \mp \frac{\Delta f}{f}\right),$$

and τ should be larger than the longitudinal energy damping time.

5.3.3 Input coupler structure and conditioning

Two types of input coupler have been used for SC cavities: coaxial antenna type couplers at CERN, DESY and KEK, and waveguide couplers at CEBAF and Cornell University. Coaxial type couplers have the following characteristics that differ from those of waveguide couplers.

- Coupler size is determined by the desired RF power level and not by the frequency.
- Variable coupling strength is achieved by changing the penetration of the inner conductor into the beam pipe.
- Bias voltage is available between the inner and outer conductor to suppress multipacting.
- Complex cooling system of inner and outer conductors.

In either case, it is important to reduce the heat load to the He bath due to RF loss as well as conduction loss through the couplers.

Figure. 5.10 shows a 152D coaxial antenna coupler used in KEKB SC cavities. The 12-mm penetration of the inner conductor into the beam pipe gives a Q_{ext} of 7×10^4. The 1-cm-thick alumina ceramic window is located ~1.75λ away from the cavity side, so that the ceramic is not positioned at the maximum field of standing waves at both on and off resonance. Three monitoring ports, attached beside the ceramic for monitoring the vacuum pressure, emitted electrons and arcing, are used as an interlock system protecting the ceramic. The outer conductor, which is made of stainless steel with 30-μm copper plating, is cooled by an ~8-l/min He gas flow. The inner conductor is cooled by water. In a test at room temperature, this power coupler achieved 800 kW of traveling wave[35].

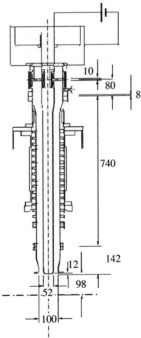

Figure 5.10: 152D coaxial input coupler for KEKB.

The most important issue for input couplers is the conditioning. If the ceramic is cracked, a large amount of air penetrate the cavity with many dust particles and lowers the cavity performance, or, more seriously, causes all the liquid He in the bath to boil. Cracks are caused mostly by discharging. Thus careful in situ conditioning is given after installation into the accelerator, in order to eliminate electron emitter sites or condensed gas in the coupler. In KEKB, 300-kW standing-wave conditioning is always given before cooling the cavity, applying a bias voltage between the inner and outer conductors; during conditioning various multipacting levels are sought by changing the DC bias voltage within ±2 kV.

5.4   Operation of KEKB SC cavities

A B-meson factory, KEKB, has been in operation since December 1998. KEKB is an asymmetric electron-positron collider using 8-GeV (HER) and 3.5-GeV (LER) storage rings. To achieve a goal luminosity of 10^{34} cm^{-2}s^{-1}, 1.1-A and 2.6-A beams of 5000 bunches are stored in HER and LER, respectively[30].

The RF system of HER is a combination of SC and NC (ARES) cavities. The SC cavities are in the NIKKO straight section, because of the existing 6.5-kW LHe refrigerator. The SC crab cavities are to be installed on both sides of the collision point in the TSUKUBA section. The KEKB rings are shown in Fig. 5.11. The designed gap voltage per cavity is 1.5 MV for SC and 0.5 MV for ARES. The RF power delivered to the beam can be shared by giving an offset to the RF phase between SC and ARES cavities.

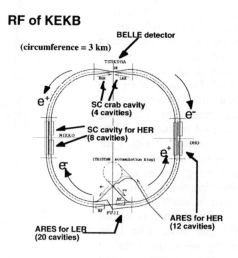

Figure 5.11: The layout of KEKB rings.

5.4.1 Cavity modules

A sectional view of the SC module is in shown Fig. 5.12. The cavity cells were made of 2.5-mm-thick Nb sheets with *RRR* of 200. After grinding the welding seams on the inner surface, the cavities were treated as follows: 1) electropolished by 80 μm, 2) degassed at 700°C for 1.5 hr, 3) electropolished by 15 μm and rinsed with 3 ppm ozonized ultra pure water (OUR) [36].

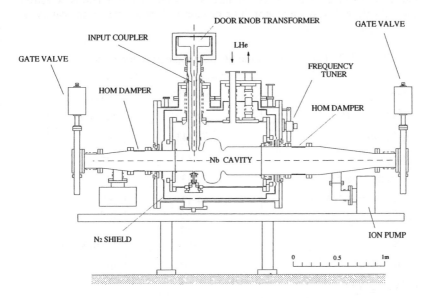

Figure 5.12: SC module for KEKB

The completed cavities were tested in a vertical cryostat before assembling. In spite of the operation voltage of 1.5 MV (6 MV/m), target voltage was set at 2.5 MV (10 MV/m) to obtain enough margin for stable operation. Two of the cavities achieved a maximum accelerating gradient of 11 MV/m and 19 MV/m on the first cold test. The limitation of 19 MV/m, which corresponds to a surface peak field of 35 MV/m and a magnetic field of 750 Oe, is not a breakdown but an available RF power. Another two cavities did not reach 10 MV/m on the first cooldown. One of them was damaged by a discharge at 14 MV/m during the first cold test, and the other was limited by a defect on the equator seam. They were repaired by grinding and additional polishing.

The inner conductor and the ceramic window of the input coupler were conditioned up to 800 kW in traveling-wave mode. Then, a set of the inner and outer conductors was conditioned to 300 kW under a reflection of 130 kW. These conditioning processes were given at a test stand before assembling. The HOM dampers were tested up to 5 kW and 7 kW for SBP and LBP, respectively, then baked at 150°C for about one month to reduce the outgassing rate.

The cryostat has a 12-mm-thick iron vacuum vessel to shield it from the geomagnetic field, and has no other shielding. The measured residual magnetic field at the center of the He vessel is 50 mG, which is due mainly to penetration from the openings for beam pipes on both ends.

After assembling, modules were tested one by one at a horizontal cold test stand, and moved to the tunnel. Each module was connected to a 1-MW klystron, being isolated with a 1-MW circulator. The geometrical parameters of the cavity are given in Table 5.3.

Table 5.3. Parameters of the SC damped cavity for KEKB

Frequency	508.887	MHz
R/Q $(R = V^2/P)$	93	Ω/cavity
Gap length	0.243	m
Diameter of iris	220	mm
Geometrical factor	253	Ω
E_{acc}/V_c (L = 0.243m) E_{sp}/V_c E_{sp}/E_{acc} H_{sp}/E_{acc}	4.11 7.65 1.84 40.3	m^{-1} m^{-1} G/(MV/m)
Total length	3701	mm
Volume of LHe	290	liter
Static loss	30	Watts
Q_{ext} of input coupler	5.3-7.7 $\times 10^4$	
HOM damper size SBP LBP	Ferrite(IB-004) 4t $\times$ 220ϕ $\times$ 120 4t $\times$ 300ϕ $\times$ 150	 mm mm
HOM Max. R (TM$_{011}$) Max. R (TM$_{110}$)	 2 37	 kΩ kΩ/m
Loss factor at $\sigma = 4$mm Cavity with tapers Ferrite dampers Total	1.2 0.3 $\times$ 2 1.8	V/pC V/pC V/pC

At every cooling down, the same procedure for cavity conditioning is used:

- Conditioning of the input coupler before cooling. RF power to 300 kW is given to the coupler for the various bias voltages up to ±2 kV between the inner and outer conductor, as mentioned above.
- Cooling down the cavity within 3 days.
- Conditioning of the cooled cavity.
- Measurement of the Q at 2.0 and 2.5 MV with the LHe consumption rate.

These conditioning processes are made under full reflection condition. For cooled cavities, conditioning with CW power is given first. Then a pulse conditioning process is carried out on the cavity with a voltage less than 2.5 MV, where a small RF pulse with a repetition rate of 100 Hz (duty cycle < 10%) is added to the CW so that the peak RF voltage reaches >3 MV. This pulse conditioning can prevent the discharge from growing into an avalanche, so that the conditioning can be continued without breaks due to vacuum burst or breakdown. Finally, the resonance phase is scanned by ±30° changing the cavity frequency so as to move the standing wave in the coupler for an additional conditioning.

5.4.2 Performance and cavities

Since its commissioning, the total RF voltage of HER has been kept at 8 MV. The four SC damped cavities supplied a voltage of 6 MV, sometimes between 4.8 MV and 8 MV as the occasion demanded. No current limitation caused by the cavities and no other problem in the cavities has been observed up to a current of 0.52 A, with few RF trips so far. This current will be pushed up to the design current of 1.1 A by adding the next four cavities in the summer of 2000.

Figure. 5.13 shows the coupler power at high power operation, where the phase offset between the SC and NC cavities of HER was set so that the SC cavities took a large part of the beam loading. Each input coupler supplied RF power of 400 kW to the cavity, and delivered 1.4 MW to the beam in total, which was 140% of the design value. Even in such a high power test, the cavities did not trip and no additional conditioning was required. The maximum HOM power absorbed by the dampers reached 2.5 kW for each cavity module and 10 kW in total so far, showing no difference in the HOM damping characteristic among all modules.

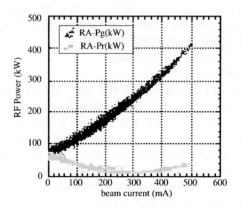

Figure 5.13: RF power vs. beam intensity of one of the KEKB cavities (RA).

One of the most important factors for such stable operation may be the vacuum around the SC cavity section. The cavity vacuum is protected by beam ducts equipped with a series of NEG pumping units to prevent residual gas condensation in the ring. The vacuum pressure of the SC cavities is shown in Fig. 5.14.

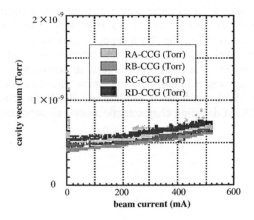

Figure 5.14: Vacuum pressure vs. beam of KEKB cavities.

6 Summary

In KEKB, four SC damped cavities have successfully accelerated a 0.52-A beam without any problem in the cavity. The intensity will be pushed up to 1.1 A with the addition of the next four cavities, scheduled for the summer of 2000. Now we can conclude that a new application field of high current accelerators, has been opened to SC cavities. And we expect to learn what should be done to proceed to the next step in the coming 1-A operation of KEKB.

Development of a SC cavity has a history of four decades. In the early stage of this period, the SC cavity was called as the dream accelerator, and nowadays it becomes the real and one of the most powerful apparatus for particle accelerators. In a storage ring the SC cavity system of CERN produces the total accelerating voltage of 3.5 GV and the SC systems of KEKB and CESR store the beam of more than 500 mA and will achieve 1 A soon. Application of SC cavities is getting wide, and R&D for a SC linear collider and proton linacs are in progress as the next step. These are the results of the many efforts for fundamental researches described in references [18]. Again I would like to emphasize these references and a nice textbook of " RF Superconductivity for Accelerators" [22] for your challenge.

392

References

1. H. A. Schwettman, et al., Proc. of 5th Int. Conf. On High Energy Accel., Frascati, 1965, p.690.
2. J. P. Turner and N. T. Viet, Appl. Phys. Lett., 16, 1970, p333.
3. B. Hillenbrand et al., IEEE Trans. Magn. Mag-13, 1977, p491.
4. U. Klein and D. Proch, Wuppertal, Nov., 1978, WU B 78-31.
5. K. W. Shepard, Proc. of the 2nd Workshop on RF superconductivity, CERN, Switzerland, 1984, p.9.
6. S. Takeuchi, et al., Proc of the 8th Workshop on RF superconductivity, LNL-INFN, Italy, 1997, LNL-INFN(Rep) 133/98, p.237.
7. H. Padamsee, IEEE Trans. Magn., 21, 1977, p.1007.
8. P. Kneisel, J. of Less-Common Met., 139 ,1988, p.179.
9. G. Muller, Proc. of the 3rd Workshop on RF superconductivity, ANL, U.S.A., 1987, ANL-PHY-88-1, p.331.
10. Y. Kojima, et al., Proc. of the 4th Workshop on RF superconductivity, KEK, Japan, 1989, KEK Report 89-21, January 1990, A, p.89.
11. B. H. Wiik, Proc. of the 8th Workshop on RF superconductivity, LNL-INFN, Italy, 1997, LNL-INFN(Rep) 133/98, p.54.
12. S. Belomestnykh, et al., Proc. of the 7th European Particle Accelerator Conference, Vienna, Austria, 26-30 June, 2000.
13. T. Furuya, et al., Proc. of the 9th Workshop on RF superconductivity, LANL, U.S.A., 1999.
14. Homepage of LHC project, http://LHC.web.cern.ch/lhc/.
15. C. Pagani, et.al., Proc of the 8th Workshop on RF superconductivity, Italy, 1997, LNL-INFN(Rep) 133/98, p36.
16. A. Citron, et al., Nucl. Instrum. Methods, 164, p.31, 1979.
17. K. Hosoyama, et al., Proc of the 8th Workshop on RF superconductivity, Italy, 1997, LNL-INFN(Rep) 133/98, p.547.
18. Places and Proceedings of the Internal Workshop on RF superconductivity.
    1) Karlsruhe, edited by M. Kuntze, KfK 3019, Germany, Jul. 2-4, 1980.
    2) CERN, edited by L. Lengeler, Switzerland, Jul. 23-27, 1984.
    3) ANL, edited by K. shepard, ANL-PHY-88-1, U.S.A., Sep. 14-18, 1987.
    4) KEK, edited by Y. Kojima, KEK Report 89-21, Japan, Aug. 14-18, 1989.
    5) DESY, edited by D. Proch, DESY-M-92-01, Germany, Aug. 19-24, 1991.
    6) J.Lab., edited by R. Sundelin, CEBAF, U.S.A, Oct. 4-8, 1993.
    7) CEA/Saclay, edited by B. Bonin, France, Oct. 17-20, 1995.
    8) INFN, edited by V. Palmieri, A. Lombardi, LNL-INFN(rep) 133/98, Italy, Oct. 6-10, 1997.
    9) LANL, edited by B. Rusnak, LA-13782-C Conference, Nov. 1-5, 1999.
19. Proc. of CERN Accelerator School, edited by S. Turner, CERN 89-04, 10 March 1989, Hamburg, Germany, 10 May – 3 June, 1988.
20. Proc. of CERN Accelerator School, edited by S. Turner, CERN 92-03, 11 June 1992, Oxford, U.K., 3-10 April, 1991.

21. Proc. of the Joint US-CERN-Japan International School on Frontiers of Accelerator Technology, edited by S.I. Kurokawa, M. Month and S. Turner, Hayama, Japan, 9-18 Sept., 1996, World Scienctific Publishing Co. Pte. Ltd.
22. H. Padamsee, J. Knobloch and T. Hays, RF Superconductivity for Accelerators, John Wiley & Sons, Inc.
23. A.A. Abrikosov, L.P. Gor'kov and I.M. Khalatnikov, Sov. Phys. JETP 8, 1959, p.182.
24. B. Bonin and W. Roth, Proc. of the 5th Workshop on RF superconductivity, DESY, Germany, 1991, DESY-M-92-01, p.210.
25. R. H. Fowler and L. Nordheim, Proc. R. Soc. Lond. A, Math. Phys. Sci., 119, 1928, p.173.
26. T. Furuya, et al., Proc. of the 3rd Workshop on RF superconductivity, ANL, U.S.A., 1987, ANL-PHY-88-1, p.95.
27. K. Saito, et al., Proc. of the 4th Workshop on RF superconductivity, KEK, Japan, 1989, KEK Report 89-21, January 1990, p.635.
28. W. Singer, et al., Proc. of the 8th Workshop on RF superconductivity, LNL-INFN, Italy, 1997, LNL-INFN(Rep) 133/98, p.850.
29. Design Report of SLAC-PEPII, SLAC-372, 1991.
30. KEK B-Factory Design Report, KEK Report 95-7, August 1995.
31. T. Kageyama, Proc. of the 15th Linear Acc. Meeting in Japan, 1990.
32. H. Padamsee, et al., Proc. of the 5th Workshop on RF superconductivity, DESY, Germany, 1991, p.138
33. T. Furuya, et al., Proc. of the 7th Workshop on RF superconductivity, CEA-Saclay, France, 1995, p.729.
34. T. Tajima, thesis, KEK Report 2000-10, September 2000.
35. S. Mitsunobu, et al., Proc. of the 7th Workshop on RF superconductivity, CEA-Saclay, France, 1995, p.735.
36. K. Asano, KEK Preprint 93-216, March 1994.

CRYOGENIC SYSTEMS

KENJI HOSOYAMA

KEK, High Energy Accelerator Research Organization
1-1 Oho, Tsukuba, Ibaraki,305-0801 Japan
E-mail: kenji.hosoyama@kek.jp

In this lecture we discuss the principle of method of cooling to a very low temperature, i.e. cryogenic. The "gas molecular model" will be introduced to explain the mechanism cooling by the expansion engine and the Joule-Thomson expansion valve. These two expansion processes are normally used in helium refrigeration systems to cool the process gas to cryogenic temperature. The reverse Carnot cycle will be discussed in detail as an ideal refrigeration cycle. First the fundamental process of liquefaction and refrigeration cycles will be discussed, and then the practical helium refrigeration system. The process flow of the system and the key components; –compressor, expander, and heat exchanger– will be discussed. As an example of an actual refrigeration system, we will use the cryogenic system for the KEKB superconducting RF cavity. We will also discuss the liquid helium distribution system, which is very important, especially for the cryogenic systems used in accelerator applications.

1   Principles of Cooling and Fundamental Cooling Cycle

2   Expansion engine, Joule-Thomson expansion, kinetic molecular theory, and enthalpy

3   Liquefaction Systems

4   Refrigeration Systems

5   Practical helium liquefier/refrigeration system

6   Cryogenic System for TRISTAN Superconducting RF Cavity

1 Principles of Cooling and Fundamental Cooling Cycle

1-1 Superconductivity and cooling by liquid helium

We use water for cooling a hot iron bar, and in the same way we use boiling liquid nitrogen and liquid helium at 1 bar, for cooling the thermal shield of the cryostats to about 80 K and the superconducting magnets and cavities to about 4 K, respectively.

Figure 1-1 shows the characteristics of the superconductors NbTi and Nb_3Sn, which are now widely used for superconducting magnets. In this lecture I will not cover the details of superconducting magnets. Please refer to the lecture notes of Maury Tigner and Martin Wilson in this book. The critical temperatures of NbTi and

Nb₃Sn are about 10 K and 18 K respectively. For superconducting RF cavities pure
niobium (Nb) is used, and its critical temperature is 9.25 K. The details of
superconducting cavities are not covered in this lecture; readers are advised to read
the lecture note by Takaaki Furuya in this book. The cooling temperature ranges of
liquid helium and liquid hydrogen are also shown in Fig. 1-1. Because the
temperature range of liquid hydrogen, i.e. 15-25 K, is too high for cooling the
superconducting devices discussed above, liquid helium must be used for this
purpose.

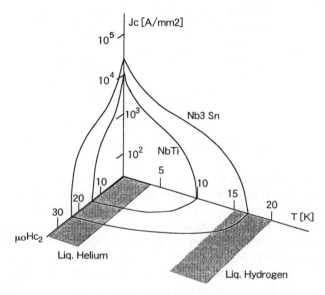

Fig. 1-1   Characteristics of NbTi and Nb₃Sn. The shadowed area shows the
cooling temperature ranges of liquid helium and hydrogen. Please
refer to the lecture notes of Tigner and Wilson to understand the exact
meaning of this figure.

1-2 Production of cooling power

The simplest method of cooling is to utilize the vaporization heat of the refrigerants, i.e. the refrigerant gas is liquefied by compression and then vaporized under low pressure for cooling; this method is used for home refrigerators and air conditioners. Figure 1-2 shows a principle of this refrigeration system.

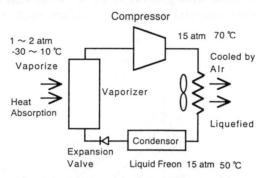

Fig. 1-2 Cooling principle of home refrigerator.

1-2-1 Critical point and triple point

Freon gases, which are widely used in home refrigerators and air conditioners, are easily liquefied by compression at room temperature. Can other gases be liquefied by compression at room temperature? Many attempts have been made to liquefy air, nitrogen and oxygen by compression up to very high pressure at room temperature, but it was found to be impossible; therefore these gases are called permanent gases.

The Irish chemist T. Andrews discovered the concept of "critical temperature" during his famous experiment, measuring the dependence of pressure P on volume v on different isotherms for carbon dioxide. He found that to liquefy a gas by compression its temperature must be kept below some point so called "critical temperature." We will discuss this Andrews' experiment here.

Figure 1-3 shows the P vs. v-curves of CO_2 at various temperatures. CO_2 gas with volume v is compressed isothermally, i.e. keeping the temperature T constant. The pressure P of the gas is increased by reducing the volume v by compression. For example at 10°C, from the point P_1 in the figure, the pressure stays constant under further compression until it starts to increase rapidly at the point P_2. P_1 is the point where liquefaction starts, while P_2 is that where all the gas is liquefied. The constant pressure between P_1 and P_2 is called the saturation vapor pressure.

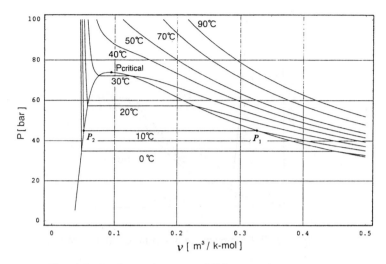

Fig. 1-3 Isothermal curves of CO_2 at various temperatures:
the famous Andrew's experiment.

The flat part of the curve corresponds to the liquefaction of the gas. At a temperature
of 20°C, the range of the flat part decreases, and at 50°C, there is no flat part at all.
The flat part diminishes gradually with increasing temperature and finally reaches
zero at a point, $P_{critical}$, called the critical point. The pressure and the temperature at
this point are called the critical pressure P_c and the critical temperature T_c. For CO_2,
T_c is 31°C and P_c is 7.4 MPa.

If we lower the temperature, we can draw a similar graph for equilibrium
between solid and vapor like that between liquid and vapor. There is obviously one
such line that is the boundary between the liquid-vapor region and the solid-vapor
region. This line is associated with the triple point. The meaning of the triple point
becomes clear if we draw the P-T diagram for a substance such as water shown in Fig.
1-4. Starting from the critical point, the vaporization curve (this line is the interface
between liquid and vapor) extends toward the triple point, from which the
sublimation curve (solid and vapor interface) and fusion curve (liquid and solid
interface) start. At the triple point, all three phases, vapor, liquid, and solid coexist.

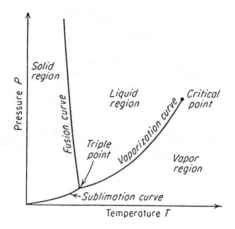

Fig. 1-4 *P-T* diagram for a substance such as water.

Figure 1-5 shows the saturated vapor pressure curves for various gases as a function of temperature. The high and low pressure ends of the curves represent the critical points and triple points of the gases, respectively. Reducing the pressure of the liquid causes the temperature to decrease and it finally reaches the freezing point. Figure 1-6 shows the cooling temperature range of various liquids from triple point to critical point.

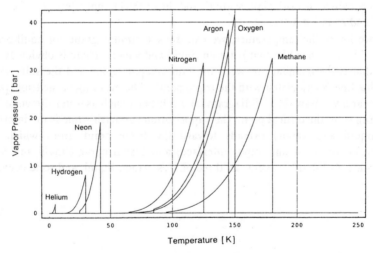

Fig. 1-5 Saturated vapor pressure curves for various gases.

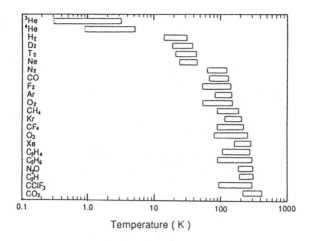

Fig. 1-6 Cooling temperature range of various liquids
(from triple point to critical point).

1-2-2 Liquefaction by cascade method

Figure 1-7 shows the saturation vapor pressure for two different kinds of gas, A and B, as a function of the temperature. The liquid temperature of gas A liquefied at pressure P_1 is T_1. If we reduce the pressure of liquid A to P_2, the temperature decreases to T_2. Because T_2 is lower than the critical temperature of gas B, we can liquefy gas B by cooling its compressed gas at P_3 by boiling liquid A. By this cascade method we can liquefy a gas step by step in a simple way, i.e. by cooling and compressing. If there is an overlapping temperature region in the saturated vapor pressure curves in Fig. 1-5 and 1-6, we can use this simple and direct method to liquefy the gas. But unfortunately there is no overlapping liquid temperature region between nitrogen and hydrogen or between hydrogen and helium. This means we need new methods to liquefy hydrogen gas and helium gas.

1-2-3 Liquefaction of hydrogen and helium by Joule-Thomson expansion

It took a long time to overcome this gap, but the first gap, between nitrogen and hydrogen, was overcome in 1891 by James Dewar using Joule-Thomson expansion of hydrogen gas pre-cooled by liquid nitrogen. The second gap was finally conquered by Kamerlingh-Onnes in 1908. Figure 1-8 is a flow diagram of his helium liquefier, in which helium gas pre-cooled by sub-atmospheric pressure liquid hydrogen was expanded at a Joule-Thomson valve and became liquefied.

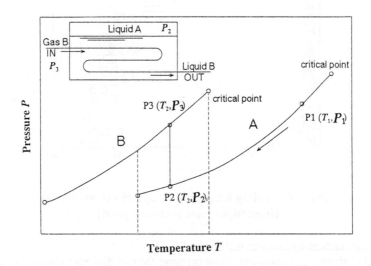

Fig. 1-7 Liquefaction by cascade method.

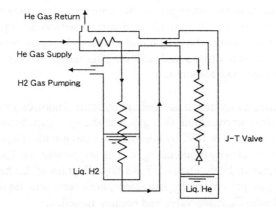

Fig. 1-8 Schematic flow of the first helium liquefaction by Kamerlingh-Onnes.

2 Expansion engine, Joule-Thomson expansion, kinetic molecular theory, and enthalpy

As is shown in Sec. 1-2-3, Joule-Thomson expansion plays an important role to cool a gas to the cryogenic temperature. In addition to it, there is another important process, namely expansion by engine. In this Section, a simple kinetic molecular theory of ideal gases is first introduced, and then the mechanism of expansion by engine is explained by the use of the kinetic theory. Second, we discuss the Joule-Thomson expansion from a molecular view point, taking intermolecular forces into account. Before discussing the Joule-Thomson expansion thermodynamically, we will introduce a quantity called the enthalpy and learn its basic features. Enthalpy is of vital importance to understand liquefiers and refrigerators. Finally after a brief review of entropy, the Joule-Thomson expansion will be discussed by using a T-s diagram.

2-1 Simple kinetic molecular theory of ideal gases

The gas in a cubic $(L \times L \times L)$ consists of a large number of molecules with various velocities moving around in the box, bombarding the wall and bouncing back. A single molecule moves around as shown in Fig. 2-1.

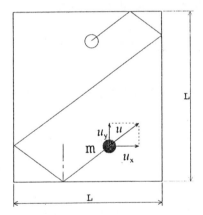

Fig. 2-1 Simple kinetic model for ideal gas.

In the ideal gas model we consider a point molecule of mas m with no volume. A molecule with velocity $u = (u_x, u_y, u_z)$ bombards the bottom of the box and bounces back with the same kinetic energy because no energy is exchanged between the wall and the molecule; momentum is exchanged by the collision. The momentum transfer to the wall is

$$\Delta p = 2mu_y, \tag{2-1}$$

and the number of collisions per unit time is $u_y/2L$. The force on the wall from this one molecule is calculated by momentum change per unit time as

$$F_y = \Delta p \times \frac{u_y}{2L} = 2mu_y\frac{u_y}{2L} = m\frac{u_y^2}{L}. \tag{2-2}$$

Then the pressure on the wall from one molecule is

$$P = \frac{F_y}{L^2} = m\frac{u_y^2}{L^3} = \frac{mu_y^2}{V}. \tag{2-3}$$

The pressure on the wall from N molecules is

$$P = \sum_i \frac{(mu_y^2)_i}{V} = \frac{m}{V}\sum_i (u_y^2)_i = \frac{m}{V}N\overline{u_y^2}, \tag{2-4}$$

The pressure can be expressed by the velocity u by using the following relations of the molecules' velocity:

$$u^2 = u_x^2 + u_y^2 + u_z^2, \quad u_x^2 = u_y^2 = u_z^2, \quad \overline{u^2} = 3\overline{u_y^2}, \tag{2-5}$$

$$P = \frac{1}{3V}Nm\overline{u^2}. \tag{2-6}$$

Form Eq. (2-6) we get

$$PV = \frac{1}{3}Nm\overline{u^2}. \tag{2-7}$$

Since the average kinetic energy of a molecule is

$$\overline{ke} = \frac{1}{2}m\overline{u^2}. \tag{2-8}$$

Equation (2-7) is expressed as

$$PV = \frac{2}{3} n N_A \overline{ke} , \qquad (2-9)$$

where $N = n N_A$, and n is the number of moles, and N_A is the Avogadro's number. On the other hand, from thermodynamics the equation of state for an ideal gas is

$$PV = nRT , \qquad (2-10)$$

where R is the gas constant. By comparing Eqs (2-9) and (2-10) we get the following equation:

$$N_A \overline{ke} = \frac{3}{2} RT . \qquad (2-11)$$

The average kinetic energy of the molecule is, therefore,

$$\overline{ke} = \frac{3}{2} \frac{R}{N_A} T = \frac{3}{2} k_B T , \qquad (2-12)$$

where k_B is Boltzmann's constant.

404

2-2 Cooling by an expansion engine

We will discuss the mechanism of cooling a gas by an expansion engine using the simple kinetic molecular theory of gases discussed above. Figure 2-2 shows the expansion process. By this expansion process, the kinetic energy of the gas molecule decreases, and this leads to the cooling of the gas.

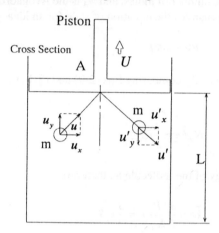

Fig. 2-2 Cooling by an expansion engine: simple kinetic model.

When a piston moves upward with velocity U, a molecule with velocity u hits the piston surface and bounds back with velocity u'. Usually the velocity of the piston U is much smaller than that of the molecule. The y-component of the velocity is decreased by the bombardment, but tangential component of the velocity is not changed:

$$u'_x = u_x, \tag{2-13}$$

$$u'_y = u_y - 2U. \tag{2-14}$$

The square of velocity u', the velocity of the rebounding molecule, is calculated from Eqs. (2-13) and (2-14):

$$u'^2 = u'^2_x + u^2_y = u^2_x + \left(u_y - 2U\right)^2 = u^2_x + u^2_y - 4Uv_y + 4U^2. \tag{2-15}$$

The bombardment causes a change in the molecule's kinetic energy, Δke:

$$\Delta ke = \frac{m}{2}\left(u'^2 - u^2\right) = \frac{m}{2}\left(-4Uu_y + 4U^2\right) \cong -2mUu_y . \qquad (2\text{-}16)$$

During expansion the molecule losses kinetic energy $\Delta ke = -2mUu_y$; on the contrary during compression it gains kinetic energy Δke. Because the molecule moves around in the cylinder and hits the piston many times, the energy change of the molecule in time τ is calculated as a product of the change in kinetic energy per hit times, the number of hits per unit time, and times τ:

$$\Delta E = \Delta ke \times \frac{u_y}{2L}\tau = -mu_y^2 \frac{1}{L}U\tau = -mu_y^2 \frac{dl}{L}, \qquad (2\text{-}17)$$

where dl is the distance of the piston movement. The total energy change dE_{kin} is calculated by summing over all molecules and averaging by time:

$$dE_{kin} = N\overline{\Delta E} = -Nm\overline{u_y^2}\frac{dl}{L} = -Nm\frac{\overline{u^2}}{3}\frac{dl}{L} = -\frac{2}{3}E_{kin}\frac{dV}{V}, \qquad (2\text{-}18)$$

$$\frac{dE_{kin}}{E_{kin}} = -\frac{2}{3}\frac{dV}{V}, \qquad (2\text{-}19)$$

where we assume $E = E_{kin}$; this correspond to a monatomic molecule.

The pressure of an ideal gas from kinetic molecular theory was discussed in the previous sub-section; from Eq. (2-9) we get

$$PV = \frac{2}{3}E_{kin} . \qquad (2\text{-}20)$$

By differentiating Eq. (2-20) and then dividing by Eq. (2-20), we get

$$Vdp + pdV = \frac{2}{3}dE_{kin}, \qquad (2\text{-}21)$$

$$\frac{dP}{P} + \frac{dV}{V} = \frac{2}{3}\frac{dE_{kin}}{PV} = \frac{dE_{kin}}{E_{kin}} = -\frac{2}{3}\frac{dV}{V}, \qquad (2\text{-}22)$$

$$\frac{dP}{P} + \frac{5}{3}\frac{dV}{V} = 0 . \qquad (2\text{-}23)$$

By integrating Eq. (2-23) we get

$$PV^{\frac{5}{3}} = const.$$ (2-24)

2-3 Cooling by Joule-Thomson expansion

In order to understand the mechanism of the Joule-Thomson expansion, we should take into account molecular size, intermolecular distance and intermolecular forces for various gases at cryogenic temperature. Figure 2-3 shows the intermolecular Lennard-Jones potential of various gases. The vertical axis is the potential measured by temperature [K], i.e. potential energy divided by Boltzman's constant k_B, and the horizontal axis is intermolecular distance [nm]. The potential has a repulsive center core and an attractive region produced by van der Waals forces between the molecules. From these intermolecular potential curves we can estimate the size of the molecules to be about 0.2 nm for helium gas and about 0.35 nm for nitrogen gas. The attractive potential depth of helium gas is very small (about 10 K) compared to those of nitrogen and other gases (about 90 K). These attractive potentials cause the molecules to stick together and form the liquid state. The weak intermolecular bonding of helium gas molecules in the liquid state can be destroyed by thermal agitation of about 10 K.

If we consider 1 mol of helium gas at 1 bar and at 273 K, then its volume is 22.4 L and the number of molecules in this volume is Avogadro's number N_A (6.03×10^{23}). If we assume, for simplicity, that the gas is in a cubic box with dimensions L^3 and all the molecules are located on a cubic lattice, then the intermolecular distance is calculated as follows. The length L of the cubic box and the number of molecules along this length of 1 mol helium gas are

$$L = \sqrt[3]{22.4 \times 1000} = 28.2\, cm, \quad N_{line} = \sqrt[3]{6.03 \times 10^{23}} = 0.845 \times 10^8.$$ (2-25)

The intermolecular distance D becomes:

$$D = L / N_{line} = 28.2 / 0.845 \times 10^8 = 33.4 \times 10^{-8}\, cm = 3.34\ nm.$$ (2-26)

This intermolecular distance is more than 10 times as large as the size of a helium gas molecule. This means that the assumption of the ideal gas model, i.e. that the molecule is a volume-less point, appears to be a good approximation at 1 bar and 273 K.

On the other hand, if the helium gas is compressed to 20 bar and cooled down to about 10 K, D becomes 0.4 nm. In this case, the intermolecular distance is also almost the same as the molecular size (see Fig. 2-3). Then the ideal gas model is no longer a good assumption, and our model must take into account the finite volume effect of the molecule and the attractive potential effect.

The Dutch physicist J.D. van der Waals proposed the phenomenological equation, now called the van der Waals equation, for real gases which took into account molecular size and intermolecular force:

$$\left(P + \frac{a}{v^2}\right)(v - b) = RT. \tag{2-27}$$

or

$$P = \frac{RT}{(v - b)} - \frac{a}{v^2}, \tag{2-28}$$

where a and b are constants that characterize the individual properties of the gas and express the effects of the intermolecular attractive force and the volume of the molecule respectively.

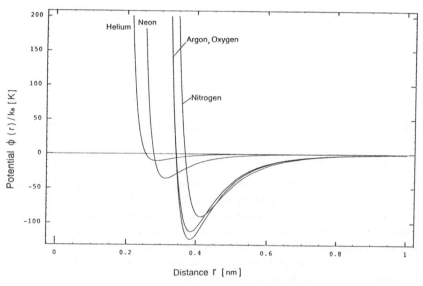

Fig. 2-3 Intermolecular Lennard-Jones potential of various gases.

Figure 2-4 shows P-v diagram of the equation of state for a van der Waals gas, where the equations of state is expressed by reduced volume v/v_c on the horizontal axis and reduced pressure P/P_c on the vertical axis as the parameters of reduced temperatures T/T_c, where v_c, P_c and T_c are the critical volume, pressure, and temperature of the gas. If we recall the isothermal P-v diagram of CO_2 gas shown in Fig. 1-3, we see that these phenomenological equations express behavior of the real gases qualitatively well.

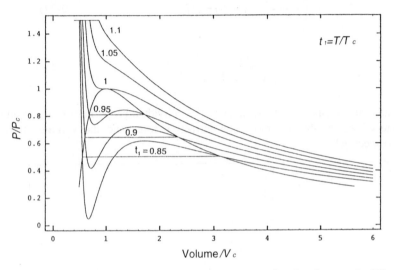

Fig. 2-4   Isothermal P-v diagram of the state equation for the van der Waals gas.

2-4   Cooing by Joule-Thomson expansion: Microscopic point of view

Here we will discuss the mechanism of cooling by Joule-Thomson expansion intuitively, from the microscopic point view. Figure 2-5 shows the intermolecular potential for real gases. If the gas is compressed to high pressure P_A, the gas molecules are pushed together and the mean intermolecular distance r becomes small, which corresponds to point A in Fig. 2-5. The molecules interact with each other via the intermolecular potential. The total energy of a gas molecule is expressed as the sum of kinetic energy T_{kinA} and intermolecular interaction U_A:

$$E_A = T_{kinA} + U_A .$$
(2-29)

If this high pressure gas at point A expands adiabatically without any work to or from the outside to a lower pressure corresponding to state B, the total energy of the molecule in state A and state B must be the same, because there is no energy transfer between the molecule and the outside. Then,

$$E_A = E_B \,, \tag{2-30}$$

$$E_A = T_{kinA} + U_A = T_{kinB} + U_B = E_B \,, \tag{2-31}$$

$$T_{kinB} - T_{kinA} = U_A - U_B = - \Delta U_0 \,. \tag{2-32}$$

We get following relation:

$$T_{kinB} < T_{kinA} \,. \tag{2-33}$$

By this expansion the mean intermolecular distance changes from that in state A to that in state B, and the kinetic energy of the gas molecule decreases. The kinetic energy of the gas is proportional to the temperature in the kinetic gas model. This expansion is called the Joule-Thomson expansion. With Joule-Thomson expansion from point A to B the temperature of the gas molecules decreases; on the other hand, if we start from the point C in Fig. 2-5, and expand the gas to the point A, the temperature of gas increases.

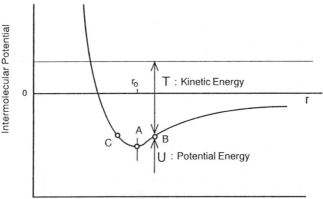

Fig. 2-5   Cooling by Joule-Thomson expansion: microscopic stand point of view.

2-5 Why we use enthalpy for the process calculation (I)

Before we discuss the Joule-Thomson expansion thermodynamically, we should fully understand the characteristics of the enthalpy. Enthalpy is important in various thermodynamic calculations and is defined as the sum of the internal energy of a system, U, and the product of the system pressure P by the volume V:

$$H = U + PV .$$ (2-34)

Usually, we discuss the internal energy, the enthalpy, the volume, the heat, etc. per unit mas (these quantities per unit mas are called the specific internal energy, the specific enthalpy, the specific volume, ...) and use small letters to denote them.

$$h = u + Pv .$$ (2-35)

The first law of thermodynamics is expressed by

$$dq = du + Pdv ,$$ (2-36)

where q is the heat. This equation is modified as follows:

$$dq = du + d(Pv) - vdP ,$$
$$= d(u + Pv) - vdP = dh - vdP .$$ (2-37)

For an isobaric process, i.e. P = constant, we get

$$dq_p = dh ,$$ (2-38)

which means that heat flow dq to or from the gas corresponds to the enthalpy change dh.

Consider a heat exchanger that has a high pressure flow channel 1 to 2 and a low pressure return channel 3 to 4, as shown in Fig. 2-6. The enthalpy change of the high pressure stream from h_1 to h_2, which corresponds to the heat transfer q from the high pressure gas stream to the low pressure gas stream, is equal to the enthalpy change from h_3 to h_4.

We can express the enthalpy as a function of P and T :

$$h = f(P,T) .$$ (2-39)

The differentiation of h is

$$dh = \left(\frac{\partial h}{\partial T}\right)_p dT + \left(\frac{\partial h}{\partial T}\right)_T dP = c_p dT + \left(\frac{\partial h}{\partial P}\right)_T dP , \qquad (2\text{-}40)$$

where C_p is specific heat at constant pressure. For an isobaric process,

$$dq_p = dh = c_p dT . \qquad (2\text{-}41)$$

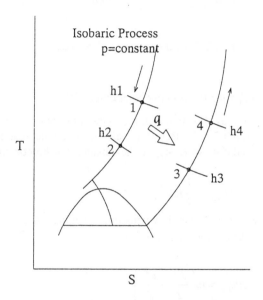

Fig. 2-6 Heat transfer at heat exchanger.

The difference between C_v (specific heat at constant volume) and C_p is easily understand from Fig. 2-7. In the first case the volume of the gas is kept constant during heating and amount of heat ΔQ_v is used to increase the internal energy of the gas. In the second case the volume must be changed to keep the pressure P constant during heating and we need additional heating ΔW to lift the weight and get $\Delta H = \Delta Q_p$.

412

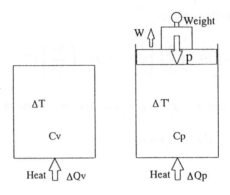

Fig. 2-7 Comparison of C_v and C_p.

2-6   Why we use enthalpy for the process calculation (II)

Consider the flow of a gas as shown in Fig. 2-8. Gas at pressure P_1 enters this system through 1 and exits through 2 at P_2. Between 1 and 2, we assume that there is a device which absorbs heat Q and does work W_{net} to the outside. The kinetic energy of the gas in unit flow time is

$$E_{kin} = \frac{1}{2} \dot{m} \mathcal{V}^2 ,$$

(2-42)

where $\dot{m}$ is mass flow rate and $\mathcal{V}$ is flow velocity. Then the kinetic energy of the gas flow into the system in unit time, ΔE_{kin}, is

$$\Delta E_{kin} = \dot{m} \left(\frac{\mathcal{V}_2^2}{2} - \frac{\mathcal{V}_1^2}{2} \right).$$

(2-43)

The work performed by the gas flow to the system at point 1, W_1, and from system to the gas stream at point 2, W_2, per unit time are calculated as:

$$W_1 = P_1 A_1 x_1 = P_1 V_1 = P_1 \dot{m} v_1 ,$$

(2-44)

$$W_2 = P_2 A_2 x_2 = P_2 V_2 = P_2 \dot{m} v_2 ,$$

(2-45)

where V_1 and V_2 are the volumes of the gas passing into point 1 and point 2 of the conduit per unit time and A_1, and A_2 are the cross section at the points 1 and 2.

The specific volume of gas at points 1 and 2 is given by v_1 and v_2. The net work done by the gas in the system by displacement L_{disp} is calculated from Eqs. (2-44) and (2-45):

$$L_{disp} = W_2 - W_1 = \dot{m}(P_2 v_2 - P_1 v_1).$$ (2-46)

The difference between the potential energy of the gas at point 1 and at point 2 is

$$L_{pot} = \dot{m}g(z_2 - z_1).$$ (2-47)

The work done by the gas system to the outside per unit time L_{12} is calculated from Eqs. (2-43), (2-46), and (2-47):

$$L_{12} = \dot{m}(P_2 v_2 - P_1 v_1) + \dot{m}\left(\frac{v_2^2}{2} - \frac{v_1^2}{2}\right) + \dot{m}g(z_2 - z_1) + W_{net}.$$ (2-48)

Substituting the relationship for L_{12} in to the equation of the first low of thermodynamics (energy conservation law) $dQ = dU + dW$,

$$Q = (U_2 - U_1) + \dot{m}(P_2 v_2 - P_1 v_1) + \dot{m}\left(\frac{v_2^2}{2} - \frac{v_1^2}{2}\right)$$
$$+ \dot{m}g(z_2 - z_1) + W_{net}.$$ (2-49)

Dividing both side of Eq. (2-49) by $\dot{m}$ gives

$$q = (u_2 - u_1) + (P_2 v_2 - P_1 v_1) + \dot{m}\left(\frac{v_2^2}{2} - \frac{v_1^2}{2}\right)$$
$$+ g(z_2 - z_1) + w_{net},$$ (2-50)

where q and w_{net} are the specific heat and work respectively. Eq. (2-50) can be expressed using enthalpy h:

$$q = (h_2 - h_1) + \left(\frac{v_2^2}{2} - \frac{v_1^2}{2}\right) + g(z_2 - z_1) + w_{net}.$$ (2-51)

414

In the case where the gas velocities to and from the system do not change, i.e. $\mathcal{V}_1 = \mathcal{V}_2$, and there is no effect of gravity, i.e. $z_1 = z_2$,

$$q = h_2 - h_1 + w_{net},\qquad(2\text{-}52)$$

or

$$q - w_{net} = h_2 - h_1.\qquad(2\text{-}53)$$

This formula will be used frequently in Sec. 3 and 4 when we study liquefiers and refrigerators. If $w_{net} = 0$, the input heat q to the gas causes a change in the enthalpy of the gas: $q = h_2 - h_1$.

In another case, where $q = 0$, i.e. an adiabatic process, and $w_{net} = 0$, the change of the enthalpy causes the change of velocity. This case corresponds to causing high speed gas flow by throttling the cross-section at the nozzle of turbo-expander.

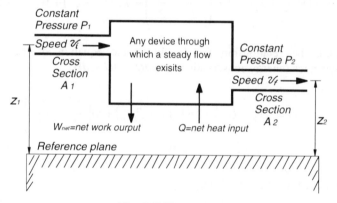

Fig. 2-8 Flow process.

2-7 Cooling by Joule-Thomson expansion: Thermodynamic point of view

We have discussed Joule-Thomson expansion from the microscopic point of view in Sec. 2-4. Now we are ready to discuss the cooling effect of Joule-Thomson expansion from the thermodynamic point of view.

A gas with pressure P_1 and volume V_1 at the left side of a porous plug S seeps through the plug to the right, to pressure P_2 and the volume V_2, as shown in Fig. 2-9. This is an expansion process: $P_1 > P_2$. In this process the Piston 1 and Piston 2 push and pull to keep the pressure on both sides constant at P_1 and at P_2. The gas extruded through the porous plug has no velocity gain by this expansion process. From the conservation of energy the work done to the gas, W, is equal to the increase of the internal energy U_2-U_1, because there is no heat flow to the gas during this process:

$$W = \int P_1 dV - \int P_2 dV = P_1 V_1 - P_2 V_2 \text{ ,} \qquad (2\text{-}54)$$

$$U_2 - U_1 = P_1 V_1 - P_2 V_2 \text{ .} \qquad (2\text{-}55)$$

From this equation, we get following equation;

$$H_2 = U_2 + P_2 V_2 = U_1 + P_1 V_1 = H_1 \text{ .} \qquad (2\text{-}56)$$

This means that in this process of expansion through a porous plug (i.e. Joule-Thomson valve) the sum of the internal energy and the product of the pressure and volume of the gas, namely the enthalpy H, is conserved.

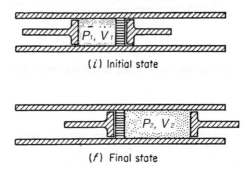

(i) Initial state

(f) Final state

Fig. 2-9 Expansion through a porous plug.

Figure 2-10 shows isenthalpic curves for nitrogen. The locus of the maxima of the isenthalpic curves, is known as the inversion curve and shown in the figure as a dotted curve. The region inside the inversion curve is called the region of cooling, since in this region if the pressure drops the temperature also drops. Figure 2-11 shows the inversion curves for helium, hydrogen, and nitrogen, and Figure 2-12 the enlarged one for helium. Note that in these figures, the horizontal axis shows T and the vertical axis P for convenience. As is shown above, we must start the Joule-Thomson expansion from inside the curves to get cooling effects. The following can be seen from the figures: (1) in the case of nitrogen gas, we can get a cooling effect at room temperature; (2) for hydrogen gas, we must cool the gas to about 200 K to get a cooling effect, and (3) for helium gas, we must go down to about 40 K.

Heat and Thermodynamics

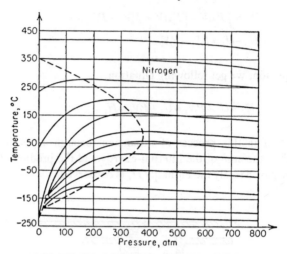

Fig. 2-10   Isenthalpic curves and inversion curve for nitrogen.

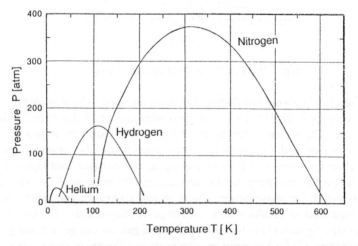

Fig. 2-11   Inversion curves of Joule-Thomson effect for helium,
hydrogen and nitrogen gases.

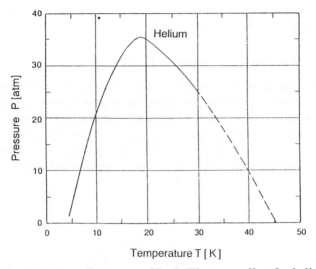

Fig. 2-12 Inversion curve of Joule-Thomson effect for helium gas.

2-8 T-s diagram

Entropy S is the one of the most important physical value in thermodynamics. In this lecture I do not have any room to explain the details of the entropy and advise you to consult one of the text books shown at the end of this note.

Only the difference of the entropy between two states has the meaning. For each infinitesimal amount of heat that enters a system during an infinitesimal portion of a reversible process, there is an equation,

$$\frac{dQ_R}{T} = dS .$$
(2-57)

It follows therefore that the total amount of heat transferred in a reversible process is given by,

$$Q_R =_R \int_i^f T ds .$$
(2-58)

The letter R means that these foumulas are only valid for reversible processes. This integral can be interpreted graphically as the area under a curve on a diagram in which T is plotted along the vertical axis and S along the horizontal axis. Usually we use the specific entropy instead of the entropy, and the diagram is called the T-s diagram. This diagram is widely used.

2-9 Cooling by Joule-Thomson expansion on T-s diagram

Figure 2-13 shows the isobar and isenthalpic curves for a gas in a T-s diagram. The Joule-Thomson expansion processes are expressed by isenthalpic curves where the enthalpy stays constant. When expansion starts at (b) in the figure from the pressure P_1 to P_2, the gas tempareture decreases by ΔT_b, whereas when expansion starts from (a) the temperature increases. Figure 2-14 shows the T-s diagram for helium gas at low temperature. Effectiveness of the Joule-Thomson expansion is evident from the figure.

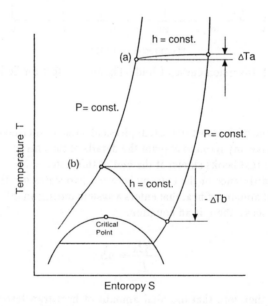

Fig. 2-13 Cooling by Joule-Thomson expansion: thermodynamic point of view.

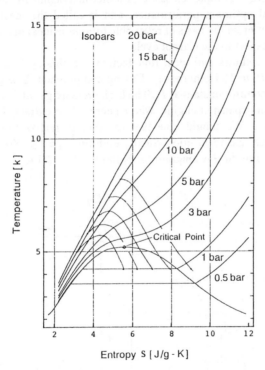

Fig. 2-14 Cooling by Joule-Thomson expansion fo helium gas. Lines connecting isobaric lines are isenthalpic lines for Joul-Thomson expansion.

3 Liquefaction Systems

3-1 *Ideal liquefaction system*

So far we have discussed two methods of cooling a gas: expansion by an engine and expansion by a Joule-Thomson valve. Here we will discuss the liquefier.

It is our aim to liquefy a unit quantity of gas with the least possible expenditure of energy. A liquefier based on a thermodynamically reversible process has the theoretical optimum performance, and is called the ideal liquefier. The reason why a liquefier on the basis of reversible process is ideal one will be discussed in Sec. 4-1 with respect to the Carnot cycle.

Figure 3-1 shows and ideal liquefaction. In this cycle the process gas is compressed from P_1 to P_2 isothermally. During compression, heat Q is extracted to keep the gas temperature constant at T_1. The high pressure gas at P_2 is expanded to P_1 isentropically by an expansion engine, where energy W_e is extracted to the outside of the system as work. We should note that in order to realize this cycle we must compress the process gas to an impractically high pressure P_2. You can understand this fact by examining the T-s diagram for helium in Fig. 2-14.

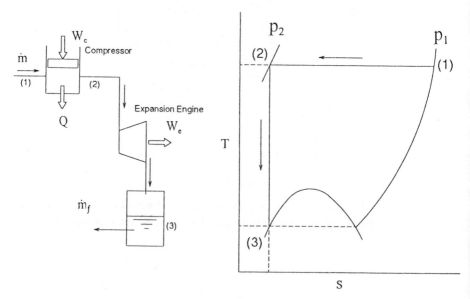

Fig. 3-1 Ideal liquefier system: flow and the cycle on T-s diagram.

This system is considered to be a flow process discussed in Sec. 2-6. By using Eq. (2-53), we can write,

$$Q_{net} - W_{net} = \sum_{outlet} \dot{m}h - \sum_{inlet} \dot{m}h, \qquad (3\text{-}1)$$

where Q_{net} is the net heat transfer to or from the system (heat transferred to the system is considered positive), and W_{net} is the net work done on or by the system (work done by the system is considered positive).

Taking into account the fact that all gas is liquefied ($m_f = m_1$),

$$Q - W_c + W_e = \dot{m}_f h_f - \dot{m}h_1 = -\dot{m}(h_1 - h_f). \qquad (3\text{-}2)$$

The heat Q produced in isothermal compression is

$$Q = \dot{m}T_1(s_2 - s_1) = -\dot{m}T_1(s_f - s_1). \qquad (3\text{-}3)$$

As an ideal case, we assume that we can utilize the power W_e produced at the expansion engine, the net power of the liquefier, $W_i = W_c\text{-}W_e$, can be calculated as follows.

$$-W_i = -W_c + W_e = -\dot{m}(h_1 - h_f) - \dot{m}T_1(s_f - s_1). \qquad (3\text{-}4)$$

By Eq.(3-4), we can calculate the required power for liquefaction of N_2, H_2 and He gases by the ideal requefier as listed in Table 3-1. We set the high temperature a orbitarily to 300 K. It is interesting to find that almost the same power, 200 Wh/L, is necessary to liquefy these gases per liter.

Table 3-1.

	T (K)	(Wh/L)
N_2	77.4	173
H_2	20.4	231
He	4.2	236

3-2 Practical liquefaction system (1): Simple Linde-Hampson cycle (Joule-Thomson cycle)

Since the ideal liquefaction system is not at all practical, a cycle that adopts Joule-Thomson expansion is usually used. Figure 3-2 shows a flow diagram and a cycle on a T-s diagram of a simple Linde-Hampson liquefier. The process gas is isothermally compressed from P_1 to P_2, and then sent to the heat exchanger, where the gas is cooled by the return cold gas to temperature T_3. The gas is then expanded at a Joule-Thomson valve. Some of the gas is liquefied there. The liquefied gas is extracted to the outside and the cold gas is sent back to the compressor through the heat exchanger.

By applying Eq.(3-1) to a system consisting of the heat exchanger HX, the Joule-Thomson valve, and the liquid receiver, and rembering that no heat and work are exchanged between this system and the outside, we can write

$$0 = \dot{m}h_2 - \left(\dot{m} - \dot{m}_f\right)h_1 - \dot{m}_f h_f, \tag{3-5}$$

then we get

$$\dot{m}(h_1 - h_2) = \dot{m}_f (h_1 - h_f). \tag{3-6}$$

From this equation, we can get the liquid yield y, i.e. the fraction of the gas flow that is liquefied:

$$\frac{\dot{m}_f}{\dot{m}} = y = \frac{h_1 - h_2}{h_1 - h_f}, \tag{3-7}$$

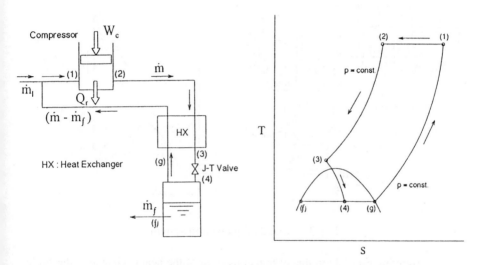

Fig. 3-2 Simple Linde-Hampson liquefaction system.

3-3 Practical liquefaction system (2): Claude cycle

Figure 3-3 shows a flow diagram and a cycle on a *T-s* diagram of a Claude cycle liquefier. In this system some of the compressed gas is sent to an expansion engine and the energy of the gas is extracted outside as work W_e, and cooling power is produced there. Another part of the gas is sent to a Joule-Thomson valve after it is cooled by heat exchangers which are cooled by return cold gas from the Joule-Thomson expansion and the expansion engine. The liquefied gas is extracted to the outside and the cold gas is sent back to the compressor through the heat exchangers.

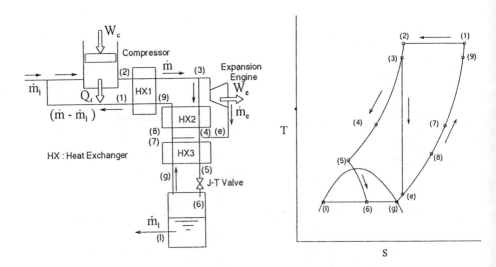

Fig. 3-3 Claude liquefaction system.

Applying Eq.(3-1) to the system consisting of the heat exchangers, the Joule-Thomson valve, and the liquid receiver, like the former section,

$$0 = \left(\dot{m} - \dot{m}_f\right)h_1 + \dot{m}_f h_f - \dot{m}_e h_e - \dot{m}h_2 - \dot{m}_e h_3 . \qquad (3\text{-}8)$$

If we define the fraction of the total flow that passed through the expander as x,

$$x = \dot{m}_e / \dot{m} . \qquad (3\text{-}9)$$

The liquefaction rate y can be obtained as

$$y = \frac{\dot{m}_f}{\dot{m}} = \frac{h_1 - h_2}{h_1 - h_f} + x\frac{h_3 - h_e}{h_1 - h_f}.$$ (3-10)

The first term of the right hand side of Eq.(3-10) is the same as Eq.(3-7), and the second term represents the improvement in performance over the simple Linde-Hampson system. The necessary net work W_{net} is expressed as,

$$-W_{net}/\dot{m} = -W_c/\dot{m} - W_e/\dot{m}.$$ (3-11)

Applying the Eq.(3-1) to the expander, we obtain the expander work expression,

$$W_e = \dot{m}_e(h_3 - h_e).$$ (3-12)

Applying Eq.(3-1) to the compressor, we get

$$-W_e - Q_r = H_1 - H_2,$$ (3-13)

and since Q_r is the heat transferred to the outside during the isothemal compression from 1 to 2,

$$Q_r = T_1(S_1 - S_2).$$ (3-14)

Combining Eqs. (3-13) and (3-14), we get

$$-W/\dot{m} = [T_1(s_1 - s_2) - (h_1 - h_2)] - x(h_3 - h_e).$$ (3-15)

The last term in the equation is the reduction in energy requirement due to utilization of the expander work output.

4 Refrigeration Systems

Refrigeration systems have the same components and thermodynamic cycles as the corresponding liquefaction systems. The difference between them is that the liquid produced in a refrigeration system is evaporated, whereas the liquid produced in the liquefaction system is extracted and utilized in some way outside the system. We will discuss this difference in detail here.

4-1 Ideal refrigeration system

4-1-1 Ideal heat engine; Carnot cycle

We first discuss the Carnot cycle, which was introduced by Sadi Carnot in studying the problem of how to build the most efficient engine. Figure 4-1 shows a Carnot cycle engine and heat reservoirs. The cycle consists of reversible isothermal and reversible adiabatic compression and expansion processes. By reversing the cycle direction, the Carnot cycle can be used for a thermodynamically ideal refrigerator.

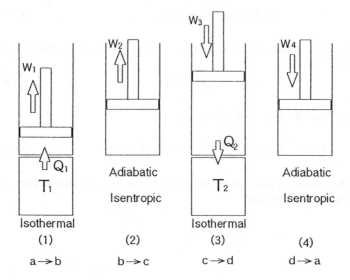

Fig. 4-1 Operation principle of ideal heat engine: Carnot cycle.

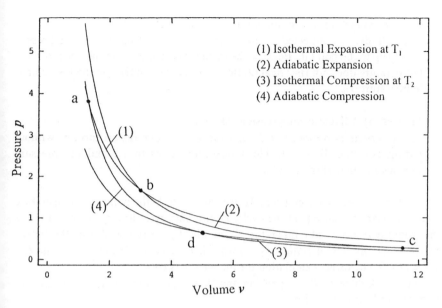

Fig. 4-2 Process of Carnot cycle shown on P-v diagram.

To simplify the discussion, we assume an ideal gas in a cylinder equipped with a frictionless piston, and two heat reservoirs at high temperature T_1 and low temperature T_2. Figure 4-2 is a plot of gas pressure P vs. gas volume v for the Carnot cycle. The cycle proceeds as follows.

Process (1) Isothermal expansion: (a $\rightarrow$ b)
The gas is heated and at the same time expanded slowly, while it is in contact with the heat reservoir at T_1. During this isothermal expansion the pressure falls as the volume increases until it stops at the point b. At the same time, a certain amount of heat Q_1 flows into the gas from the reservoir to keep the gas temperature at T_1.

Process (2) Adiabatic expansion: (b $\rightarrow$ c)
The heat reservoir is taken away and expansion continues adiabatically. The gas continues to expand to the point c, and the temperature falls to T_2 since there is no longer any heat entering.

Process (3) Isothermal compression: (c → d)
The cylinder is connected to the reservoir at T_2 and the gas is compressed slowly up to the point d. Because the cylinder is connected to the reservoir, the temperature does not rise; heat Q_2 flows from the cylinder into the reservoir at temperature T_2.

Process (4) Adiabatic compression: (d → a)
The reservoir is removed at T_2 and the gas is compressed further, without letting any heat flow out. The temperature rises to T_1, and the pressure increases to that at the point a.

If we carry out each step properly, we can return to the point a at temperature T_1, where we started, and repeat the cycle. During one cycle heat Q_1 is put in to the engine from the reservoir at temperature T_1, and heat Q_2 is removed from the engine to the heat reservoir at temperature T_2 and work W is extracted. We can determine Q_1, Q_2, ($W = Q_1-Q_2$) by direct calculation for the ideal gas case.

The work done by the expansion and that required to compress the gas are given by the following equations:

$$W_{a \to b} = \int_a^b PdV \; , \qquad W_{c \to d} = \int_c^d PdV .$$ (4-1)

In an ideal gas, the energy of each molecule depends only on the temperature; this means that in the isothermal process the internal energy u is the same. From the first law of thermodynamics, all the work done by the gas during expansion, $W_{a \to b}$ and the work given to the gas during compression $W_{c \to d}$ are equal to heat flow Q_1 taken from the reservoir at T_1 and Q_2 given to the reservoir at T_2 respectively:

$$Q_1 = \int_a^b Nk_BT_1 \frac{dV}{V} = Nk_BT_1 \, ln\frac{V_b}{V_a} .$$ (4-2)

$$Q_2 = \int_c^d Nk_BT_2 \frac{dV}{V} = Nk_BT_2 \, ln\frac{V_d}{V_c} .$$ (4-3)

In the adiabatic process of an ideal gas PV^γ is a constant. This equation is modified in the following way. Since the points a and b are at the ends of the isothermal process at temperatures T_1 and the points c and d at the ends of the isothermal process at temperature T_2, PV in the equation can be replaced by temperatures T_1 and T_2 in the following way:

$$PV^\gamma = (PV)V^{\gamma-1} \to TV^{\gamma-1} = constant, \qquad (4\text{-}4)$$

and we get

$$T_1V_b^{\gamma-1} = T_2V_c^{\gamma-1}, \quad T_1V_a^{\gamma-1} = T_2V_d^{\gamma-1}. \qquad (4\text{-}5)$$

If we divide the first of these equations by the second one in (4-5), we find that V_b/V_a is equal V_c/V_d, therefore, the ln's in Eqs. (4-3) and (4-4) are equal, and we get

$$\frac{Q_1}{T_1} = \frac{Q_2}{T_2}. \qquad (4\text{-}6)$$

This is the relation we are seeking. Although we have used the properties of an ideal gas for the calculation, the same result can be obtained for any reversible engine!

4-1-2 Efficiency of a reversible engine

Figure 4-3 shows an ideal heat engine schematically. The engine gets heat Q_1 from the heat reservoir at high temperature T_1 and drains heat Q_2 to the reservoir at low temperature T_2 and does work W to outside. Figure 4-4 shows the Carnot cycle on P-v and T-s diagrams. From the law of conservation of energy, $W = Q_1 - Q_2$, we get,

$$Q_1 / T_1 = Q_2 / T_2$$

$$W = Q_1\left(1 - \frac{T_2}{T_1}\right). \qquad (4\text{-}7)$$

The efficiency of an engine, i.e. the amount of work we get from a given amount of heat, becomes

$$Efficiency = \frac{W}{Q_1} = \frac{T_1 - T_2}{T_1}. \qquad (4\text{-}8)$$

The efficiency is proportional to the difference between the temperatures $T_1 - T_2$ at which the engine runs divided by the higher temperature T_1.

430

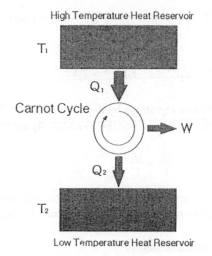

High Temperature Heat Reservoir

T_1

Q_1

Carnot Cycle

W

Q_2

T_2

Low Temperature Heat Reservoir

Fig. 4-3   Energy flow in an ideal heat engine: Carnot cycle.

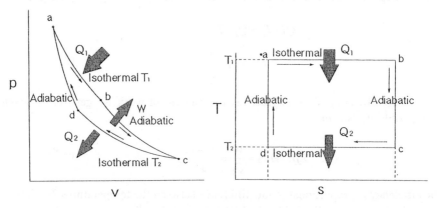

Fig. 4-4   Ideal heat engine: Carnot cycle on P-v and T-s diagrams.

4-1-3 Refrigeration cycle; Reverse Carnot cycle

The Carnot cycle discussed above is reversible and therefore can be used for refrigeration. Figure 4-5 shows this cycle. By giving work W to the engine, heat Q_2 can be absorbed from the heat reservoir at low temperature T_2, and then heat Q_1 (= Q_2

+ W) is drained to the heat reservoir at high temperature T_1. This reverse Carnot cycle is shown in Fig. 4-6 on P-v and T-s diagrams.

The coefficient of performance, COP, for a refrigerator is the ratio of cooling power Q_2 to required power W. From the above discussion of the Carnot cycle we get the COP for an ideal refrigerator as

$$COP = \frac{Q_2}{W} = \frac{T_2}{T_1 - T_2}. \qquad (4\text{-}9)$$

Table 4-1 shows the COP's of ideal and practical refrigerators for cooling to liquid nitrogen, liquid hydrogen and liquid helium temperatures. The ratio of a COP of a practical refrigerator to that of an ideal one is called the % Carnot.

In the case of an ideal refrigerator working at 300 K, to get a cooling power Q_2 of 1 W at 77 K and at 4.2 K, we must spend power W of about 3 W and 70 W respectively.

The % Carnot of widely used helium refrigerators is about 10% for a medium sized refrigerator (300 W at 4.4 K) and about 30% for a large refrigerator (10 kW at 4.4 K).

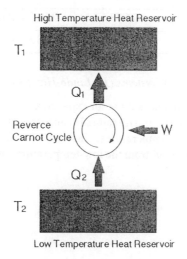

Fig. 4-5 Ideal refrigerator: reverse Carnot cycle.

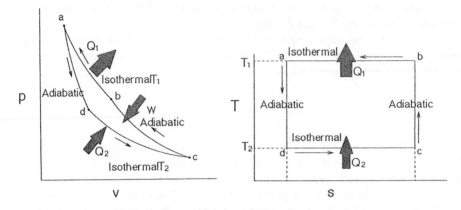

Fig. 4-6 Reverse Carnot cycle on *P-v* and *T-s* diagrams.

Table 4-1

	T (K)	COF (Ideal)	COF (Practical)	% Carnot
N_2	77.4	0.35	0.1 – 0.16	28 – 45
H_2	20.4	0.073	0.013 – 0.025	17 – 34
He	4.2	0.014	0.0013 – 0.0058	9.3 – 30

4-2 Practical refrigeration systems (1): Linde-Hampson refrigerator

Refrigeration systems that do not use an expansion engine, as shown in Fig. 4-7, are classified as Joule-Thomson refrigerators because they depend on the Joule-Thomson effect to produce low temperatures. Instead of withdrawing liquid from the refrigerator, heat is absorbed from the low-temperature source to evaporate the liquid.

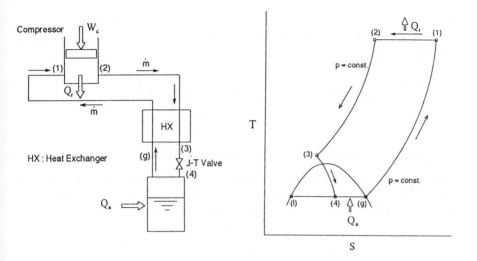

Fig. 4-7 Simple Linde-Hampson refrigerator.

Applying Eq. (3-1) to a system consisting of the heat exchanger, the Joule-Thomson expansion valves, and the evaporator, then

$$Q_a = \dot{m}\left(h'_1 - h_2\right),$$ (4-10)

where h'_1 is the actual enthalpy of the fluid leaving the heat exchanger at the warm end. The heat exchanger effectiveness can be defined by

$$\varepsilon = \frac{h'_1 - h_g}{h_1 - h_g}.$$ (4-11)

From Eqs.(4-10) and (4-11), we get

$$\frac{Q_a}{\dot{m}} = \left(h_1 - h_2\right) - \left(1 - \varepsilon\right)\left(h_1 - h_g\right).$$ (4-12)

The work requirement for the system is given by

$$\frac{-W}{\dot{m}} = \frac{T_2\left(s_1 - s_2\right) - \left(h_1 - h_2\right)}{\eta_{c,o}},$$ (4-13)

where $\eta_{c,o}$ is the overall efficiency of the compressor.

From the definition of performance we find the *COP* for the Linde-Hampson refrigerator to be

$$COP = \frac{-Q_a}{W} = \frac{\eta_{c,o}\left[\ (h_1 - h_2) - (1-\varepsilon)(h_1 - h_g)\ \right]}{T_2(s_1 - s_2) - (h_1 - h_2)} \ . \qquad (4\text{-}14)$$

4-3 Practical refrigeration systems (2): Claude refrigerator

A schematic of a Claude refrigerator is shown in Fig. 4-8. If we apply the Eq. (3-1) to the three heat exchangers, the expansion valve, and the evaporator as a unit, we obtain the following relation for the heat absorbed by the refrigerant:

$$\frac{Q_a}{\dot{m}} = (h'_1 - h_2) + x(h_3 - h'_e), \qquad (4\text{-}15)$$

where $x = \dot{m}_e / \dot{m}$ is expander mass-flow-rate ratio, $\dot{m}_e$ mass flow rate through the expander, and $\dot{m}$ mass flow rare through the compressor. The refrigeration effect may be written in terms of the expander adiabatic efficiency as follows:

$$\frac{Q_a}{\dot{m}} = (h'_1 - h_2) + x\eta_{ad}(h_3 - h_e), \qquad (4\text{-}16)$$

where the expander adiabatic efficiency is defined as

$$\eta_{ad} = \frac{(h_3 - h'_e)}{(h_3 - h_e)} \ . \qquad (4\text{-}17)$$

The net work requirement, assuming that the expander work is utilized to help compress the gas, is given by

$$\frac{-W_{net}}{\dot{m}} = \frac{\left[\ T_2(s_1 - s_2) - (h_1 - h_2)\ \right]}{\eta_{c,o} - x\eta_{e,m}\eta_{ad}(h_3 - h_e)} \ , \qquad (4\text{-}18)$$

where $\eta_{c,o}$ is the overall efficiency of the compressor and $\eta_{e,m}$ is the mechanical efficiency of the expander.

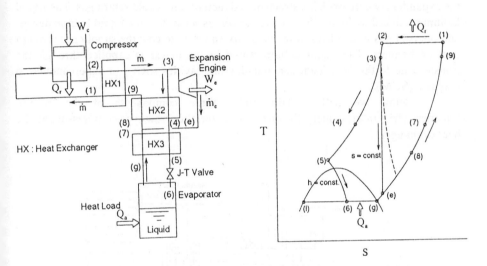

Fig. 4-8 Claude refrigeration system.

5 Practical helium liquefier/refrigeration system

5-1 Typical cryogenic system for superconductiong magnet and cavity

Figure 5-1 shows a typical helium liquefier/refrigeration system. The helium gas is compressed from 1 bar to about 16 bar by a compressor unit. The helium gas is, after cooled by water, sent to an oil separator, where oil contamination in the gas from the compression process is removed, and then sent to a cold box, which contains heat exchanger, piping and valves under vacuum insulation. At the cold box, the high pressure supply helium gas is cooled by cold return gas through heat exchanger and expanded at a Joule-Thomson valve. Part of the higher pressure gas is expanded in the expanders, where work is extracted, and then becomes cold return gas. The liquid helium produced by Joule-Thomson expansion is separated in a liquid helium dewar, and the cold gas is sent back to the cold box and used to cool the supply gas through heat exchangers. The liquid helium stored in the dewar is supplied through the transfer line to the heat load: superconducting cavities (SCC) or superconducting magnets (SCM).

Below we will explain major three components of the helium liquefier/refrigerator, namely, the turbo-expander, the screw-type compressor and the heat exchanger.

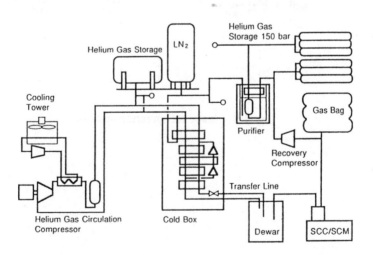

Fig. 5-1 Typical helium liquefier/refrigerator system.

5-1-1 Gas-bearing turbo-expander

Historically, reciprocating expansion engines were used in the early stage of cryogenic systems. George Claude of France developed liquefiers with reciprocating expanders to liquefy air. Later Dr. Samuel C. Collins of MIT developed a helium liquefier with high-efficiency reciprocating expanders. The first effective turbo-expander was designed by P. Kapitza for an air separation plant. Commercially available helium liquefiers with high performance turbo-expanders have been manufactured by BOC in England, Air Liquide in France, and Sulzer in Switzerland, and have been supplied to universities and research institutes all over the world. In Japan, Hitachi and Kobe Steel have developed helium liquefiers with turbo-expanders and supplied them to universities and research institutes in Japan.

Figure 5-2 shows a turbo-expander made by Hitachi, used for helium refrigerator of KEK superconducting RF cavities. High pressure gas is throttled by the nozzle and its enthalpy is changed to kinetic energy, i.e. it forms a high speed gas stream and then is injected to the turbine wheel. The momentum of the gas is transferred to the wheel and this makes the turbine wheel turn. The gas in the turbine wheel is further expanded and this imparts rotating force to the wheel. Then the expanded gas is sent to the defuser, and there the velocity is reduced. The work extracted at the turbine wheel is used to compress the gas at the brake compressor, and the heat in this gas flow is absorbed by the cooling water. The high speed rotating part of the turbine, i.e. the turbine wheel and shaft and brake compressor, are supported by gas bearings: radial and thrust bearings.

5-1-2 Oil-flooded screw-type compressor

Screw-type compressors are widely used for helium liquefier/refrigerator systems. The compressor consists of two cylindrical rotors with multiple helical lobes or grooves, as shown in Fig. 5-3. Oil is injected into the rotors for lubrication and sealing and hence the term "oil-flooded screw compressor." The injected oil also cools the gas during compression. The oil that contaminates the gas during compression must be removed before sending the gas to the refrigerator cold box; otherwise the oil in the gas stream will stick to the surfaces of heat exchangers and piping and cause deterioration of the heat-exchanger performance and plugging of the pipe. Multistage, usually four-stage, oil filters are used to remove the oil contamination in the gas flow.

438

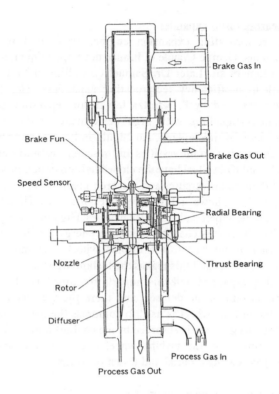

Brake Gas In

Brake Fun

Brake Gas Out

Speed Sensor

Radial Bearing

Nozzle

Thrust Bearing

Rotor

Diffuser

Process Gas In

Process Gas Out

Fig. 5-2   Gas bearing turbo-expander used in Hitachi 8 kW at 4.4 K helium refrigerator for TRISTAN superconducting RF cavities.

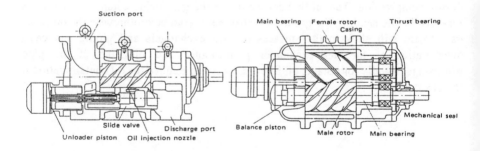

Suction port

Main bearing   Female rotor   Thrust bearing

Casing

Slide valve

Discharge port

Balance piston   Male rotor   Main bearing

Unloader piston   Oil injection nozzle

Mechanical seal

Fig. 5-3   Oil-flood screw type compressor.

5-1-3 Heat exchanger

The heat exchanger is also a key component in high performance liquefier/
refrigeration systems. There are many different heat exchanger configurations.
Aluminum-plate fin heat exchangers are usually used for practical helium
refrigeration systems. Typical structures are shown in Fig. 5-4. The heat exchanger
surface is formed by dip brazing aluminum plates to the corrugated aluminum
exchanger surface. The heat exchangers are assembled by brazing.

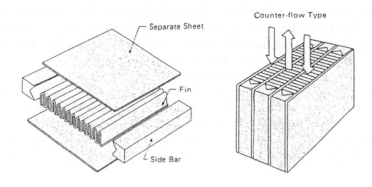

Fig. 5-4 Aluminum plate-fin heat exchangers.

5-2 Refrigerator and liquefier modes

Usually a helium liquefier that is designed and used to liquefy helium gas can also
operate as a refrigerator. If a heater used to heat the liquid helium is turned on during
operation, the heat balance and temperature distribution of the system will be
changed and the liquefaction rate will decrease by some amount. As the heater power
increases, the liquefaction rate decreases further and reaches zero. If the
liquefier/refrigerator operates at zero heater power the system is said to operate in
liquefier mode; if the system operates at maximum heater power with no liquefaction,
it is said to operate in refrigeration mode. Figure 5-5 shows the heater power, i.e.
cooling power, curve of a helium liquefier vs. liquefaction rate.

The liquefaction rate is maximum at no heat input, and it decreases with
increasing heater power almost linearly, and finally reaches zero at maximum heater
power, i.e. maximum refrigeration power. As a rule of thumb, because of the
thermodynamic properties of helium gas, a liquefier with a 100-L/hr liquefaction
capacity has a cooling power of about 300 W at 4.4 K. This means that a liquefaction
rate of 1 L/hr corresponds to a cooling power of about 3 W. On the other hand, the
cooling capacity by vaporization of liquid helium at 1 L/hr is about 0.7 W. This big

difference between cooling powers of these modes can be understood from the following: In the liquefaction mode we can use only the vaporization heat of the liquid helium, but in refrigeration mode we can also use the enthalpy of the helium gas.

Here we will discuss the difference between the liquefier and refrigeration modes. Figure 5-6 shows the temperature dependence of the enthalpies of helium and nitrogen. The enthalpy gaps at 4.2 K and 77 K correspond to the latent heat (vaporization heat) of helium and nitrogen respectively. The latent heat of helium is very small compared to that of nitrogen because the attractive force between its molecules is weak, but helium has as large enthalpy dependence on temperature as nitrogen and has a large cooling power. But usually for cooling superconducting magnets and cavities, the small latent heat of helium is utilized. We must consume about 1 L/hr liquid helium to get about 0.7 W cooling power.

Figure 5-7 shows the difference between the liquefier mode and the refrigerator mode. In the liquefier mode the vaporized gas which was used for cooling the magnet/cavity is sent back to the compressor suction as warm gas (through path Ⓐ), and the cooling power of the vaporized cold gas is not recovered by the system. In the refrigerator mode the vaporized cold gas is returned to the cold end of the refrigerator (through path Ⓑ) and the cooling power of the gas is recovered.

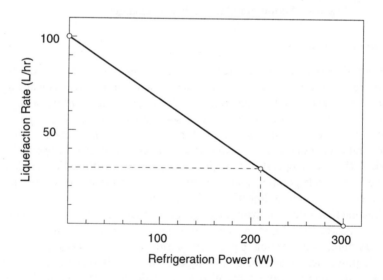

Fig. 5-5 Liquefaction rate vs. cooling power of helium liquefier/refrigerator.

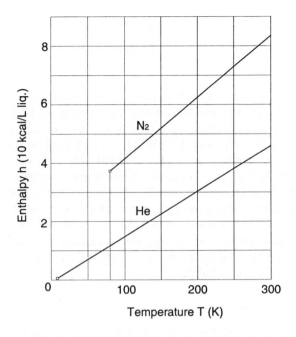

Fig. 5-6 Temperature dependence of the enthalpies of the helum and nitrogen.

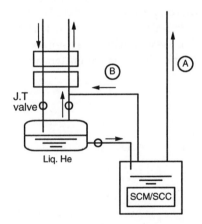

Fig. 5-7 Difference between the liquefier and the refrigerator modes.
In the former case, the return gas goes to the compressor through
path Ⓐ, while in the latter case the gas returns to the refrigerator.

6 Cryogenic System for TRISTAN Superconducting RF Cavity

6-1 Introduction

As an example of an operating helium refrigeration system, we will discuss a large cryogenic system for the superconducting RF cavities at KEK. A large-scale cryogenic system with a large helium refrigerator (8-kW cooling capacity at 4.4 K), and a liquid nitrogen refrigerator (6.5-kW at 80 K) for the 80K radiation shield, were designed and constructed at KEK in 1984 for cooling the TRISTAN superconducting RF cavities. Thirty-two 5-cell superconducting RF accelerating cavities (508 MHz) in 16 cryostats were installed in the TRISTAN electron-positron collider to upgrade the electron and positron beam energy from 27 GeV to 32 GeV. Operation of the system was started in 1988 and continued until the end of the TRISTAN physics run in 1995. During the cumulative operating time of about 38,000 hours in these 7 years, the superconducting RF cavities and the cryogenic system worked stably and reliably.

6-2 Cryogenic overview

Figure 6-1 is a schematic flow diagram of the cryogenic system for TRISTAN superconducting RF cavity. The cryogenic system consists of a large helium refrigeration system with a helium gas compressor unit, cold boxes and a liquid helium dewar, 16 cryomodules for superconducting RF cavities, a liquid nitrogen circulation system with a compressor unit and cold boxes, a large-scale liquid helium and liquid nitrogen distribution system with a long transfer line, a helium gas recovery and purification system, helium gas storage vessels, and a large capacity liquid nitrogen storage tank. The main parameters are shown in Table 6-1.

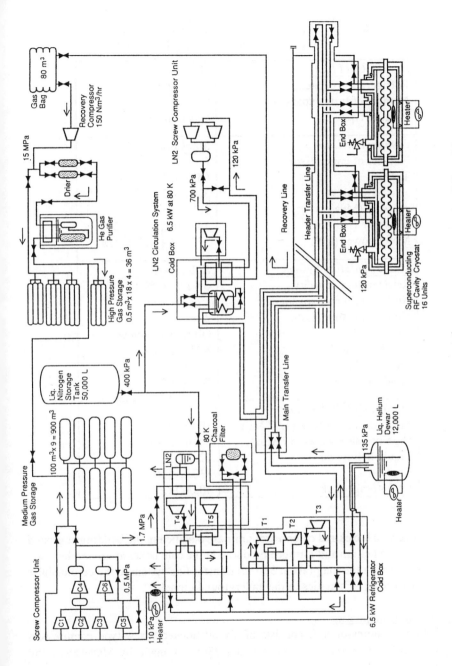

Fig. 6-1 Cryogenic system for TRSITAN superconducting RF cavities.

Table 6-1. Main parameters of cryogenic system of
TRISTAN superconducting RF cavities.

He Ref. Cold Box	Hitachi
Refrigeration capacity	8 kW at 4.4 K
Number of turboexpanders	5
Bearing type	Gas bearing
Compressor Unit	Maekawa
Number of compressor	6
Type	Oil flooded screw
Mycom	320L × 3 + 320S × 1
pressure (Mpa)	1.9/0.4/0.105
Flow rate (Nm³/hr)	14,600/2,300/12,200
Motor power (kW)	373 × 3 + 127 + 423
LN2 Circulation System	Kobe Steel
Refrigeration capacity	6.5 kW at 80 K
Number of turboexpanders	1 gas bearing
Number of compressor	2
Type	Oil flooded screw
Kobe steel	KTS75A × 2
Pressure (Mpa)	0.7/0.12
Flow rate (Nm³/hr)	700 × 2
Storage and Buffer Tank	
Medium pressor tank (1.97 MP)	100 m³ × 9
High pressure vessel (15 Mpa)	0.5 m³ × 18 × 4
Liquid helium dewar	12,000 L
Cryostat	830 L × 16
LN2 storage tank	50,000 L

The cold boxes and compressor units are installed in the building at ground level. The liquid helium and nitrogen produced at the cold boxes at ground level are transferred through the long (about 380-m) transfer line system to the cryomodules of superconducting RF cavities in the underground tunnel.

6-2-1 Helium compressor

The helium compressor unit consists of six oil-flooded screw-type compressors (Mycom 320L × 3, 320S × 1, 250L × 1 and 250S × 1 made by Mayekawa) and four-stage oil filters. We adopted oil-flooded screw-type compressors to attain reliability for long-term operation. The total required electric power for this compressor unit is 2.6 MW including 130 kW for cooling water and oil pumps. The process helium gas is compressed by a two-stage compressor unit from 0.1 MPa to 1.6 MPa and then sent to a cold box. The isothermal efficiency (i.e. the ratio to ideal isothermal compression) of the compressor unit is about 53%. To remove the oil contaminated in compressed helium gas we use four-stage oil filters as shown in Fig. 6-2.

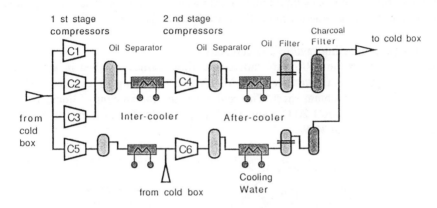

Fig. 6-2 Oil-filters of screw-type compressor unit for KEK 8-kW helium
refrigeration system.

6-2-2 Cold boxes

The vertical cold boxes (4 m in diameter × 6 m in height, and 2 m in diameter × 3 m in height) have aluminum-plate fin heat exchangers, five turbo-expanders (T1 to T5) and control valves. A supercritical turbo-expander T3 was installed just before the Joule-Thomson valve to increase the cooling capacity from 4-kW to 8- kW[7]. A cooling capacity of 8 kW at 4.4 K could be obtained by this system without liquid nitrogen pre-cooling. The turbo-expanders T4 and T5 for 80K pre-cooling are installed in an auxiliary cold box. An 80 K charcoal filter is also installed in another auxiliary cold box to purify the process helium gas during cool-down of the system. After Joule-Thomson expansion the liquid helium is separated from the return gas and stored in a 12,000-L dewar. The liquid helium level in the dewar is automatically controlled by the electric heater in it.

446

6-2-3 Distribution system and transfer line

Liquid helium produced in a 12,000-L liquid helium dewar is distributed to the 16 cryostats installed in the ~200-m-long straight section of the underground TRISTAN tunnel through an ~ 200-m-long supply line of the multi-channel transfer line, and vaporized cold helium gas at the cryostats is returned through the return line of the same multi-channel transfer line to the cold box. Figure 6-3 shows the structure of the multi-channel transfer line. Helium channels which supply liquid helium and return gas, are guarded by an 80 K radiation shield, and the aluminum radiation pipe is cooled by liquid nitrogen flow and return lines. The cooling liquid nitrogen circulation system will be discussed in detail later.

6-2-4 Superconducting cavities and cryostats

Figure 6-4 shows the TRISTAN superconducting cavities installed in a cryostat. Each cryostat contains two 5-cell 508-MHz superconducting RF cavities made of 2-mm-thick pure niobium sheet. The cold mass and the amount of liquid helium stored in a cryostat are 1,200 kg and 830 L respectively. The heat load to 4.4 K of each cryostat is about 30 W.

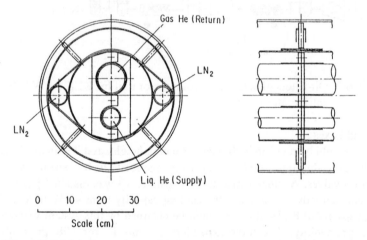

Fig. 6-3 Multi-channel transfer line for helium and liquid nitrogen.

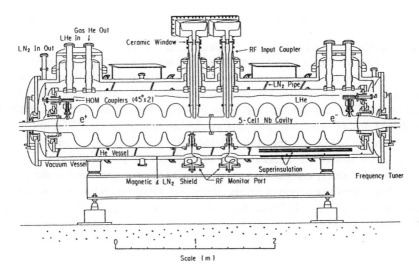

Fig. 6-4 Superconducting RF cavities for TRISTAN

6-2-5 Helium gas handling

The total amount of liquid helium handled in the whole system during steady-state operation is about 16,500 L including about 2500 L of liquid helium in the 12000- L dewar.

For storing helium gas during shutdown of the system, nine 100-m³ medium-pressure (1.9-MPa) helium gas storage tanks are connected to the system. An off-line helium gas recovery and purification system consists of a 5-stage air-cooled oil-lubricated reciprocating compressor (15 MPa, 150 Nm³/hr), high-pressure low-temperature purifiers (80 K, 15 MPa, 150 Nm³/hr), and high-pressure storage vessels (4 × 1350 Nm³).

The whole cryogenic system is controlled by means of a process control computer system.

6-2-6 Liquid nitrogen circulation system

We adopted the liquid nitrogen circulation system with a turbo-expander for the 80 K thermal shield of 16 cryostats to ease the control of 16 parallel cooling paths and to reduce the liquid nitrogen consumption. The flow diagram of the liquid nitrogen circulation system and its main parameters are shown in Fig. 6-1 and Table 6-1 respectively.

448

6-3 Operation of the cryogenic system

6-3-1 Design heat load of the superconducting RF cavities

Figure 6-5 shows the total heat load of the KEK superconducting cavity refrigeration system as a function of the operating accelerating field, E_{acc} of the cavities. The total design heat load of about 4 kW consists of about 1 kW static heat loss of the 16 cryostats and the transfer line and about 3 kW RF loss of the superconducting RF cavities at design condition: accelerating electric field E_{acc} = 5 MV/m at Q_o = 1 × 10^9. As shown in Fig. 6-5, the cavity RF loss is proportional to E_{acc}^2 and $1/Q_o$. In our refrigeration system we decided on a design cooling power of 6 kW at 4.4 K including 50% safety margin, because we worried about degradation of the Q_o value over a long operating period of several years. The superconducting RF cavities were operated at about E_{acc} = 4 MV/m at Q_o = 1.5 × 10^9 on average for about 7 years. This RF loss corresponds to about half the design value. We had enough safety margins for the operation of the cryogenic system.

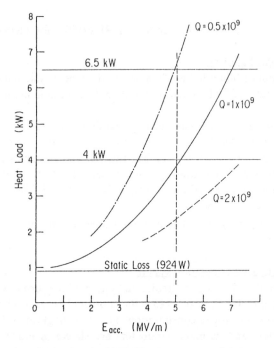

Fig. 6-5   Total heat load of TRISTAN superconducting cavities as a function of the operational accelerating field E_{acc}.

6-3-2 Cool-down and steady-state operation

The 32×5-cell superconducting RF cavities in 16 cryostats were cooled from room temperature down to about 150 K in parallel with about 80 K helium gas precooled by liquid nitrogen in the helium refrigerator, and then the turbo-expanders were started for further cool-down. After the cavities reached 5 K, we stopped the supply of cold helium gas to the cryostats and started liquefaction in the helium dewar, and then sent the liquid helium in the dewar to the cryostats to fill them with liquid helium. It takes 3 days to cool down from room temperature to liquid helium temperature and 1 day to fill the 16 cryostats with liquid helium. The cool-down speed of the cavities, about 10 K per hour, is limited by a requirement regarding the cryomodules side, i.e. not to cause leaks at the seals of the cryostat due to thermal shrinkage during cool-down. Before cool-down of the cavities, about 8,000-L liquid helium was liquefied in the 12,000-L dewar.

The multi-channel header transfer line has 16 connection boxes along with the cryostats, and the cryostats are connected to the boxes by sub-transfer lines (helium and nitrogen supply and return lines). Figure 6-6 shows the flow diagram of the helium and liquid nitrogen to the cavity cryostat. The flows of the liquid helium and nitrogen to the cryostats are controlled by the valves at connection boxes. The liquid helium levels in the cryostats are automatically controlled by liquid helium supply valves at the connection boxes.

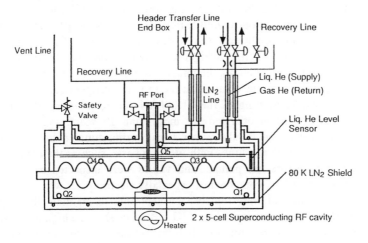

Fig. 6-6 Cryostat for SCC and distribution system for liquid helium and nitrogen.

The heat load of each cryostat increases from about 30 W to about 120 W in about 2 minutes and remaining for about 1.5 hours, according to the operation pattern of TRISTAN, i.e. injection, acceleration, and collision modes. The electric heaters in the cryostats are used to compensate the large change of RF loss in the cryostats and to minimize pressure fluctuations. These compensation heaters in the cryostats during the long operation period eliminate the need to control the return valves.

6-4 Cryogenic system for KEKB superconducting cavities

After the TRISTAN project was over, construction of a high luminosity double-ring 8 × 3.5-GeV asymmetric electron-positron collider, the KEK B-Factory (KEKB)[2], was started in 1994 as a five-year project in the existing TRISTAN tunnel. Installation of eight superconducting single-cell higher-order-mode damped acceleration RF cavities[3] in the high energy ring was proposed, and four cavities in four cryostats were installed in the summer of 1998 and started operation at the end of 1998. The other four superconducting RF cavities will be installed in the summer of 2000. The existing cryogenic system, which had been designed, constructed and operated for the TRISTAN superconducting cavities, has been reused[8,9] for cooling the KEKB superconducting RF cavities.

Installation of four superconducting RF deflectors, so-called crab cavity[4], in the KEKB intersection region has also been proposed to eliminate the possibility of beam-beam instability due to the finite crossing angle scheme of KEKB. For cooling the crab cavities we want to adopt a satellite refrigeration schemes[9], in which about 350 L/hr of liquid helium produced at the existing large helium refrigerator (central refrigerator) is sent to two satellite refrigerators about 1 km away from the central refrigerator. With this liquid helium, the total cooling power of about 800 W needed for the four crab cavities will be obtained at the satellite refrigerators.

After the commissioning of the TRISTAN superconducting RF cavity in 1984, the cryogenic system operated for more than 10 years. Inspections and maintenance were carried out regularly once a year during summer shut-down of the system. During this period machines with wearing parts, such as the compressors, including air compressors, oil pumps, and valves were inspected carefully and repaired if necessary. Many components, especially electric parts, wore out during the long running time and were replaced by new ones to maintain the required performance. Before the commissioning of KEKB we washed the high pressure line of the first heat exchanger for the helium refrigerator to remove the oil contamination accumulated during 10 years of operation. We also overhauled the turbo-expanders, the screw compressors and the electric motors. The cooling water towers for the helium gas compressor unit, the electric motors of oil pumps, relay switches, and sequencers of helium compressors were replaced by new ones during shut-down of the cryogenic system. The electropneumatic converters of valves at header connection boxes in the KEKB tunnel, which were damaged by high radiation doses, were also replaced by new ones.

6-4-1 High performance sub-transfer lines

We installed four KEKB HOM damped single-cell acceleration superconducting RF cavities in the high energy ring in the straight section of the Nikko experimental area in the summer of 1998. We also installed the sub-transfer lines that connect the cryostats for KEKB superconducting cavities and the existing header transfer line used at TRISTAN.

Figure 6-7 shows the layout of the KEKB superconducting cavities in the tunnel. Because we had to use the existing header transfer line, designed for TRISTAN superconducting cavities, the locations of the connection boxes were not optimized for KEKB superconducting cavities. We needed long sub-transfer lines for connection of the KEKB superconducting cavities and connection boxes. We developed a new type of high performance sub-transfer line with an 80 K liquid nitrogen thermal shield and a bayonet-type mid-joint connection[10]. The length of this transfer line can be easily adjusted by changing the angle of this mid-joint part. Figures 6-8 and 6-9 show the cross-sections and mid-bayonet joint of the sub-transfer line. The heat leak from the room temperature vacuum jacket is intercepted by the 80 K thermal shield made from aluminum alloy by extrusion and conducted into the liquid nitrogen in the stainless steel piping. The helium pipes and the 80 K shield are wrapped by 10 and 30 layers respectively of double-sided aluminized 12-μm-thick polyester film with polyester inter layers net.

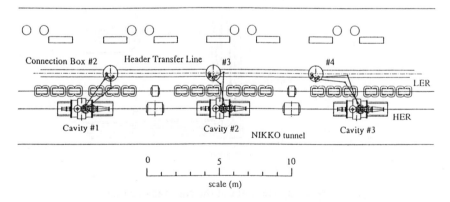

Fig. 6-7 Layout of the KEKB superconducting cavities in the tunnel.

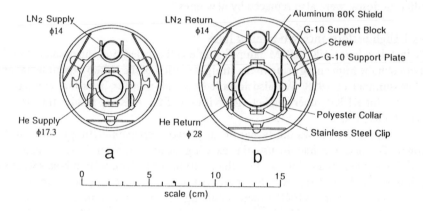

Fig. 6-8 Cross-sections of the sub-transfer line.

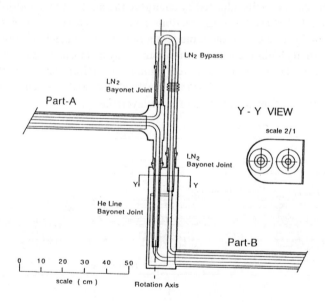

Fig. 6-9 Mid-bayonet joint of the sub-transfer line.

6-4-2 Satellite refrigeration system for KEKB crab cavities

We are planning to install four KEKB superconducting crab cavities with a total heat load of about 800 W in the Tsukuba straight section in 2003. For cooling the four crab cavities we want to adopt a so-called satellite refrigeration system like the Tevatron refrigeration system at FNAL[12,13], which is also similar to the liquid nitrogen circulation system discussed above. The reasons for adopting the satellite refrigeration scheme for the KEKB crab cavity cryogenic system are as follows.

1) We can use the surplus cooling power, about 5 kW at 4.4 K, of the existing cryogenic system, which is now operating for KEKB superconducting accelerator cavities, and we can also use its helium gas handling facility.

2) If the four crab cavities were connected directly to the Nikko cryogenic system, to keep the pressure drop of return cold gas small we would have to construct a high performance transfer line with large size piping for it.

3) In the satellite refrigeration cooling scheme, only a one-way supply helium transfer line is required, and the return helium gas is sent back to the central cryogenic system as room temperature gas. The heat loss and the construction cost of this long transfer line can be reduced and the system operation flexibility will be increased by this scheme.

Figure 6-10 shows the flow in the satellite system for two KEKB superconducting crab cavities. The satellite refrigerator with no turbo-expander can produce about 400 W at 4.4 K cooling power by using about 150 L/hr liquid helium supplied from the central refrigerator, i.e. the existing cryogenic system at Nikko. This means that all the cooling power required at the satellite refrigeration systems in Tsukuba can be produced by the large refrigerator, which has a high efficiency, i.e. a %Carnot of about 20%, and very high reliability. This satellite refrigerator scheme is very attractive compared to an independent refrigeration system, even if the heat load of about 200 W (0.2 W/m) from the 1-km-long transfer line is taken into account, because the required helium gas flow rate is less than half that of an independent helium refrigerator, since it has no turbo-expander, and its system configuration is very simple. The liquid nitrogen used in this system can be supplied from the Nikko cryogenic system through 80 K radiation shielded piping of a 1-km-long transfer line.

454

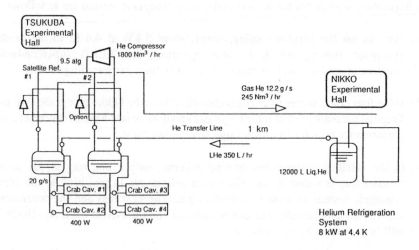

Fig. 6-10 Satellite refrigeration system for KEKB superconducting crab cavities.

Text Books

1. Randall F. Barron, Cryogenic Systems, Oxford University Press, New York ; Clarendon Press, Oxford (1985).
2. Russell B. Scott, Cryogenic Engineering, van Nostrand, Princeton, New Jersey (1966).
3. Steven W. Van Sciver, Helium Cryogenics, Plenum Press, New York and London (1986).
4. Richard P. Feynman, Robert B. Leighton and Matthew Sands, The Feynmann's Lecture on Physics, Addison-Wesley, Reading, Chapters 39, 44, and 45, MA, (1965).
5. V.A. Kirillin, V.V. Sychev, and A.E. Sheindlin; Engineering Thermodynamics, Mir Publishers, Moscow (1976).
6. K. Mendelssohn, The Quest for Absolute Zero, Weidenfeld and Nicolson (1966).

Reference

1. Y. Kimura, TRISTAN project and KEK activities, Proceedings of the XIII International conference on high energy accelerators, Novosibirsk, U.S.S.R., (1986).
2. KEKB B-Factory Design Report, KEK Report 95-7 (1995).
3. T. Furuya et al., Beam test of a superconducting damped cavity for KEKB, Proceeding of Particle Accelerator Conference, Vancouver (1997).
4. K. Hosoyama et al., Crab cavity for KEKB, Proceedings of the 7th Workshop on superconductivity (1995).
5. K. Hara et al., Cryogenic system for TRISTAN superconducting RF cavity, Advances in Cryogenic Engineering, Vol.33, Plenum Press, New York, p.615 (1988).
6. K. Hosoyama et al., Cryogenic system for the TRISTAN superconducting RF cavities: performance test and present status, Advances in Cryogenic Engineering, Vol.35, Plenum Press, New York, p.933 (1990).
7. K. Hosoyama et al., Cryogenic system for TRISTAN superconducting RF cavities: upgrading and present status, Advances in Cryogenic Engineering, Vol.37, Plenum Press, New York , p,683 (1992).
8. A. Kabe et al.,Cryogenic system for KEKB superconducting RF cavity, Advances in Cryogenic Engineering, Vol.37, Plenum Press, New York (1992).

9. K. Hosoyama et al., Design and performance of KEKB superconducting cavities and its cryogenic system, Advances in Cryogenic Engineering, Vol.43, Plenum Press, New York, p.123 (1998).
10. K. Hosoyama et al., Development of a high performance transfer line system, Advances in Cryogenic Engineering, Vol.45, Plenum Press, New York (1992).
11. K. Hosoyama et al., Cryogenic system for TRISTAN superconducting RF cavities, Description and Operating Experience, ICEC16 / ICMC Proceedings, Elsevier Science, New York, p183 (1992) .
12. C. Rode et al., Energy Doubler refrigeration system, in: "IEEE Trans. Nucl.Sci., Vol.NS-27,No.3, p.1328 (1977).
13. P.C.Vander Arend, Helium refrigeration system for Fermilab energy doubler, Advances in Cryogenic Engineering, Vol.23, PlenumPress,New York, p.420 (1977).

HISTORY OF HIGH ENERGY ACCELERATORS IN JAPAN

KEN KIKUCHI

Director General, JSPS

1. Establishment of the Institute for Nuclear Study, University of Tokyo

The first cyclotron in Japan was completed in 1938 by Y. Nishina's group at the Institute for Physical and Chemical Research (RIKEN=理化学研究所), and in the early 1940's four cyclotrons (2 at RIKEN, 1 in Osaka and 1 in Kyoto) were available for experimental nuclear physics research. In particular, outstanding research activities were carried out at RIKEN and Osaka University, where excellent young researchers from Tokyo and Kyoto joined in.

After World War II, in the fall of 1945, all cyclotrons were destroyed by the General Head Quarters of the Occupying Forces (GHQ). The two RIKEN cyclotrons were disassembled and thrown into Tokyo Bay and the other two were thrown into Osaka Bay. The GHQ prohibited all basic and applied research on the development of atomic energy. Fearing Japanese reprisals against the US, the GHQ also prohibited the presentation of a well-known popular play "忠臣蔵 (Chusin-Gura)", in which retainers avenged actions against their liege lord. The news that cyclotrons in Japan were destroyed by the US Occupying Forces disturbed US scientists, who expressed their sharp criticism of the harsh acts and policy of the GHQ.

In 1952, the GHQ lifted the prohibition following the advice of E. O. Lawrence, the well-known inventor of the cyclotron, who visited Japan and was impressed by the dreadful research conditions at universities. In the following year, when the Peace Treaty was finally concluded between the Governments of Japan and the Allied Powers, the RIKEN, Kyoto and Osaka groups started discussions about the reconstruction of their cyclotrons, under the guidance of the Specific Committee for Nuclear Physics (原子核特別委員会, Currently the National Committee for Particle and Nuclear Physics) of the Science Council of Japan (JSC, 日本学術会議). The Science Council of Japan was established in the Prime Minister's Office in 1949, on the advice of the GHQ, to serve as a representative organization for scientists and to give advice or recommendations to the Prime Minister on important scientific issues. The Committee discussed the reconstruction plans and gave the first priority to the Osaka cyclotron as the project to be funded by a budget from MONBUSHO (Ministry of Education, currently Ministry of Education, Science, Sports and Culture). The RIKEN and Kyoto groups also started to reconstruct small cyclotrons with funds from MITI (Ministry of International Trade and Industry) and private industries.

However, the nuclear physics group in the Tokyo area, not being satisfied with a relatively small cyclotron (26-inch-diameter magnet) at RIKEN, started a campaign to construct a large cyclotron in the Tokyo area, including the possibility of constructing a synchrocyclotron capable of producing pions; after lively discussions, they submitted a bill to the Specific Committee for Nuclear Physics, the Science Council of Japan. The Committee, after lengthy arguments over few years, summarized the opinions of particle and nuclear physicists and made a proposal to JSC to establish a new national center in this field. The JSC, accepting the proposal from the Committee, discussed the issue and in 1953 gave the Prime Minister a recommendation on the establishment of a new research institute in the field of nuclear physics. The new institute was recommended to be a national center open to users from all universities (for joint use) providing a synchrocyclotron with an energy of several tens of MeV and used for cosmic ray research as well as theoretical research. It was also recommended that, since the immediate construction of a weak-focusing synchrotron with an energy over 1 GeV was impossible because of its high construction cost, a preparatory study should be initiated taking account of new technology such as the strong focusing principle invented by E.D. Courant, et. al. in order to realize a 1-GeV-class accelerator in the near future. In response to the recommendation from the JSC, MONBUSHO started to investigate, in consultation with the JSC and relevant scientists, how to establish and how to operate the new institute. They finally concluded that the new institute for joint use should be affiliated with the University of Tokyo, following the model of the Institute for Fundamental Physics (Yukawa Hall), Kyoto University, which had been established in 1952. The preparatory working group, organized to conduct a design study of the synchrocyclotron, fixed a final conceptional design of an FF and FM dual cyclotron capable of providing proton beams in a variable energy range of 7-14 MeV with a fixed frequency (FF) rf system and a 50-MeV proton beam with a frequency-modulated (FM) rf system. The new institute was named the Institute for Nuclear Study (INS) and, in summer 1954, a part of the experimental farm at Tanashi was chosen as the site. Construction started in late fall, prior to the formal establishment of INS in 1955. The FF cyclotron was completed at the end of 1957 and commissioned in early 1958, and the FM cyclotron was completed in 1960 and experiments started the next year. In the late 1950's, this FF and FM cyclotron facilitated the production of various remarkable experimental results by its unique feature of variable energy, and played an important role in advancing experimental nuclear physics research in Japan to the level of that in the rest of the world. The first experiments carried out by FF cyclotron were:

1. Proton-proton and proton-deuteron scattering
2. Deuteron pick-up (p, d) reactions
3. Elastic and inelastic scattering of protons by medium nuclei

The working group, organized in INS to conduct preparatory studies on high energy accelerators, reached a conclusion after R&D to propose the construction of a 750-MeV electron synchrotron as the first step toward experimental high energy physics. This proposal was approved immediately by MONBUSHO and construction started in 1956. The first 750-MeV electron beam was obtained at the end of 1961, more than a year behind schedule. On the same day, we learned that the Cambridge Electron Accelerator in the US had produced an intense 5-GeV electron beam, and we realized that Japan was far behind the world in the field of experimental high energy physics. The original purpose of the electron synchrotron was to acquire the fundamental techniques needed for a high energy accelerator and almost nothing was prepared for particle experiments. The next year was spent on constructions of an experimental hall and experimental facilities. In early 1963, particle experiments were initiated by the INS in-house group and five university groups (Tohoku, Tokyo, Nagoya, Kyoto and Hiroshima). Typical experiments were as follows:

1. Double pion production by high energy gamma rays
2. Überall effect of γ-rays
3. Polarization of recoil proton from neutral pion production
4. Single pion production
5. Proton Compton scattering

It is well known that any accelerated charged particle emits electron-magnetic waves when they are bent in a dipole magnet. In 1947, this phenomenon was actually observed in an electron synchrotron; hence the name "Synchrotron (Orbital) Radiation" (SR). When the INS electron synchrotron became operational, experiments with SR were already being carried out at several advanced electron synchrotrons, and some Japanese scientists participated in SR experiments at European laboratories, mostly at DESY in Germany. They were quite anxious to use the new electron synchrotron for experiments using SR. However their desire was not easily accepted by high energy physicists, who were reluctant to share the electron beam as well as the budget with the SR group. As the intensity and stability of the electron beam improved, the SR group was allowed to share the beam as parasite users. The group grew up gradually and, in 1967, organized the INS-SOR group.

The energy of the INS synchrotron was increased to 1.3 GeV in 1967, and the machine operated for more than 30 years until its formal shut-down in September 1999. Despite its low energy, the INS synchrotron made a great contribution to advancing experimental high energy physics in Japan as well as to training young scientists in this field.

2. Establishment of the National Laboratory for High Energy Physics (KEK) and Related Institutes: the Institute for Cosmic-Ray Research (ICRR) and the Research Center for Nuclear Physics (RCNP)

Concurrently with the construction of the 750-MeV electron synchrotron at the Institute for Nuclear Study, scientists took a step toward the next high energy accelerator. Encouraged by a statement by the INS Director, S. Kikuchi, at a Science Policy Committee meeting that it was time to think about the next high energy accelerator, a volunteer group of high energy physicists, headed by T. Kitagaki, started to investigate the next accelerator to be constructed after the INS electron synchrotron. The tasks to which they directed most effort were to decide the type and energy of the next accelerator and to survey an appropriate site for it. At first two types of accelerators were considered: a 1-GeV proton linear accelerator and a high intensity 10-GeV-class proton synchrotron. Kitagaki's group put more emphasis on the high intensity synchrotron, whereas the high-energy group of INS headed by H. Kumagai strongly insisted on the linear accelerator. In either case, the next accelerator was supposed not only to be an experimental facility but also to have the capability for use as an injector for a larger synchrotron which scientists wished to construct in the future. In particular, the linear accelerator was taken into consideration because of its merit as an injector. On the other hand, the proton synchrotron was considered to be more useful and appropriate for various experiments. This issue was controversial and the opinions of scientists were divided on this point. However, in the course of time, arguments converged to put more emphasis on experimental utility and versatility and to choose a high intensity proton synchrotron as the next accelerator.

Meanwhile, the Kitagaki group made a wide survey to find a site for the new accelerator. They first selected 13 candidates from the original 80 proposed, using the following criteria:

1. Sites in areas without heavy snow, which would hinder construction in winter,

  2. Sites without large population nearby, and
  3. Sites, which were flat.

Next, each of these 13 candidates was investigated in a thorough field survey by the official working group organized in 1964. The group finally selected two candidates, the Tsukuba area and Kuroiso area, taking account of various geographical and social conditions. In 1967, these two areas were more thoroughly examined by borings and the Tsukuba area was chosen.

The Kitagaki group, supported by the experimental high energy physics community, proposed the construction of a high intensity proton synchrotron to the Specific Committee for Nuclear Physics of JSC. In the Committee, controversies arose because of the high construction cost in comparison with available resources for academic research. In fact an estimated construction cost of 20 billion yen was almost ten times as large as the total amount of the research fund (Grants-in-aid) under MONBUSHO in 1960. Moreover the experimental high energy physics group was a small fraction in the community of particle and nuclear physics that hardly carried a great weight in the Committee. The majority of the Committee consisted of theoretical, nuclear and cosmic-ray physicists who were either against or reluctant to launch such a large project so soon.

In the course of several years of arguments in the Committee, the strong desire of experimental high energy physicists gradually became acceptable to scientists in other fields; in particular, some theoretical physicists who had visited CERN or US high energy physics laboratories emphasized the necessity for a high energy accelerator in Japan. At last, the Committee, having watched the completion of the 750-MeV electron synchrotron at INS, submitted a proposal for constructing a proton synchrotron to JSC. In 1962 the general assembly of JSC approved the proposal and made a recommendation to the Prime Minister "On the realization of the future plan in the field of nuclear physics", which included the construction of a 12-GeV high intensity proton synchrotron and future plans for nuclear and cosmic-ray physics.

In response to the recommendation from JSC, MONBUSHO gave the commission of drafting a practical plan to the Council of National Universities and Institutes (研究所協議会), which concluded that the following three projects should be launched immediately:

  1. The basic study for a high energy accelerator,
  2. construction of the next accelerator for nuclear physics, and

3. equipping universities to strengthen cosmic-ray and theoretical research activities.

The next year the Council advised MONBUSHO that all of these projects should be initiated in fiscal 1964 and, when the high energy accelerator project was approved, a new type of national institute should be established to conduct research with the high energy accelerator. On the basis of Council's conclusion, the first preparatory research fund of 100 million yen was allocated to INS in 1964 and an official working group was organized, chaired by S. Tomonaga. For the seven years from 1964 until the establishment of KEK in 1971, a total of 1.785 billion yen was allocated for preparatory studies for a proton synchrotron and experimental facilities. The working group was composed of two sub-groups, the construction group and physics group. The construction group consisted of two subgroups, one, headed by H. Kumagai, in charge of the accelerator and the other, headed by I. Miura, in charge of basic experimental infrastructures such as beam lines and a bubble chamber and R&D for them. The physics group was in charge of R&D for experimental equipment and of promoting experimental programs, which included particle experiments using cosmic rays and the INS electron synchrotron, and bubble chamber film analysis, in collaborations with CERN and US laboratories.

Shortly after the accelerator design study was started, a big argument again took place among scientists about the type and the energy of proton synchrotron. When the 12-GeV high intensity proton synchrotron was chosen as the next accelerator, emphasis was on its beam intensity as it would be used as an injector for a future accelerator. However, the future prospective was still vague and several technical difficulties were apparent in a rapid-cycling proton synchrotron. Taking a broad view of high energy accelerators in the world, most of the high energy physicists began to think a conventional type of proton synchrotron with a higher energy would be more feasible and a better choice, and after much discussion it was concluded that a proton synchrotron with an energy 30-40 GeV would be much more appropriate as the next accelerator, provided that the construction budget could be expanded to about 27 billion yen from the original 20 billion yen. The change in the type of accelerator caused some changes in the organization of the working group. The chairperson of the working group, S. Tomonaga, gave way to S. Hayakawa as acting chairperson; and the head of the accelerator, H. Kumagai, was replaced by S. Suwa, who was a member of the FF/FM cyclotron team and had been working at Minnesota and Argonne since 1960. T. Nishikawa, who was at BNL, also joined the working group, at Suwa's request.

In MONBUSHO, this issue was continuously discussed in a new organization, "Council for the Promotion of Science (学術奨励審議会)", which was organized after the dissolution of the Council of National Universities and Institutes. Reflecting the opinions of the working group at INS and leading scientists in this field, the new Council reported to MONBUSHO in 1965 that a 35-GeV proton synchrotron would be better choice as the next accelerator in order to formulate a strategy for experimental high energy physics in Japan and to raise the level of Japanese high energy physicists so as to be able to participate in international cooperative experiments in the near future. The Council also reported that, without a national center for experimental high energy physics as a domestic base, immediate participation in an international organization such as CERN would be hardly possible and less cost-effective. They added that comparison with "big science" under other ministries showed that the annual budget of 5-6 billion yen needed for this project would be comparable to the 1965 annual budget of the Atomic Energy Research Establishment, or to the space science program if the planned satellite be included; and if an expected amount of 190 billion yen estimated by JSC for the total amount of research resources in 1970 under MONBUSHO, which did not include personnel, was correct, the high energy physics project would not cause a serious imbalance between different field of science.

In 1967 MONBUSHO reorganized the Council for the Promotion of Science into a new organization "the Science Council (学術審議会)", in order to strengthen its capability in formulating policies for promoting scientific research. The new Council continued discussions on the system and management of the new Institute for Particle Physics (素粒子研究所) and on the scientific merit of the project, and summarized its conclusions on the organization of the new institute, which resulted in a substantial part of KEK. An entirely new feature was that the new institute was directly supervised by MONBUSHO but its staff members were to have the same status as in national universities; this was considered to be a key point to assure the academic freedom and autonomy of the new institute. This new category of institute is currently called "Inter-University Research Institute (大学共同機関)".

The last and most crucial problem for MONBUSHO was the funding source for the project. During the 1960's the budget for research did not increase as had been expected, and in fact the total amount of Grants-in Aid under MONBUSHO in 1967 was about 4 billion yen. Under the circumstances, some Council members still had strong objections to allocating an annual budget of 5 billion yen to high energy physics, and Council could not reach any conclusion after long arguments. Finally, at the end of 1968, some members of the Council, including K. Husimi, with the agreement of MONBUSHO, proposed the launching of a project with a quarter scale of the original budget, i.e. the construction of an 8-GeV proton synchrotron instead of a 40-GeV one. This proposal was approved by the Council, but it excited controversies among particle, nuclear and cosmic ray physicists on two points: (1) the usefulness and the scientific merit of an 8-GeV proton synchrotron and (2) the omission of future plans for nuclear and cosmic ray physics. At first, the high energy experimental physics group was confused, but it soon united to accept the Council's conclusion, regarding the 8-GeV machine as an important milestone for high energy physics in Japan, and it drew up a concrete plan with a construction cost of 8 billion yen for the new high energy physics laboratory. Nuclear and cosmic-ray physicists, who were at first strongly against the conclusion, finally agreed to the new plan under the condition that research facilities for nuclear and cosmic-ray physics would be reinforced as soon as possible.

In 1970, construction of the preinjector house started on the Tsukuba site, and on April 1, 1971, the National Laboratory for High Energy Physics (KEK) was established as the first Inter-University Research Institute under MONBUSHO. In later years, the Research Center for Nuclear Physics (RCNP) at Osaka University, and the Institute for Cosmic-Ray Research (ICRR) at the University of Tokyo were established as university-affiliated institutes for joint use. In the 1980's, a ring cyclotron was constructed at RCNP, which, together with a ring cyclotron at RIKEN, has been playing an important role in promoting nuclear physics experiments. Meanwhile, the Kamiokande project (a large water-tank neutrino detector) was launched at ICRR, which achieved an outstanding outcome in the discovery of neutrinos from Super Nova 1987A.

3. 12-GeV Proton Synchrotron at KEK (KEK-PS)

In the design of the 8-GeV proton synchrotron, high energy physicists had two opinions on the injection system to the main synchrotron:

(1) Preinjector + linear accelerator
(2) Preinjector + linear accelerator + booster synchrotron.

The use of a smaller synchrotron as an injector for a large ring, which was originally proposed by T. Kitagaki in 1961 and was named a cascade system, is now adopted in many high energy accelerators, but in the late 1960s it was not so popular and used only by the Tevatron at Fermilab. From a technical point of view, mode (1) was considered to be easier, but mode (2) has the outstanding feature that the booster could be used for a pulsed neutron source and other applications. In response to the earnest of neutron users, mode (2) was chosen and the 8-GeV proton synchrotron was designed to consist of:

750-KeV   Cockcroft-Walton perinjector
20-MeV   Alvarez-type linear accelerator
500-MeV   Booster synchrotron
8-GeV   Main synchrotron

Construction started in 1970, and the first 8-GeV beam was obtained in March 1975, as scheduled. After a year the energy was raised to 12 GeV; this was made feasible by R&D on the high field magnetic steel used in fabricating the main ring magnets. In later years, another Cockcroft-Walton was constructed for a polarized proton source and an accelerating tank was added to the linear accelerator to increase its energy to 40 MeV.

4. Booster Utilization Facility (BUF)

When a booster was adopted as an intermediate injector to the 12-GeV proton synchrotron, it was naturally considered that 500-MeV beams from the booster were available for various applications. Among the possible capabilities for the use of 500-MeV beams, three facilities were constructed:

(1) A pulsed neutron source (KENS),
(2) A pion and muon experiment facility (BOOM), and
(3) A medical facility for cancer therapy.

This research complex was named Booster Utilization Facility (BUF). The booster produces 20 pulses/second of 500-MeV beam, of which 9 pulses are injected to the main ring during its one cycle of about 2 seconds. Then about 3/4 of the 500-MeV beam can be used in the BUF. It should be noted that KENS was one of the first pulsed neutron sources and broke new ground in the field of material and bio-sciences. Since then KENS and BOOM have yielded a number of remarkable results over the past 20 years. The medical facility, operated by the Medical School, University of Tsukuba, uses proton beams for cancer therapy and has been found to be very beneficial. Based on the success in BUF, several accelerators were constructed in Japan as dedicated machines for cancer therapy.

5. Synchrotron Radiation Research Facility (Photon Factory)

The INS-SOR group organized to use synchrotron radiation from the INS electron synchrotron was not satisfied with parasitic use and proposed the construction of a storage ring dedicated to SR research. The proposal was approved in 1972 and an additional 300-MeV storage ring was constructed at the INS electron synchrotron. This SR facility operated as a part of the Institute for Solid State Physics, University of Tokyo, until its shutdown in 1999.

In fall 1973, at the Annual Meeting of the Physical Society of Japan in Kyusyu, SR users held a large symposium to discuss future plans for SR research in Japan and decided to start a campaign for the establishment of a new laboratory for SR research, which would include construction of a 2-GeV-class storage ring as SR source. Soon a conceptual design was finished and the total construction cost was estimated to be about 20 billion yen. However finding an appropriate site for the new accelerator was very difficult, and getting approval from the Government for establishing a new laboratory was not easy. Finally, it was decided to construct a new SR facility as a part of KEK. In this decision, it was also taken into consideration that an electron accelerator could be used as an injector for a future electron-position collider to be constructed on the KEK site. The proposed accelerator consisted of

A 2.5-GeV linear accelerator (400 m) and
A 2.5-GeV storage ring (180 m in circumference).

With strong support from KEK, this was accepted by the Government and construction began in 1978. The first stored beam was obtained in 1981 and after elaborate tuning of the accelerator and test experiments, the commissioning started in 1982 with several completed beam lines. Within a few years, ingenious improvements were made to stablize the beam and to make the beam emmittance smaller. Since then the facility, known as the Proton Factory, has been successfully operated as one of the top-class SR machines in the world.

6. Electron-Position Colliding Accelerator (TRISTAN)

As a natural development in high energy physics, a future plan was considered while the 12-GeV proton synchrotron was under construction. One simple possible plan was to construct a hadron collider by using the 12-GeV synchrotron as an injector. In fact, a design study of a hadron collider was carried out in the early 1970's by a group headed by T. Nishikawa. In this plan, in addition to superconducting hadron rings, a conventional magnet ring was to be constructed to store electron/positron beams as well. Because the tunnel for the planned accelerator will accommodate two superconducting ring for hadrons and one conventional ring for electron/positron beams, the plan was named "TRISTAN," i.e. Three-Ring Intersecting Storage Accelerator in Nippon. However, the maximum possible beam energy of the hadron ring that could be constructed on the KEK site was estimated to be about 800 GeV, by using superconducting magnets, and this did not exceed the energy of the Tevatron, then under construction. Taking this into consideration, KEK changed their plan, putting emphasis on an electron-positron collider. One of the reasons was that neither PETRA at DESY nor PEP at SLAC had verified the existence of the top quark, thought to exist in their energy region. The TRISTAN accelerator consisted of two rings,

An 8-GeV accumulation ring (AR) and
A 30-GeV colliding ring (Main Ring).

The electron linear accelerator of the Photon Factory was used as the injector, and a new 200-MeV linear accelerator was built to produce the positron beam. The total construction cost was estimated to be 85 billion yen (45 billion yen for the accelerator and two detectors, and 40 billion yen for the tunnel and housings).

This was the largest and costliest project that MONBUSHO had ever funded. After reviews and discussions by the Science Council of MONBUSHO, it was approved as a five-year plan, starting in 1982. Construction proceeded as scheduled, and the first collision was observed in December 1986. Three experiments, VENUS, TOPAZ and AMY, started taking data in March the next year and steadily accumulated data. It was unfortunate that TRISTAN was not able to discover the top quark, whose mass was far beyond 30 GeV. However, TRISTAN accumulated precise data on electron-positron collisions in the center-of-mass energy region 48-65 GeV and brought high energy physics in Japan to the top level of high energy physics in the world. It should be noted that groups from China, Korea and the US, jointly with a Japanese group, formed the experimental group AMY. Experimental runs at TRISTAN ended in 1995 and the ring was converted into a new ring for KEKB (B-Factory).

7. Japanese Commitment to ICFA, SSC and LHC

In the 1960s, organized international cooperation in high energy physics was initiated under memoranda between CERN and the Soviet Union (Dubna and Serpukhov). Reflecting the increasing necessity of international cooperation in this field, the first international meeting to discuss future perspectives on high energy physics was held in Riga in 1967, with representatives from the US, the Western Europe (CERN member states) and the Soviet Union (Dubna member states), followed by meetings in Semmering (1968), in Tbilisi (1968) and in Morges (1971). After a 4-year-hiatus, a meeting on "Perspective in High Energy Physics" was held in New Orleans in March 1975, in which Y. Yamaguchi and T.

Nishikawa participated as the first representatives from Japan. The main purpose of the series of meetings was to discuss a framework for international cooperation to use regional large facilities and to construct and operate a future large accelerator. At this meeting, this accelerator was named VBA (Very Big Accelerator), and it was decided to organize an International Committee for Future Accelerator (ICFA). The first ICFA organizing committee meeting was held at CERN in October 1975, and the second at Serpukhov in May 1976, at which the VBA was defined as a fixed-target proton machine with energy above 10 TeV or an electron-positron collider with energy above 100 GeV. In 1976 the ICFA became an official committee under C13 (Committee for Particles and Fields) of the International Union of Pure and Applied Physics. At the ICFA meeting held in Hamburg in August 1977, the task of ICFA was formulated and the quota of representatives from each region was decided (three members each from the US, CERN member states and the Soviet Union; and one each from Dubna member states and Japan). In 1980 ICFA set guidelines for joint use of large accelerators in the world. However any practical plan for the VBA was not easily reached, though there had been a vague consensus about its scale.

In May 1983 the High Energy Physics Panel in the US proposed the construction of a 20-TeV Superconducting Super Collider (SSC). Meanwhile CERN started to build a 50-GeV electron-positron collider (LEP = Large Electron-Positron Ring). The Japanese high energy physics group started to participate in the joint R&D efforts on superconducting magnets and experimental detectors for SSC in the framework of the Implementing Arrangement for US-Japan Cooperation in High Energy Physics signed by MONBUSHO and the US DOE in 1979. Meanwhile the University of Tokyo Group, which had been carrying out an experiment at DESY, decided to join the experimental collaboration OPAL at LEP, which became available for experiments in 1989. In parallel with experiments at LEP, CERN started to prepare for the construction of 7-8 TeV hadron collider (LHC = Large Hadron Collider). Controversies arose when the Japanese government was officially asked to participate in the construction of SSC. The funding needed for participation was enormous compared to funding that Japan had ever allocated to one academic project. Negotiations and discussions between the two governments went on until the US decided to terminate the SSC project. The Japanese high energy physics group, which was planning to participate in experimentation at the SSC, was forced to change its plans. Finally its significant part organized a new group to participate in the ATLAS experiment at LHC, having merged with the University of Tokyo group that had been carrying out an experiment at LEP. Being moved by the earnest desire of scientists, the Japanese MONBUSHO contributed funds to the construction of LHC and the ATLAS detector.

8. B-Factory (KEKB)

On the other hand, the major part of Japanese high energy physicists, who were doing experiments at TRISTAN, pursued the next accelerator plan on the KEK site. An electron-positron linear collider was considered to be one of the most important high energy accelerators in the next generation, and KEK, with SLAC in the US and DESY in Germany, had put great efforts into R&D in the frame work of US-Japan cooperation in the field of high energy physics. However a number of technical problems remained in realizing an electron-positron linear collider in the energy range of several hundred GeV. Therefore most of the world's high energy physics communities think that the linear collider should become a reality after having observed experimental results at LHC.

Under these circumstances, in 1991 KEK proposed the construction of a new electron-positron collider KEKB (B-Factory) to search for CP violation, which would be expected to be observed in decays of B-mesons. Though the circumference is too large to optimize the luminosity, the TRISTAN tunnel was used to accommodate two new rings for asymmetric collision of 8-GeV electron-beam and 3.5-GeV positron beam. At almost the same time, a B-Factory plan was proposed at SLAC, using the PEP ring. Both proposals were approved in fiscal 1994 and construction of two B-Factories started, at KEK and SLAC, as five-year projects. Both were commissioned last year and began to produce experimental results.

Addendum (by Shin-ichi Kurokawa, October 2001)

The KEKB was completed in November 1998 and after half-a-year long commissioning without the detector, physics experiment with Belle, the experimental detector of KEKB, started in June 1999. The improvement of performance of KEKB has been quite smooth and by July 2001 it recorded a peak luminosity is $4.5 \times 10^{33} cm^{-2}sec^{-1}$. This is the highest luminosity ever achieved by any types of colliders. By July 2001 Belle has logged 33.1/fb.

At SLAC, another B-Factory called PEP-II has been running with its detector named BaBar. Both Belle and BaBar experiments have observed matter-antimatter asymmetry (CP-violation) in the decays of B mesons in July 2001. This observation is the first observation of CP-violation for particles other than K mesons; it marks a major experimental breakthrough in a search that has been going on almost 40 years since the discovery of CP-violation in K mesons in 1964. These results are published in the August 27, 2001 issue of Physical Review Letters.

The observed asymmetries by BELLE and BaBar are:

$A_{cp} = 0.99 \pm 0.14 \pm 0.06$ (BELLE)
$A_{cp} = 0.59 \pm 0.14 \pm 0.05$ (BaBar)

The statistical significance of the Belle measurement corresponds to the possibility of no CP -violation smaller than one part in 10 million, whereas that of BaBar smaller that one part in 30,000.

MAKING SUPERCONDUCTING SOLENOID MAGNET

KENJI HOSOYAMA, KAZUFUMI HARA, ATSUSHI KABE, YUJI KOJIMA,
YOSHIYUKI MORITA, HIROTAKA NAKAI, SHIN-ICH KUROKAWA,

High Energy Accelerator Research Organization,
Tsukuba, Ibaraki, Japan

AKIO SATO

National Research Institute for Metal
Tsukuba, Ibaraki, Japan

KOH AGATUMA

Electrotechnical Laboratory
Tsukuba, Ibaraki, Japan

HAO YAODOU, HAN SHIWEN

Institute of High Energy Physics
Beijing, China

1 Introduction

We held a hands-on training course on superconducting magnets for the participants in this accelerator school who had no or little experience with superconducting and related technologies. The experience obtained in this course will be valuable to help beginners in this field to understand the total concept of superconducting magnets and cyogenics. Figure 1 shows the time schedule for the one-week course. In this hands-on course, the design, winding, and assembly of four superconducting solenoid magnets were carried out, and all the magnets were cooled down and excited. Because it takes a long time to wind a coil, we had to spend most of the time on winding, i.e. except for the lectures on designing the magnet and handling the liquid helium we allocated almost all the time to coil winds.

472

The total number of participants in the magnet course was about 45, including those who attended only to this course. They were divided into four groups (A, B, C, D) of about eleven. The winding and fabrication of four magnets were carried out by each group in parallel in the same experimental hall. Figure 2 charts the progress of winding by each group. The groups were faurther divided into sub-groups of three. The sub-groups took turns the coil. The persons not attending the coil winding used their time in designing the coil and attending lectures on computer calculations of magnetic field.

As this kind of hands-on training course on superconducting magnets had never been tried before in an accelerator school, it seemed ambitious to hold such a

	Nov. 29 Mon.	Nov. 30 Tue.	Dec. 1 Wed.	Dec. 2 Thu.	Dec. 3 Fri.
9:00	Magnet Course Guidance	Measurement	Handling of Liq. He and Liq.N2 / Magnet Excitation	Meeting	Meeting
10:00				Winding Mesurement System Set Up	System Check
	Magnet Design	Winding Magnet Design Exercises	Winding Magnet Design Exercises	Winding Mesurement System Set Up	
11:00	Winding Demonstration	Magnet Design Exercises	Winding Magnet Design Exercises	Winding Mesurement System Set Up	Cool Down and Excitation Test #1
12:00	Lunch	Lunch	Lunch	Lunch	Lunch
13:00	Winding Magnet Design Exercises	Winding Magnet Design Exercises	Winding Mesurement System Set Up	Winding Mesurement System Set Up	
14:00	Winding Magnet Design Exercises	Winding Magnet Design Exercises	Winding Mesurement System Set Up	Winding Mesurement System Set Up	
15:00	Winding Magnet Design Exercises	Winding Magnet Design Exercises	Winding Mesurement System Set Up	Magnet Set Up System Check	Cool Down and Excitation Test #2
16:00	Winding Magnet Design Exercises	Winding Magnet Design Exercises	Winding Mesurement System Set Up		
17:00	Winding	Winding	Winding	Winding	
18:00	Winding	Winding	Winding	Winding	

Fig. 1 Time schedule of hands-on training for magnet course.

473

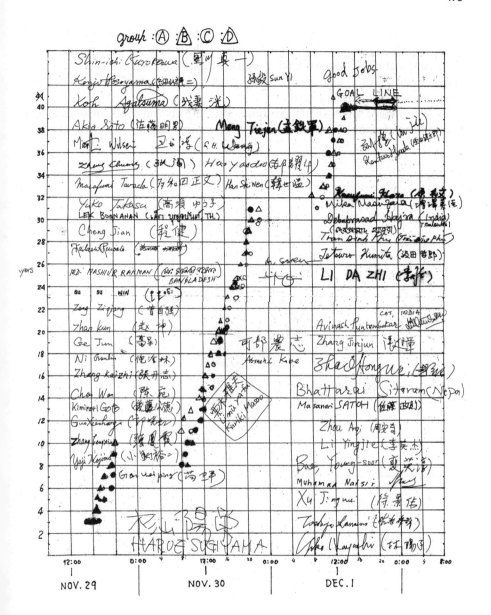

Fig. 2 Chart of winding progress by each group.

large-scale hands-on course in a short time and in IHEP in Beijing where no cryogenic facility was available. Fortunately we had made several magnets of the same design at KEK, and had held a similar hands-on training course twice at KEK summer schools in Tsukuba, as "cryogenic technology courses" for training young scientists and engineers in Japan.

We prepared four winding machines and all parts and tools for fabrication of four superconducting solenoid magnets; i.e. superconducting wires, bobbins for

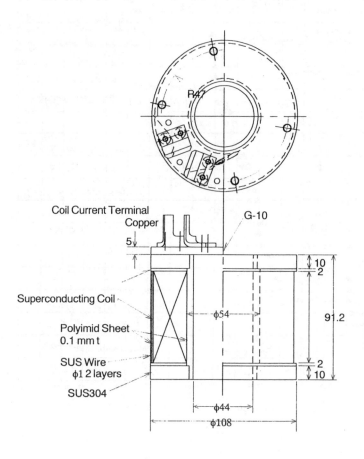

Fig. 3 Dimensions and structure of superconducting solenoid coil.

superconducting solenoid coils, etc. We also prepared two vertical 200-mm-diameter cryostats, two DC power supplies (160 A x 6 V) for excitation tests of the magnets, and two measurement systems. All these were prepared and checked at KEK in Tsukuba and sent to IHEP in Beijing by ship, and returned to KEK after the hands-on training was over.

Specifications of the 7-T superconducting solenoid coil

Many types of superconducting magnets are used in accelerators. We decided on a small solenoid magnet for our hands-on training course, because this magnet is easy to fabricate using fine superconducting wire. We could use a high performance fine multi-filament NbTi superconducting wire 0.68 mm in diameter, and 6-μm-diameter filament with a copper-super ratio of 1.8. The wire is electrically insulated by 0.1-mm-thick polyvinylformar varnish. This superconducting wire was manufactured by Furukawa for the outer coil cable of the SSC (Superconducting Super Collider) dipole magnet.

Figure 3 shows the dimensions and structure of a superconducting solenoid magnet fabricated in our course. The coil inner and outer diameters are 54 mm and 108 mm respectively and its length is 67 mm. The coil has about 3800 turns with

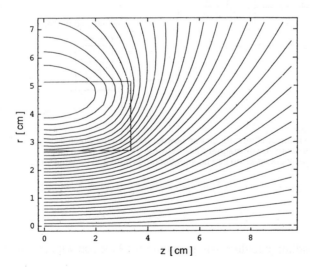

Fig. 4 Computer plot of the magnetic flux lines of the solenoid magnet wound in the hands-on training course.

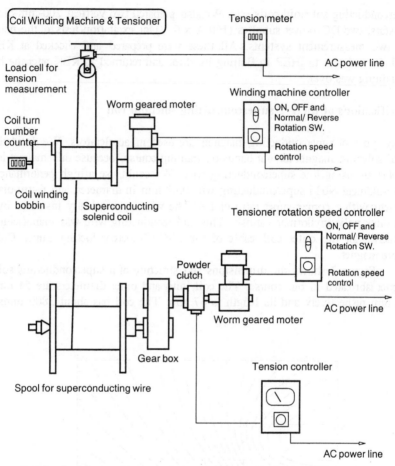

Powder clutch :
The torque is transmitted through iron powder between the clutch disks, and
its torque is controlled by magnetic flux between the disks. When the
magnetic flux increases, the iron powder becomes hard and transmit torque
increase.

Fig. 5 Coil winding procedure using winding machine and wire tensioner.

40 layers, and the self-inductance L is 0.64 H. This magnet could produce 7 T of central field at an operating current of about 150 A at 4.2 K.

We developed a computer code for this course, which could calculate the magnetic field distribution and stored energy of a solenoid magnet numerically using Mathematica. Figure 4 shows the result. We prepared six notebook-type computers (Machintosh G3) for this magnetic field calculation and held several 30 minute lectures, in parallel with the coil winding, In which the principles of the magnetic field calculation and the use of Mathematica were explained.

Coil winding

The coil was wound on a stainless steel bobbin using a winding machine, with no impregnation of epoxy resign, and the wire tension was kept at 4 kgf by a tensioner during winding. Figure 5 shows the coil winding procedure using a winding

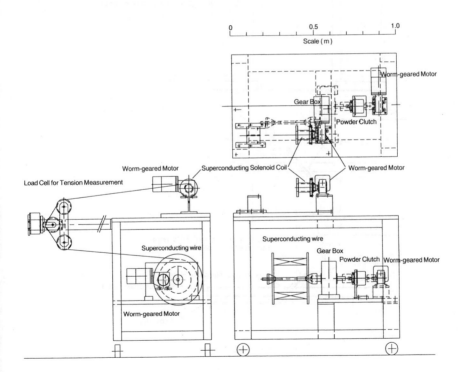

Fig. 6 Coil winding machine and tensioner.

machine and tensioner. Figure 6 is a drawing of the winding machine and tensioner.

Figure 7 shows the detail of the starting and end parts of the winding. The wire must be wound as tightly as possible because small wire movements in the coil during magnet excitation, due to the strong electromagnetic force, cause coil quenches. The weak parts of the coil winding are the two end parts of the layer, i.e. the transition parts of winding layers because void spaces in the winding occur there,

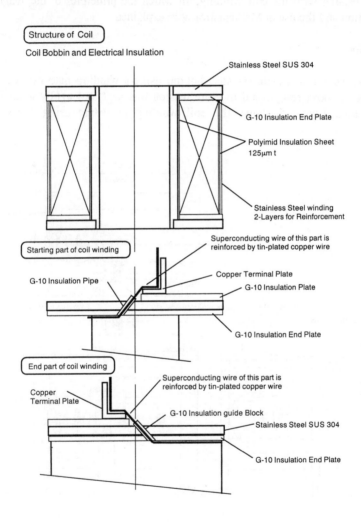

Fig. 7 Detail of the starting and end parts of the coil winding.

as shown in Fig. 8. During coil winding these void spaces must be filled by an epoxy putty called "green putty." The electric insulation between the coil winding and the bobbin is guaranteed by 0.15-mm-thick polyimid film. After the coil winding was finished, the outside of the coil was reinforced by two-layer winding of 1-mm-diameter stainless steel wire. Winding of the coil and a complete magnet are shown in Figs. 9 and 10 respectively.

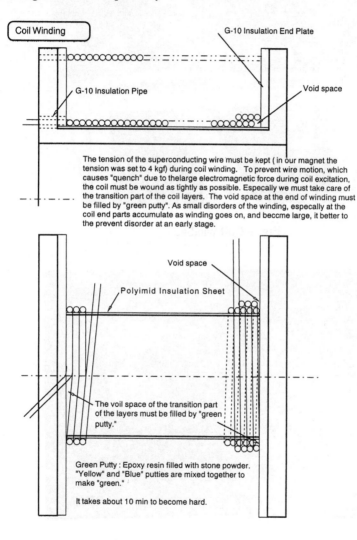

Fig. 8 Weak parts of the coil winding.

480

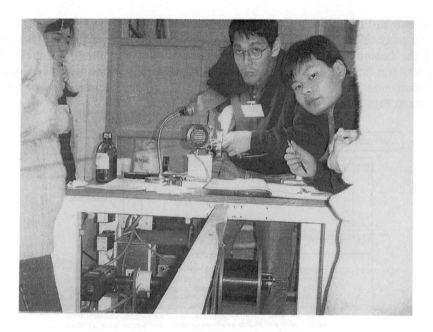

Fig. 9 Coil winding.

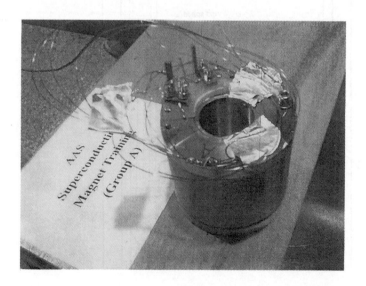

Fig. 10 A complete superconducting solenoid magnet.

Coil assembling and set up

After the coil windings were finished, these magnets were set in the top flanges of the cryostats, as shown in Fig. 11. The magnets were connected to the current leads by superconducting bus bars. The excitation current of about 150 A is supplied to the

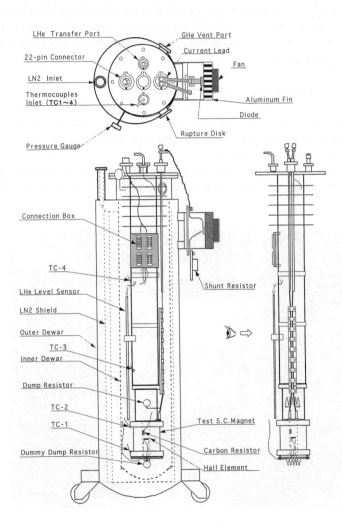

Fig. 11 A superconducting magnet installed in a vertical cryostat.

coil through these superconducting bus bars, which are made of the same superconducting wire as the coil winding. The bus bars are reinforced by 2-mm-diameter copper wire by soldering to increase the thermal stability.

A quench protection resistor of 0.165 ohm at 4.2 K is connected to the coil in parallel for quench protection of the magnet, as shown in Fig.11. This resisteor is used to bypass the excitation current when the magnet quenches and also to dump the stored energy of the magnet after cut-off of the power supply. Coil quench can be detected by measureing the coil voltage (voltage between the coils). Resistance measurements on the protection resistors at low temperature were carried out at 4.2 K using dummy resistors.

Cool-down and items measured on the magnet

We prepared two sets of measurement systems for cool-down and excitation of four magnets. The cool-down and excitation tests were carried out first on two magnets, and then on the other two magnets.

In our hands-on training course the available time for setting up the measurement system and cooling down and excitation of the magnet was limited.

Fig. 12 Measurement system.

Table 1   Items measured in the superconducting magnet test.

Resistivity of Protection Resistor
    at Room   Temperature
    at Liq. He Temperature
Decay Time of Coil Current   - Coil Current Shut-Down Test
    at   2 , 5,10 A

Magnet Excitation
    System Check
        DC Power Supply
        Excitation of Magnet Up to 75 A
    Excitation of Magnet Up to Quench
        Excitation Current
        Magnetic Field
        Coil Voltage
        Coil Temperature
        Liquid He Level
Cool-Down of Magnet
        Temperature, by Thermo-couple

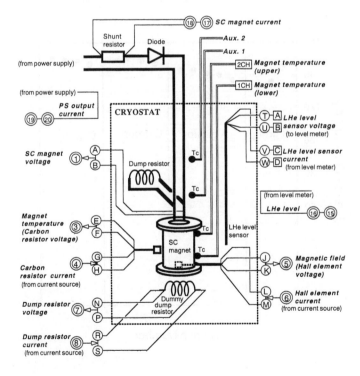

Fig. 13 Monitoring sensors to the superconducting solenoid magnet.

Therefore we selected as few measurement items as possible. These items are listed in Table 1: we measured temperatures of the magnet during cool-down using thermocouples, liquid helium level during magnet excitation, and coil current and voltage and central magnetic field by Hall probe during magnet excitation. The data taking system was simple, i.e. a 3-pen recorder for fast measurements and a 12-point dot-print recorder for slow measurements. Figure 12 shows the measurement system used in this experiment. Figure 13 is a schematic of the magnet set-up and the monitoring sensors in the cryostat. At the bottom of the cryostat we set a resistor of the same type used as dump resistor of the magnet to measure its resistance at 4.2 K.

After cool-down of the magnet was completed, we carried out current cut-off experiments at 2, 5, and 10 A and measured the decay time constants of the magnet current. By using the measured values of the dump resistor discussed above we could calculate the self-inductance L of the magnet. We found good agreement between the measured values and the calculated one.

Finally we carried out the quench tests. Figure 14 shows the results of quench tests of four magnets. The quench current is plotted as a function of quench

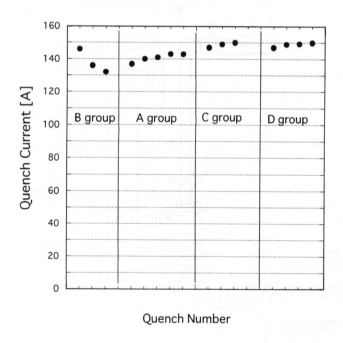

Fig. 14 Quench histories of superconducting solenoid magnets A, B, C, D.

Fig. 15 Experimental set-up of the magnet test.

number for four coils. The first quench currents are 146 A, 137 A, 147 A, 147 A for magnets A, B, C, D respectively. In the case of magnets C, D the quench current reached its short sample limit in the fourth quench. All the magnets except magnet B could exceed the central field of 7 T. We could carry out about four quench tests on average: a total of 15 quenches for the four magnets.

We prepared a total of 1000 L of liquid helium in two 500-L dewars and used a total of about 700L for the four magnet experiments. Figure 15 shows the experimental setup: a vertical cryostat, a 500-L liquid helium dewar, and a measurement system in the experimental hall in IHEP.

BASIC DESIGN OF SOLENOID SUPERCONDUCTING MAGNET FOR HANDS ON TRAINING

K. AGATSUMA

*Electrotechnical Laboratory, 1-1-4 Umezono, Tsukuba, Ibaraki,
Japan
E-mail: agatsuma@etl.go.jp*

This exercise for hands-on training on magnet design helps the students to understand the principal design procedure for air-core superconducting solenoids with a rectangular cross section and uniform current density, and also makes it possible to design a model superconducting magnet that the students themselves can wind.

1 Introduction

We start the study of solenoid superconducting magnet design with the following case study. We assume that we got an order for a superconducting magnet with the following specifications from our customers; therefore that we need to determine the magnet dimensions, and winding bobbin size, select the superconductor wire, determine the length of the wire and its cost, estimate the winding tension, and so on.

[Case A] The specifications of the magnet are as follows:
  a) The central magnetic field is 7 Tesla and the inside winding-diameter is 54 mm.
  b) The magnetic field homogeneity is 5% in a sphere of half winding-radius around the center of the magnet.

[Case B] The specifications of the magnet are as follows:
  a) The central magnetic field is 7 Tesla and the inside winding-diameter is 54 mm.
  b) Minimum winding length (cost minimum).

In these cases you use NbTi fine multi-filamentary wire of 0.7 mm diameter. The critical current of this wire is 100 A at 8.5 T, and 383 A at 4 T in liquid helium.
You should fill the blanks in the following design sheet.

Winding dimensions:

  inner winding radius; $a_1 = \underline{27}$ mm (given),

  outer winding radius; $a_2 = \underline{\hspace{2cm}}$ mm

  winding length; $2b = \underline{\hspace{2cm}}$ mm

packing factor; $\lambda = $ _____(%)
superconductor wire length; _____ m
number of the turns in a layer; _____ turns
number of layers; _____ layers

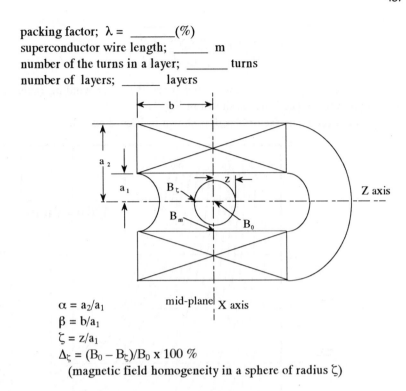

$\alpha = a_2/a_1$
$\beta = b/a_1$
$\zeta = z/a_1$
$\Delta_\zeta = (B_0 - B_\zeta)/B_0 \times 100\ \%$
　(magnetic field homogeneity in a sphere of radius ζ)

Figure 1. Cross section of air-core solenoid superconducting magnet.

2　Basic Design Theory

2.1 Uniform Current Density and Magnetic Field

These procedures are for a simple air-core solenoid with a rectangular cross section and uniform current density, as illustrated in Fig. 1. For a square-ended solenoid with a rectangular cross section, uniform current density, j, in the superconducting material, and an amount of this material per unit cross section of coil, λ (packing factor), the axial magnetic field at the center of the coil B_0 (Tesla) can be written as follows:

$$B_0 = j\ \lambda\ a_1 F(\alpha,\beta) \times 10^{-6} \qquad (1)$$

where a_1 is the inner winding radius (m), and the Fabry factor $F(\alpha,\beta)$ [1,2] is given as follows

$$F(\alpha, \beta) = \frac{4\pi}{10} \beta \, \log_e \frac{\alpha + \sqrt{\alpha^2 + \beta^2}}{1 + \sqrt{1 + \beta^2}} \qquad (2)$$

where $\alpha = a_2/a_1$, $\beta = b/a_1$, and a_2 and b are the outer winding radius and the winding length, respectively, as shown in Fig. 1.

The graph of the Fabry factor $F(\alpha,\beta)$ is shown in Figs. 2 and 3.

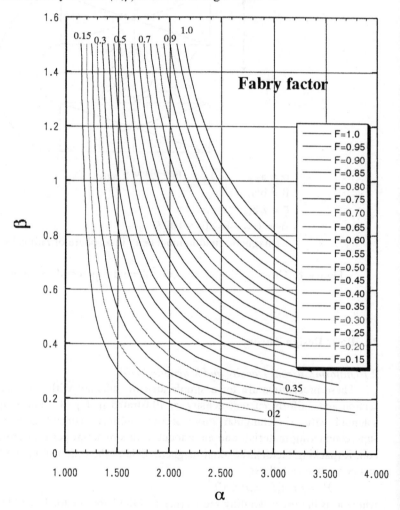

Figure 2. Fabry factor for various α and β.

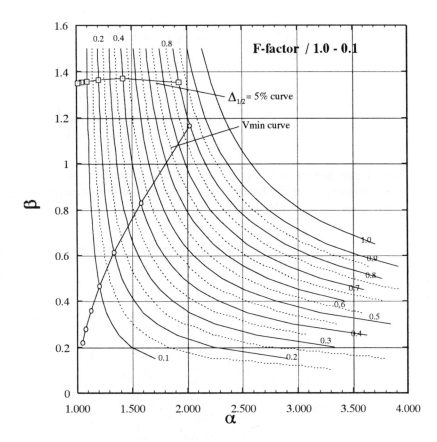

Figure 3. Fabry factor and 5% magnetic field uniformity line within the sphere of a half winding-radius.

2.2 Field Homogeneity in a Sphere of Normalized Radius ζ (z/a_1).

We consider the magnetic field homogeneity in a sphere of normalized radius ζ ($=z/a_1$) in the center of a solenoid coil (see Fig. 1). The homogeneity Δ_ζ in the sphere of radius ζ is given as follows:

$$\Delta_\zeta = (B_0 - B_\zeta)/B_0 = (F_0 - F_\zeta)/F_0 \quad (\times 100 \ \%) \tag{3}$$

where B_ζ and F_ζ represent respectively the axial magnetic field and the Fabry factor at the point ζ ($= z / a_1$) on the Z axis of the solenoid coil, as shown in Fig. 1.

$$F_\zeta = \frac{2\pi}{10}\left[(\beta-\zeta)\,\log_e\frac{\alpha+\sqrt{\alpha^2+(\beta-\zeta)^2}}{1+\sqrt{1+(\beta-\zeta)^2}}\right.$$

$$\left.+(\beta+\zeta)\,\log_e\frac{\alpha+\sqrt{\alpha^2+(\beta+\zeta)^2}}{1+\sqrt{1+(\beta+\zeta)^2}}\right] \qquad (4)$$

Typical homogeneity is $\Delta_{1/2}$ or Δ_1 for $\zeta = 1/2, 1$. (Sometimes we use $\zeta = (a_1-t)/a_1$, where t=thickness of the bobbin.) Various values of homogeneity in a sphere of radius ζ are shown in Fig. 4.

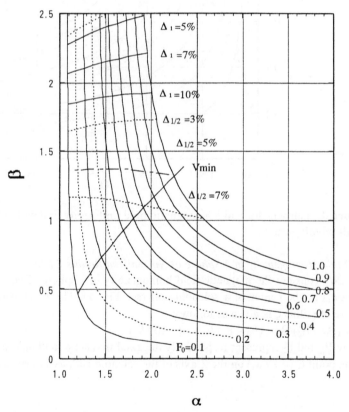

Figure 4. Magnet design curves show various Fabry factors and various magnetic field uniformity lines. The symbols $\Delta_{1/2}$ and Δ_1 are represent the magnetic field

uniformity within a sphere of a half winding radius ($\zeta=z/a_1=1/2$) and a winding radius ($\zeta= z/a_1=1$) respectively.

2.3 The Maximum Magnetic Field B_m in the Coil

The maxim magnetic field B_m in the coil is at the center of the inner winding layer (see Fig. 1), and the ratio κ (B_m/B_0) of the maximum magnetic field B_m at the winding to the central field B_0 at the center of the coil is shown in Fig. 5.

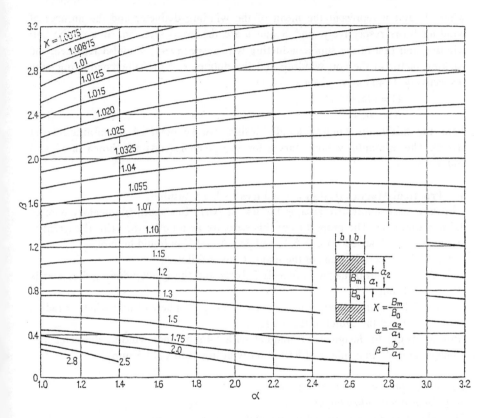

Figure 5. Ratio κ (B_m/B_0) of the maximum magnetic field B_m in the winding to the central field B_0 at the center of the solenoid coil.

2.4 Minimum Winding Length (cost minimum) Design.

For a square-ended solenoid with a rectangular cross section, shown in Fig. 1, the winding volume $v(\alpha,\beta)$ is given as follows;

$$v(\alpha,\beta) = 2b\pi(a_2^2 - a_1^2) = 2a_1^3\beta \ (\alpha^2 - 1) \tag{5}$$

so that the length of superconductor L_s required is derived as follows:

$$L_s = \pi D_n = \pi(a_1 + a_2)S\lambda/s$$

$$= \pi a_1(1+\alpha)\frac{2a_1^2\beta(\alpha^2+1)\ \lambda}{(\pi d^2/4)} = \frac{8a_1^3\beta(\alpha^2-1)\ \lambda}{d^2} \tag{6}$$

where D and n represent the mean of the winding diameter and the number of winding turns respectively, and S and s are the cross sectional area of the coil winding and that of the superconducting wire respectively. λ and d represent the packing factor in the coil winding and the diameter of the superconducting wire respectively. D and n are given as follows;

$$D = \pi(a_1 + a_2), \tag{7}$$
$$n = S\lambda/s. \tag{8}$$

Thus the minimum winding length design can be obtained by minimumizing $v(\alpha,\beta)$. The minimum volume curves for any given $F_0(\alpha,\beta)$ are shown in Figs. 3 and 4.

3 Practical Magnet Design Procedures by Case Studies.

In these case studies we assume the use of NbTi fine multi-filamentary wire of 0.7 mm diameter. The properties of this wire are as follows: the critical current is assumed to be about 100 A at 8.5 T, and 383 A at 4 T in liquid helium.

[Case A] The specifications of the magnet are as follows:
   a) The central magnetic field is 7 Tesla and the inside winding-diameter is 54 mm.
   b) The magnetic field homogeneity is 5% in a sphere of half winding-radius around the center of the magnet.

Now we begin to design a 7T magnet with 54mm inner winding-diameter and homogeneity 5% in a sphere of half inner winding-radius (13.5 mm).

The design procedures are as follows;
A-1) Determine Fabry factor.
The Fabry factor is determined from equation (1) as follows:

$$F(\alpha,\beta) = B_0 \times 10^6 / j \lambda a_1 \tag{9}$$

Here $B_0 = 7$ (T) and $a_1 = 27 \times 10^{-3}$ (m) are given. The current density of the wire j is estimated from the critical current data shown in Fig. 6. The maximum field B_m is estimated to be about 6 or 7% higher than B_0 because of the homogeneity, so that the critical current I_c is assumed to be about 160 A (at 7.5 T). Therefore j is estimated as follows:

$$j = 160 \div (\pi d^2/4) = 4.16 \times 10^8 \ (A/m^2) \tag{10}$$

The packing factor λ is about 0.9 for close pack as illustrated in the upper part on the right, and for straight pack it is about 0.79 as illustrated in the lower part on the right. In case of our design we assume a packing factor λ of about 0.8 as straight pack, so that we can get a values of the F-factor of 0.799 to substitute into equation (9).

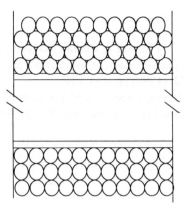

A-2) Determine the coil dimensions from Magnet Design Curves

We can find the intersection of curves for $\Delta_{1/2} = 5$ % and Fabry factor F= 0.8 is at $\alpha = 1.94$ and $\beta = 1.35$ in Fig. 3 or Fig. 4, therefore we can get $a_2 = 52.4$ (mm) and b = 36.5 (mm). The length of superconducting wire L_s required is estimated about 956 m from equation (6). Magnet road line is defined by $B_0/I = 7/160$ (T/A) $= 0.0438$ (T/A) = 438 (G/A). The maximum field B_m in winding is defined.

[Case B] The specifications of the magnet are as follows:
   c) The central magnetic field is 7 Tesla and the inside winding-diameter is
      54 mm.
   d) Minimum winding length (cost minimum).

B-1) Procedures to determine Fabry factor same as for Case A

We can find that the intersection of the curves for minimum volume (V_{min}) and Fabry factor F = 0.8 is at $\alpha = 2.03$ and $\beta = 1.17$ in Fig. 3 or Fig. 4, therefore we can get $a_2 = 54.8$ (mm) and b = 31.6 (mm). The length of superconducting wire L_s required is estimated as about 939 m from Eq. (6).

4   Discussion

Practical magnet design procedures by case studies described above, help you understand basic magnet design theory. Now you can calculate the magnet design parameters more accurately by using a computer. It is not so difficult for you to plot the Fabry factor and estimate the magnet design parameters by computer. It is a good exercise for you to make a computer program for solenoid magnet design.

494

References

1. Fabry C., Sur le Champ Magnetique au Centre d'une Bobine Cylindrique et la Construction des Bonines Galvanometers, *L'Eclairage Electrique*, vol. **17**, no. **43** (1898) pp. 133-141.
2. Fabry C., Production de Champs Magnetiques Intenses au Moyen de Bobines Sans Fer, *J. Phys.*, vol. **9** (1919) pp. 129-134.

EXPERIMENT ON A SUPERCONDUCTING CAVITY

TAKAAKI FURUYA, YUKO KIJIMA, TAKESHI TAKAHASHI
SHIN-ICHI YOSHIMOTO

KEK, 1-1 Oho, Tsukuba, Ibaraki, 305-0801 Japan
E-mail: takaaki.furuya@kek.jp

and   ZHAO SHENGCHU

IHEP, No. 19 Yu Quan Lu, Beijing 100039, China

1   Outline of the SC cavity training program

We planned an experiment using a real 1.3GHz superconducting cavity as a hands-on training program to introduce superconducting technology. A five-day period is not enough for a complete superconducting cavity experiment including the cavity fabrication and treatment processes. Therefore our program consists of introducing of basic technology for measurements on the superconducting cavity, calculation of the cavity parameters, setting the cavity in a cryostat, assembling the RF system, cooling the cavity, and making measurements. First, the students calculated the cavity parameters of a pill-box cavity analytically, such as shunt impedance, transit time factor, Q-value for the Cu wall and so on. After that, they compared these figures with the calculation results that they obtained by using the SUPERFISH code, where the students understood the input and output of SUPEFISH. Finally, the parameters of the test cavity required in this experiment were calculated out using SUPERFISH.

On the second day, the students opened the packages and set the cavities into a cryostat. At the same time they assembled the RF system with phase-lock-loop (PLL). In this RF system a double-balanced-mixer (DBM) was used for the PLL. On the third and fourth days, the students cooled the cavity to 4.2 K and measured it by both CW and pulse methods.

On the last day, the students disassembled the cavity and inspected the inside, where they could see the input and monitor couplers as well as the cavity surface, and also the indium-sealing gasket for Nb flanges.

In this experiment, the maximum cavity field was limited to around 2 MV/m, because of the available RF power source of 3 W and cavity temperature of 4.2 K, and also the limitation of radiation safety. Nevertheless, we think the students could obtain the basic knowledge and technology of high-Q cavity measurements through this experiment. (See Table 1.1)

Table 1.1. Curriculum of the cavity experiments.

DATE		A. M.	P. M.
Nov. 29	Mon.	Calculation (I) For a pill-box cavity 1) analytical calculation 2) using SUPERFISH	Calculation (II) For the 1.3-GHz test cavity 1) using SUPERFISH
Nov. 30	Tue.	Open packages Cavity setup 1) connecting cables	Set up the RF system Installing the cavity into a cryostat
Dec. 1	Wed.	Cooling the cavity 1) LN$_2$ pre-cooling 2) transfer of LHe (4.2 K)	Measurements/Discussion 1) cable correction 2) CW measurements
Dec. 2	Thurs.	Cooling the cavity 1) Transfer of LHe (4.2 K)	Measurements/Discussion 1) change coupling 2) pulse measurements 3) warm up the cavity
Dec. 3	Fri.	Open the cavity 1) Inspection of the cavity surface 2) Indium sealing	Packing the tools
Dec. 4	Sat.	Presentation	Closing

2 Calculate the test cavity using SUPERFISH (first day; 29 Nov.)

In the experiment, we have to use many parameters related to the cavity shape. The purpose of the following steps is to understand the parameters, such as shunt impedance, R/Q, transit time factor and so on, and to use the SUPERFISH code to obtain the cavity parameters for the test.

As the first step, consider the RF field of a pill-box cavity, shown in Fig. 2.1. The calculation of a pill-box can be treated analytically and will give us an idea of cavity parameters such as Q and shunt impedance R. Furthermore, it will help us to understand the output of the computer calculations.

The field components of the TM_{010} mode are given as

$$E_r = E_\theta = 0 ,$$
$$H_r = H_z = 0 ,$$
$$E_z = E_0 J_0(k_c r)\cos(\omega t) ,$$
$$H_\theta = -\frac{E_0}{Z_0} J_1(k_c r)\sin(\omega t) ,$$
$$H_z = 0 ,$$

where

$$Z_0 = \sqrt{\frac{\mu_0}{\varepsilon_0}} = 377\ \Omega .$$

The angular frequency ω of this mode is

$$f = \frac{\omega}{2\pi} = \frac{c}{2\pi} k = \frac{c}{2\pi} \frac{\rho_{01}}{a} = \frac{23.0}{2a[cm]}\ [GHz] , \quad \rho_{01} = 2.405 ,$$

which shows that the frequency of TM_{010} depends only on the cavity diameter 2a. Thus the diameter of the cavity becomes 17.69 cm for 1.3 GHz.

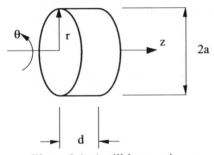

Figure 2.1: A pill-box cavity.

(Step 1) Derive the cavity parameters from these field components.

The stored energy U and the power dissipated in the cavity wall P for TM_{010} are obtained as

$$U = \frac{\varepsilon_0}{2} \int E_z^2 dv = \frac{\pi \varepsilon_0 a^2 d}{2} E_0^2 J_1^2(\rho_{01}) ,$$

$$P = \frac{R_s}{2} \int H_\theta^2 ds = \frac{\pi R_s a(a+d)}{Z_0^2} E_0^2 J_1^2(\rho_{01}) ,$$

where R_s is the surface resistance of the cavity wall.

From the definition of the Q-value, the transit time factor T, the shunt impedance R, and R/Q are described as

$$Q = \frac{\omega U}{P} = \frac{1}{R_s} \frac{1}{\dfrac{a}{d}+1} \frac{\rho_{01}}{2} Z_0 \, , \tag{2-1}$$

$$T = \frac{V_c}{dE_0} = \frac{\sin(\pi d/\lambda)}{\pi d/\lambda} \quad \left(\text{for} \quad \beta = \frac{v}{c} \approx 1 \right) , \tag{2-2}$$

$$R = \frac{V_c^2}{P} = \frac{T^2}{\lambda R_s} \left(\frac{d}{\dfrac{a}{d}+1} \right) \frac{2Z_0^2}{\rho_{01} J_1^2(\rho_{01})} \, , \tag{2-3}$$

and

$$\frac{R}{Q} = \frac{V_c^2}{\omega U} = \frac{T^2}{\lambda} d \frac{4Z_0}{\rho_{01}^2 J_1^2(\rho_{01})} \, . \tag{2-4}$$

For the pill-box with a half wave length ($d=\lambda/2$),

$$Q = 257 \frac{1}{R_s} \, , \quad R = 5.14 \times 10^4 \frac{1}{R_s} \, , \quad T = \frac{2}{\pi} = 0.637 \, ,$$

and

$$\frac{R}{Q} = 200 \ \Omega \, .$$

Thus Q and R are functions of the surface resistance R_s, where

$$R_s^{NC} = \sqrt{\frac{\omega \mu}{2\sigma}} \, ,$$

for normal-conducting cavities, and

$$R_s^{SC} = R_{res} + R_s^{BCS} \, , \quad R_s^{BCS} = A \frac{\omega^2}{T} \exp\left(-\frac{\Delta(0)}{k_B T_c} \cdot \frac{T_c}{T} \right) \, ,$$

for superconducting cavities.
On the other hand, R/Q depends only on the cavity shape and has no dependence on either cavity material or frequency.

(Step 2) Calculate the pill-box cavity using SUPERFISH.

In the cavity measurements we can obtain the RF power dissipated in the cavity walls and the Q. These two quantities give the stored energy U using the definition of Q in Eq. (2-1). Thus the peak accelerating voltage V_c can also be obtained using the stored energy U and R/Q, where R/Q is a figure of merit that relates V_c to U and depends only on the cavity shape. Further, the transit time factor T gives the peak surface electric field gradient E_0 from Eq. (2-2). To obtain these cavity shape parameters, the computer codes LALA, SUPERFISH, URMEL and MAFIA are available. In this school the SUPERFISH code is used.

(Step 3) Calculate the test cavity using SUPERFISH.

The shape of the test cavity is shown in Fig. 2.2. This shape is a 1/2.6 scale copy of the center cell designed for the KEK-TRISTAN 5-cell structure. This cavity has a spherical shape to avoid maltipacting and the end plates are tilted by 80 degrees, so that the acid or rinsing water can be dropped smoothly out of the cavity during the surface treatments.

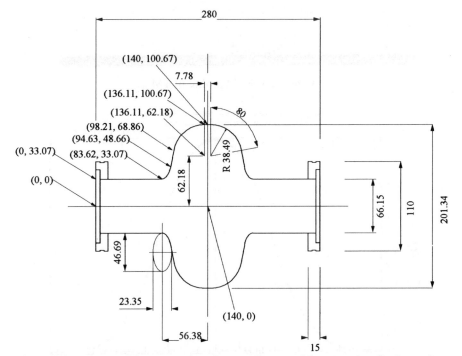

Figure 2.2: Shape of the test cavity; the coordinates of the segments for input data are shown.

The SUPERFISH code gives the profile of the electric field and its distribution along the beam axis, as shown in Figs. 2.3 and 2.4. From these results one can obtain the parameters, as shown in Table 2.1.

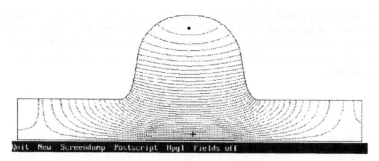

Figure 2.3: Field profile of TM_{010}.

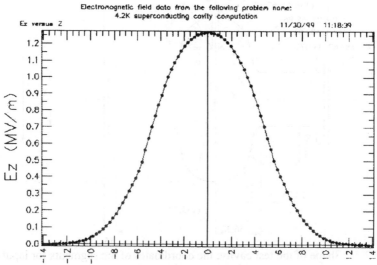

Figure 2.4: Field distribution along the beam axis.

Table 2.1: Calculated parameters of the test cavity.

parameter	
Effective gap length (L_{eff})	0.116 m ($=1/2\lambda$)
Frequency	1296.283 MHz
Transit time factor	0.580
Q for Cu wall	2.90E+04
Shunt impedance(Z_{TT})	3.64E+06 ohms
G ($=R_s{}^*Q$)	270 ohms
R/Q (Z_{TT}/L)	127 ohms
Loss factor k	0.259 V/pC
Field strength Gap voltage (V_c) Gradient ($E_{acc}=V_c/L_{eff}$) E_{sp}/E_{acc} H_{sp}/E_{acc}	11.28×SQR(PQ) V 97.52×SQR(PQ) V/m 1.81 39.6 gauss/(MV/m)

3   Experimental setup (second day; 30 Nov.)

Figure 3.1 shows the setup for this experiment. The system is simple, but it contains all the basic components except the diagnostic tools. The He vessel

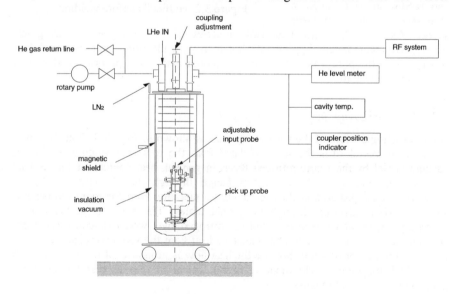

Figure 3.1: Setup of a cryostat.

contains the various sensors, such as a liquid He level meter, thermometers on the top and bottom of the cavity, and a position indicator of the adjustable input coupler.

The test cavity for this experiment has already been treated, evacuated and closed by a metal valve. But it should be emphasized that the cavity vacuum is usually connected to a pumping unit located on the room temperature side. Further, the setup should be shielded against X-rays from the cavity at high field measurements.

3.1   The test cavity

The 1.3-GHz test cavity was made from 2.5-mm pure Nb sheets of RRR=150. A pair of half-cells is formed and dipped in a HCl bath to remove the iron contaminant implanted during the forming process. Then they are welded by electron beam welding (Fig. 3.2).

The inner surface of the cavity was electropolished by 80μm with a mixture of HF and H_2SO_4 so as to make a strain-free and dust-free

Figure 3.2: Half cells before welding.

surface. After this chemical treatment, a series of rinsing processes was given; rinsing with pure water, dipping in a hot bath with ultrasonic agitation, and final rinsing with ultra-pure water. Assembling of input and output probes was done in a class-100 clean room.

3.2   Setup the cavity in a vertical cryostat

A cross-sectional view of the vertical cryostat is shown in Fig. 3.3. The 300-mm-diameter He vessel of is thermally isolated from the room-temperature side of the vacuum vessel by the liquid nitrogen layer. In general, heat sources for the He bath are (a) heat conduction from the top flange through the He vessel wall, (b) convection of He gas around the top flange, (c) radiation from the room temperature side, and (d) conduction through the signal cables. The reflectors at the top are especially important to reduce (b) and the nitrogen layer works to reduce (c). If the cryostat has no nitrogen shield, the radiative heat from the room temperature side is increased by a factor of >200, because the heat flow is proportional to $(T_1^4 - T_2^4)$. A 2-mm-thick magnetic shield made of high-μ metal in the He vessel reduces the geomagnetism to ~20 mgauss.

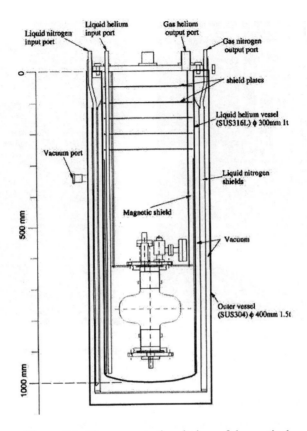

Figure 3.3: The cross-sectional view of the vertical cryostat.

Two ports are on the top flange of the cavity; one is for an input coupler and the other for evacuation. The outer conductor of the input coupler was made by SUS-bellows (Fig. 3.4) to allow changing the penetration of the inner conductor even during low temperature measurements to obtain optimum coupling. The position of the input coupler is detected using a linear potentiometer. A pick-up antenna on the bottom flange has sufficiently small coupling strength.

In usual experiments, the cavity is equipped with a mapping system of surface temperature or x-ray distribution, as shown in Fig. 3.5, in which carbon resisters and PIN photo diodes are used to detect of temperature and radiation, respectively.

Figure 3.4: Adjustable input probe.   Figure 3.5: Mapping sensors of carbon resisters and PIN diodes,

3.3   Assembling the RF system

A block diagram of the RF system used in this school is shown in Fig. 3.6. In general, the Q is obtained from the full width at half-maximum (FWHM) of the resonance Δf_{half} as

$$\frac{\Delta f_{half}}{f_0} = \frac{1}{Q_L} \ . \tag{3-1}$$

But this method cannot be used for superconducting cavities. Because of the extremely high Q of superconducting cavities, the bandwidth is too narrow for the measurements. Typically it is only ~1 Hz at 1 GHz. Therefore an RF system with a frequency-lock-loop is required in order to maintain the resonance and measure the RF power. As shown in Fig. 3.6, the phase-lock-loop (PLL) signal using a double balanced mixer (DBM) returns to the signal generator to fix the generator frequency to the cavity resonance. The RF switch driven by a function generator is used to create the pulse RF for the decay measurements. Either P_r or P_t can be used for the input signal to the DBM.

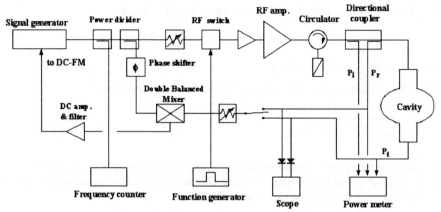

Figure 3.6: Block diagram of the RF system.

From the equivalent circuit, the impedance of the cavity is given as

$$\frac{\vec{Z}}{Z_0} = \frac{1}{1 + jQ_L \left(\dfrac{\omega}{\omega_0} - \dfrac{\omega_0}{\omega} \right)} \, , \tag{3-2}$$

where ω and ω_0 are the angular frequency of the RF generator and the cavity resonant frequency, respectively. This can be rewritten as

$$\frac{\vec{Z}}{Z_0} = e^{j\psi} \cos \psi \, , \tag{3-3}$$

by using δ and ψ, given as

$$\delta = \frac{\omega - \omega_0}{\omega_0}, \quad \tan \psi = -2Q_L \delta \, . \tag{3-4}$$

The ψ is called the tuning angle. From the real part of Eq. (3-3), one can obtain the FWHM of Eq. (3-1) for amplitude. On the other hand, the equation Eq. (3-4) gives the sensitivity of the phase ψ as a function of Q and frequency shift δ. Hence, the phase change of the cavity field against the reference position can be used for the frequency-lock-loop. From this point of view, the SC cavity seems to be a sensitive band-pass filter.

4 Measurements (third and fourth days; 1 and 2 Dec.)

4.1 Loaded Q and unloaded Q

From the definition of Q, the stored energy $U(t)$ can be written as

$$\frac{dU(t)}{dt} = -\frac{\omega}{Q}U(t).$$

In a real cavity system, the stored energy is dissipated not only on the cavity wall (P_c), but also as leakage through an input coupler (P_{rad}). Thus, when the RF power is turned off, the total power loss is described as

$$P_{tot} = P_c + P_{rad}.$$

For these losses we can define the Q values as

$$Q_L = \frac{\omega U}{P_{tot}}, \quad Q_0 = \frac{\omega U}{P_c}, \quad Q_{ext} = \frac{\omega U}{P_{rad}}, \tag{4-1}$$

and

$$\frac{dU(t)}{dt} = -P_{tot} = -\left(P_c + P_{rad}\right)$$

$$= -\omega\frac{1}{Q_L}U(t) = -\omega\left(\frac{1}{Q_0} + \frac{1}{Q_{ext}}\right)U(t), \tag{4-2}$$

where Q_L, Q_0 and Q_{ext} are loaded Q, unloaded Q and external Q. Further, by defining the coupling constant β as

$$\beta = \frac{P_{rad}}{P_c} = \frac{Q_0}{Q_{ext}}, \tag{4-3}$$

we can obtain the relation between Q_0 and Q_L as

$$Q_0 = (1+\beta)Q_L. \tag{4-4}$$

From the solution of Eq. (4-2), the decay time of the stored energy is given as

$$\tau = \frac{Q_L}{\omega}, \tag{4-5}$$

and is rewritten as

$$Q_0 = \frac{(1+\beta)\omega\tau_{1/2}}{\ln 2}, \tag{4-6}$$

using the half decay time of the energy $\tau_{1/2}$, which is easily measured from the power decay P_{rad}. To obtain the unloaded Q (Q_0), we have to know the coupling constant β.

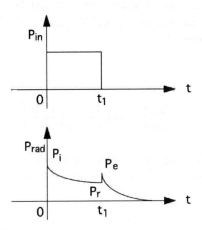

Figure 4.1: Input pulse (top) and power from the cavity (bottom).

4.2 Coupling constant β (one-port cavity)

Consider a power pulse, as shown in Fig. 4.1. As is well known, the analysis using a circuit model gives the transient solution of the cavity field, and on resonance ($\omega = \omega_0$) the power coming out from the cavity (P_{rad}) for $0 \leq t \leq t_1$ is

$$\begin{aligned}
P_{rad} &= P_{in} - (P_c + P_{field}) \\
&= P_{in} - \left(\frac{\omega_0}{Q_0} U(t) + \frac{d}{dt} U(t) \right) \\
&= \frac{1}{(1+\beta)^2} \left(1 - \beta + 2\beta e^{-\frac{\omega_0}{2Q_L}t} \right)^2 P_{in}, \tag{4-7}
\end{aligned}$$

and for $t_1 \leq t$ it is

$$P_{rad} = \frac{\omega_0}{Q_{ext}} U(t)$$

$$= \frac{4\beta^2}{(1+\beta)^2} e^{-\frac{\omega_0}{Q_L}(t-t_1)} P_{in} , \tag{4-8}$$

where P_{field} is the power spent for building up the cavity field and vanishes at the steady state of $t = t_1 \gg 2Q_L/\omega_0$. Therefore one can find four expressions for the coupling constant β using the powers P_{in}, and P_{rad} at $t = t_1 \gg 2Q_L/\omega_0$, that is, P_i, P_r, and P_{rad} just after t_1, namely P_e.

$$\beta = \frac{P_{rad}}{P_c} = \frac{P_e}{P_i - P_r} ,$$

$$\frac{P_r}{P_i} = \left(\frac{1-\beta}{1+\beta}\right)^2 , \quad \frac{P_e}{P_i} = \left(\frac{2\beta}{1+\beta}\right)^2 , \text{ and } \frac{P_r}{P_e} = \frac{1}{4}\left(\frac{1-\beta}{\beta}\right)^2 . \tag{4-9}$$

Thus, you can calculate β using one of these equations and the measured power ratio of P_i, P_r and P_e. Especially at $\beta=1$, P_r becomes zero and P_i is equal to P_e.

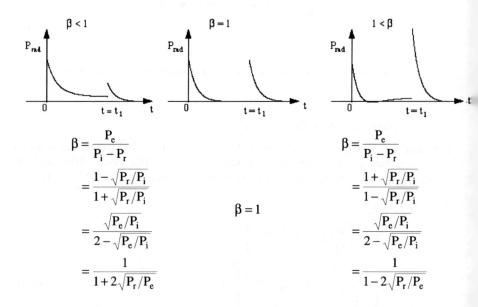

$$\beta < 1 \qquad \qquad \beta = 1 \qquad \qquad 1 < \beta$$

$$\beta = \frac{P_e}{P_i - P_r}$$

$$= \frac{1 - \sqrt{P_r/P_i}}{1 + \sqrt{P_r/P_i}}$$

$$= \frac{\sqrt{P_e/P_i}}{2 - \sqrt{P_e/P_i}}$$

$$= \frac{1}{1 + 2\sqrt{P_r/P_e}}$$

$$\beta = 1$$

$$\beta = \frac{P_e}{P_i - P_r}$$

$$= \frac{1 + \sqrt{P_r/P_i}}{1 - \sqrt{P_r/P_i}}$$

$$= \frac{\sqrt{P_e/P_i}}{2 - \sqrt{P_e/P_i}}$$

$$= \frac{1}{1 - 2\sqrt{P_r/P_e}}$$

Figure 4.2: Reflected power from the cavity for various coupling β.

4.3 Coupling constant β (two-ports-cavity)

When the cavity has another coupler of P_t, we can define another coupling constant β_t for that coupler as the ratio of P_t and P_c:

$$\beta_t = \frac{P_t}{P_c} = \frac{Q_0}{Q_t}. \tag{4-10}$$

Because of this additional power dissipation, Eq. (4-2) has to be changed to

$$\frac{dU(t)}{dt} = -\omega \frac{1}{Q_L} U(t) = -\omega \left(\frac{1}{Q_0} + \frac{1}{Q_{ext}} + \frac{1}{Q_t} \right) U(t),$$

and the unloaded Q becomes

$$Q_0 = \frac{(1 + \beta + \beta_t) \omega \tau_{1/2}}{\ln 2}. \tag{4-11}$$

4.4 Cavity voltage V_c

In Fig. 4.1, the cavity loss P_c is

$$P_c = P_i - P_r \tag{4-12}$$

Thus we can calculate the peak cavity voltage V_c using the definitions of Q and R/Q. By repeating the pulse measurements, we can obtain a series of Q values for various cavity voltages.

By the way, from Eq. (4-1) and (4-10), the stored energy is described as

$$\omega U = Q_L P_{tot} = Q_0 P_c = Q_{ext} P_{rad} = Q_t P_t. \tag{4-13}$$

On the other hand, it is clear that Q_t depends only on the geometrical size of the coupler and not on the cavity performance. Thus you never need to measure it again, you once get the Q_t at the first pulse measurements. After that, by measuring the input power, reflected power, and P_t under the steady state at each power level, you can calculate the Q_0 using Eq. (4-13), where

$$P_c = P_i - P_r - P_t. \tag{4-14}$$

Finally, it should be mentioned that all the RF powers described above are just the powers at the cavity. Thus the power measured at the directional coupler in Fig.3.6 has to be corrected using the cable attenuation between the directional coupler and the cavity. The attenuation of the cables connected to the cavity is obtained as the square root of the power ratio of forward and reflected power at the cable end using the frequency of off resonance.

5 Summary

All curriculums of the cavity experiments were smoothly carried out and we could reach our goal perfectly. Our results are summarized as follows.

- Measuring the Q increase during the cooldown to 4.2 K by using a network analyzer.
- Measuring the cable attenuation factor at 4.2 K.
- PLL using P_r and P_t; resonant frequency of 1297.92 MHz
- Measuring Q at $\beta=1$; a decay $\tau_{1/2}$ of 13 msec was observed.
- Changing the coupling β from under to over coupling and measuring the Q at $\beta=0.35\sim2.2$.
- The maximum field reached 2.5 MV/m with the cavity loss of 2.4 W.
- The Q at the maximum field was 3×10^8, which corresponds to a surface resistance of 900 nΩ. This is due to a theoretical limit of 1.3 GHz at 4.2 K.

It was very impressive that all the students always helped each other and tried to reach their goal enthusiastically, though this experiment was the first experience for most of them. Unfortunately we could not prepare a pumping unit with enough capacity to cool the cavity to 2 K. If we had had such a unit, the students would have learned much more about the surface resistance of superconducting cavities.

The success of these experiments was due in large part to the efforts and close cooperation of the students. We are deeply indebted to our friends at the Institute of Electrical Engineering for their support in supplying LHe. Finally we would like to thank all of our colleagues who gave us the opportunity for this training course.

6 Photo album

Photo 1: Dream team of SC cavity measurements.

Photo 2: Assembling the RF system.

Photo 3: Setting up the test cavity.

512

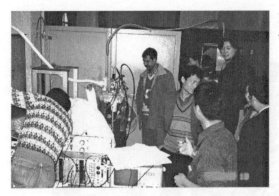

Photo 4: Cooling the cavity to 4.2 K.

(a) under coupling

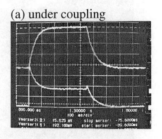

(b) β=1

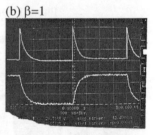

Photo.5: Q measurements and power decay at
(a) β<1, (b) β=1, and (c) β>1.

(c) over coupling

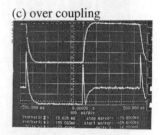

Report from group A

KURIKI Masao
KEK, Japan

BAE Young Soon
POSTECH, Korea

GARCIA Alipio
U. Phil., Philippines

GOU Weiping, GUO Xiaohong, XU Jingwei
IHEP, China

HAJRA Debaprasad
VECC, India

HU Yanle
Peking Univ., China

NAKANISHI Kota
MHI, Japan

NATSIR Muhamad
BATAN, Indonesia

SATOH Masanori
Tokyo Univ., Japan

Hands-on training on superconducting magnet technology was given at the Asian Accelerator School held in Beijing, China. Starting from the winding of superconducting wire, Group A made a superconducting magnet designed to induce 7-T magnetic field at the center of the coil. The magnet initially quenched at 6.33 T but successive tests showed progressively increasing quench fields up to 6.80 T. This magnet is very reliable and useful for practical applications.

1 Introduction

Superconducting magnets are among the most important components of high energy physics and engineering, particularly in state-of-the-art accelerator technology.

The superconducting magnet training course was a five-day hands-on program with the following objectives:

1. To provide the participants with the necessary skills for the design and fabrication of magnets.

2. To develop skills in handling various instrumentation devices in superconducting magnet testing.

The ultimate challenge of the program was the fabrication of a 7-T superconducting magnet.

We endeavored to achieve this goal through team efforts, which demanded the utmost care, coordination, and patience.

2 Making the coil

2.1 Winding

A 0.648-mm-diameter superconducting wire (Furukawa SG-99019-2) was used. The winding was done by using the winding machine specially made by Hosoyama san and his company. The tension was 4.0 kgf during the winding.

The number of turns in per layer was kept at 96 for all the layers except the outermost layer. This shows that our magnet was fabricated accurately and our winding procedure was well established. Ofcourse, the sophisticated winding machine was one of the main reason for this smooth winding.

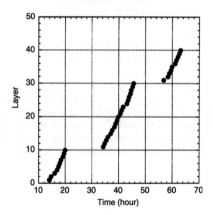

Figure 1: Winding progress

Figure 1 shows the progress of the winding. We finished the winding on the evening of the third day. The total number of turns was 3839 for 40 layers.

2.2 Specifications of our magnet

Table 1 shows the dimensions of our magnet. From these numbers, the packing factor was calculated by Eq. (1):

$$\lambda = \frac{S_C}{S_A} = \frac{\pi(a_{eff}/2)^2 N}{(R_1 - R_2)L}, \tag{1}$$

where S_C is coil area within the solenoid magnet, S_A is the cross sectional area of the solenoid magnet, and a_{eff} is the effective superconducting coil diameter under tension. The a_{eff} was found from the measured coil length, L, divided by the number of turns, i.e. $67.5/96 = 0.703$ mm. Finally, the packing factor λ was calculated to be 0.901.

Table 1: Dimensions of the magnet

Inner radius R_1 (mm)	Outer radius R_2 (mm)	Length L (mm)	Average turns per layer	Total turns N
27.0	51.5	67.5	96	3839

The magnetic field B_0 is expressed as follows:

$$B_0 = j\lambda R_1 F(\alpha, \beta)10^{-6} \text{ [T]}, \tag{2}$$

where j is the current density, and F is the fabrication factor, which is evaluated as

$$F(\alpha, \beta) = \frac{4\pi}{10}\beta \log\left[\frac{\alpha + \sqrt{\alpha^2 + \beta^2}}{1 + \sqrt{1 + \beta^2}}\right], \tag{3}$$

where $\alpha = R_2/R_1$ and $\beta = L/(2R_1)$. From the measurements, these constants were determined to be $\alpha = 1.91$ and $\beta = 1.25$, giving $F = 0.75$.

The current density required for a 7-T magnetic field is $j = 3.83 \times 10^8 \text{A/m}^2$, which corresponds to a current on the wire, $I = j \times [\pi(a_{eff}/2)^2] = 147.5$ A.

This current should be below the critical current. Normally, the magnetic field in the innermost layer, B_m, is the highest. Thus the critical current should be testedd for B_m rather than the central magnetic field, B_0.

The ratio of this highest magnetic field and the central magnetic field, $\kappa(B_m/B_0)$, is parameterized by α and β. For our coil dimension, this ratio was obtained[1] to be 1.11, i.e. the current should be less than the critical current for $7.0 \times 1.11 = 7.77$ T.

Figure 2 shows critical current as a function of magnetic field. The dashed line shows magnetic field induced by our magnet. The small circle shows our

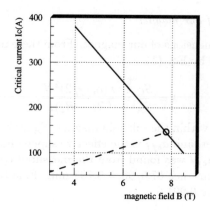

Figure 2: Critical current as a function of magnetic field. The dashed line shows the operation line of our magnet and the small circle is the target corresponding to 7.0 T at the center of the magnet.

target, corresponding to 7-T magnetic field at the center of the coil. This figure shows that our target is almost on the line of the critical current.

3 Experiments

3.1 Cooling down

After setting up the magnet in the cryostat, we poured liquid nitrogen followed by liquid helium, which took almost one and a half hours.

The impedance of the dummy protection resistor was measured to be 0.161 Ω at 4.2 K.

3.2 Decay time

The shutdown test was done to measure the decay time. The applied current was 10 A. The result is shown in Fig. 3. From this measurement, the decay time was obtained to be 0.243 sec. Assuming 0.161 Ω as the impedance of the protection resistor, the self inductance of the superconducting coil was calculated as 0.642 H.

3.3 Quench test

The quench test was done five times. The quench point on the first trial was 135.8 A, but it progressively increased, and finally, it almost saturated at

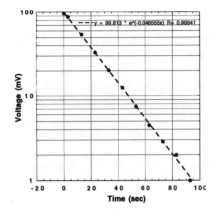

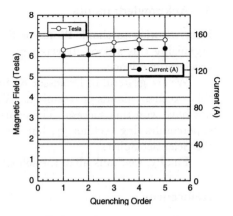

Figure 3: The result of the shutdown test. The applied current was 10A. The decay time was obtained as 0.243 sec giving 0.642 H as the self inductance of the coil

Figure 4: Coil current and Magnetic field at the quench point. The data are plotted in order of quench tests.

143.7 A. These results are summarized in Fig. 4. The highest magnetic field was 6.80 T as determined by the Hall probe.

Although the central field did not reach the design value, 7.00 T, the maximum field, $6.80 \times 1.11 = 7.55$ T was larger than that. The most important feature of our magnet was that the quench point never decreased in the iterative tests. This great performance shows that our magnet will never quench below the established quench point, and thus it is very reliable and stable. This is a very nice feature for practical applications.

4 Summary

We successfully fabricated a superconducting magnet with a packing factor of 0.901 and a maximum quench field of 6.80 T.

The magnet initially quenched at 6.33 T but successive tests showed progressively increasing quench fields indicating that the magnet can be very useful for practical applications.

This training behavior of our magnet can be attributed to the uniformity in winding at 96 turns per layer.

A higher quench field of 7.0 T could be achieved if "coil movements" could be further diminished. This could be done through fewer gaps in the winding, more uniform tension levels, and more rigid reinforcements.

5 Acknowledgments

First we wish to thank Professor S. Kurokawa and Professor Z. Chuang for organizing this school. It was very fruitful for us to study the foundation of accelerator science.

We are grateful to Professor K. Hosoyama and his company for developing this excellent hands-on training. Without their frequent advices, we would not have won this success. The course is too short to learn everything about superconducting technology, but he gave us a very very important treasure, the Scientist Spirit.

Finally, we thank all of the school participants and hope to see them again.

References

1. K.Agatsuma, *Basic Solenoid Superconducting Magnet Design*, Asian Accelerator School lecture note, 1999.

Report from Group B

Toshiya Sanami[a]

KEK, Japan : E-mail toshiya.sanami@kek.jp

Kentaro Harada

Tokyo Univ., Japan : E-mail kentaro@issp.u-tokyo.ac.jp

Kailash Ruwali

CAT, India : E-mail ruwali@cat.ernet.in

Zhao Hongwei

Chinese Academy Sci., China : E-mail zhw@ns.lzb.ac.cn

Sitaram Bhattarai

Suryodaya Secondary School, Nepal : E-mail sur@LBK.mos.com.np

Ge Jun

Qinghua Univ., China : E-mail christ@cenpok.net

Li Dazhi

IHEP, China : E-mail dazhi_li@hotmail.com

Tetsuro Kumita

Tokyo Metropolitan Univ., Japan : E-mail kumita@phys.metro-u.ac.jp

Zhang Fengxia

IHEP, China : E-mail ZhangFx@mailcity.com

We fabricated a superconducting magnet during hands-on training at the Asian Accelerator School in Beijing. In this training, we learned magnet design and winding techniques. The parameters of the magnet design were confirmed by the experiment. Our magnet successfully produced around a magnetic field of about 7 T in the excitation test up to quench, but no training effect couldn be observed.

1 Introduction

Superconducting technique is a key technology for recent high power accelerators. At the Asian Accelerator School, hands-on training was given focusing on superconducting technique. Group B fabricated a superconducting magnet to learn coil design and winding techniques. Figure 1 is a photo of the group,

---

[a] Corresponding author

with Drs. Kurokawa and Hosoyama. Group B had 9 students from Japan, China, India and Nepal. This report summarizes our work in this training.

Figure 1: Members of Group B with its leaders

2 Winding

We started winding at 13:30 on Nov. 29, and finished at 14:30 on Dec. 1. During the winding process, we paid attention to keeping constant tension on the wire (4.0 kgf/cm^2). Every time we finished winding a layer, we filled the gaps at the ends of the layer with epoxy, checked the insulation between the wire and the coil bobbin, and measured the coil diameter. The average time to wind one layer was less than 20 minutes. The total number of layers was 40 (3837 turns). Table 1 shows the person winding each layer; each one wound 4 or 5 layers. To protect the surface of the superconducting wire, we wound a final two layers consisting of stainless steal wire.

3 Dimensions of our magnet

Figure 2 shows our magnet after the winding process was complete. We measured the magnet parameters as follows:

- Inner radius : $a_1 = 27$ mm

- Outer radius : $a_2 = 51.085$ mm

Table 1: Person winding each layer

Ruwali	#1	#10	#11	#18	
Zhang	#2	#9	#20	#30	#34
Harada	#3	#12	#23	#38	
Bhattarai	#4	#14	#22	#29	#39
Ge	#5	#15	#27	#33	
Kumita	#6	#16	#24	#35	
Li	#7	#17	#25	#32	#40
Sanami	#8	#19	#26	#37	
Zhao	#13	#21	#28	#31	#36

- Length : $2b$=67.14 mm

The definition of these parameters are shown in Fig. 3.

From these parameters, the Fabry factor ($F(\alpha,\beta)$) and the packing factor (λ) were calculated to be

$$F(\alpha,\beta) = \frac{4\pi}{10}\beta log_e \frac{\alpha + \sqrt{\alpha^2 + \beta^2}}{1 + \sqrt{1 + \beta^2}} = 0.736, \tag{1}$$

$$\lambda = \frac{3837 \times 4\pi/d}{2b(a_2 - a_1)} = 0.913 \quad (91.3\%). \tag{2}$$

The ratio κ (B_m/B_0) was obtained from the plot of the relationship between κ and α, β.

- $\kappa = (B_m)/(B_0) = 1.08$

Using these parameters, the central and maximum magnetic fields were

$$B_0[T] = j\lambda a_1 F(\alpha,\beta) \times 10^{-6} = 0.0471 \times I[A], \tag{3}$$

$$B_m[T] = B_0 \times \kappa = 0.0509 \times I[A]. \tag{4}$$

For B_0=7 [T], we need B_m=7.56 [T]. From the plot of critical current vs magnetic field, the critical current for the magnet was predicated to be about 150 A.

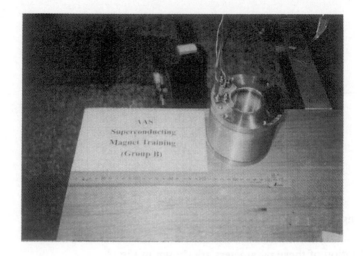

Figure 2: Our coil

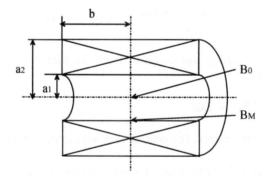

Figure 3: Cross section of solenoid superconducting magnet

4 Measurement

4.1 Setting

After the winding was finished, the magnet was mounted on a test bench and connected to sensors for monitoring magnetic field [Hall probe], temperature [TC and carbon resistor] and Li-He level [level meter]. Figure 4 shows the test bench with our magnet.

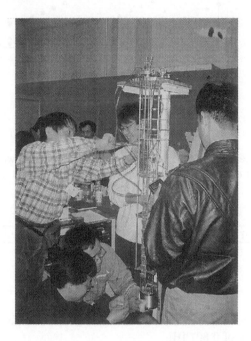

Figure 4: The magnet mounted on the test bench

The test bench was installed in a cryostat that had a high vacuum thermal insulation space and a liquid nitrogen shield. We first filled the shield with liquid nitrogen and then poured liquid helium into the inner vessel with monitoring of its level.

4.2 Inductance

After sufficient cooling time, we started decay time measurements with currents of 2, 5 and 10 A. Figure 5 shows the results of the measurements.

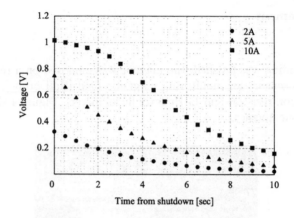

Figure 5: Results of inductance measurement

By fitting the curves in Figure 5, a decay constant was obtained. The inductance of the coil could also calculated; Table 2 shows results. inductance.

Table 2: Results of curve fitting

current [A]	time const. [1/s]	inductance [H]
2	0.265	0.574
5	0.247	0.616
10	0.242	0.629

Averaging these three results, we obtained a time constant of 0.2513 [sec^{-1}], and an inductance of 0.607 [H].

4.3 Quench test

Before the quench test, we checked the whole system at 75 A. In the quench test (Figure 6), the current applied to the magnet was gradually increased at a rate of 150 A/500 sec up to the critical current causing quench. The current and the central magnetic field were measured when quench occurred. We did three quench tests to check for the training effect; Table 3 shows the results. The central magnetic field measured using the Hall element is in good agreement with that estimated from the current: within 1%. But no training effect appeared.

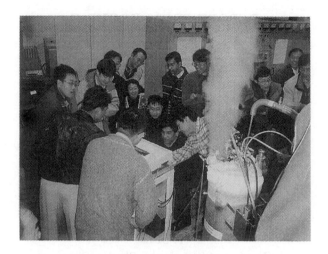

Figure 6: Quench test

Table 3: Results of quench test

Quench test (Run No.)	Hall element [mV]	Magnetic field (measured) [T]	Current [A]	Magnetic field (calculated) [T]
1	50	6.925	146	6.877
2	46	6.371	135	6.359
3	45.5	6.302	132	6.217

5 Conclusion

During this hands-on training, everyone in Group B mastered the technique of winding. The quench measurements were exciting because we were the first group to do the test. Unfortunately, no training effect could be seen. We expect the reasons may be slight movement of the wire or loose connection of the dump resister, but only God knows the real reason. The most important result is that we made eight good friends and spent a very pleasant time.

6 Acknowledgment

The authors wish to thank the organizers of this school on the staffs of JSPS, CAS, KEK and IHEP.

REPORT FROM GROUP C

TIEJUN MENG

Beijing Univ., China

GANLIN NI

Institute of High Energy Physics, China

MD. MASHIUR RAHMAN

Bangladesh Atomic Energy Comission, Bangladesh

HARUE SUGIYAMA

Nagoya Univ., Japan

YI SUN

Institure of High Energy Physics, China

YUKO TAKASU

Univ. of Tokyo, Japan

MASAFUMI TAWADA

High Enegy Accelerator Research Organization, Japan

DINH PHU TRAN

Institute of Physics, Vietnam

SU SU WIN

National Synchrotron Research Center, Thailand

ZIQANG ZENG

China Institite of Atomic Enegy, China

ANQI ZHOU

Univ. of Science and Technology of China

The Asian Accelerator School (AAS) was held at Huairou and IHEP in Beijing, China, from November 22 to December 4, 1999. The second week was devoted to the hands-on training about the superconducting magnets. Our group fabricated a superconducting solenoid magnet, and cooled it down, and measured the magnetic field strength. We succeeded in obtaining a high quality solenoid magnet with a central field of 7.17 [T] at a quench current of 149.5 [A].

1 Introduction

Superconducting technology is one of the most important technologies used to achieve advanced accelerators and detectors for high energy physics.

The Asian Accelerator School was held at Huairou and IHEP in Beijing, China, from November 22 to December 4, 1999. The first week was devoted to lectures on accelerator baisics and the following week to hands-on training in superconducting technologies: superconducting cavities and superconducting magnets. The purpose of this hands-on training was to provide experience in superconducting technology to the students who were not familiar with it.

In the hands-on training course, the group spent three days learning the design of superconducting solenoid magnets and fabricating one, and the following two days setting up experimental equipment, cooling the magnet down and doing the excitation test.

2 Design and fabrication

The specifications of a solenoid magnet are as follows.

1. Central magnetic field is 7 [T].
2. Inside winding diameter (a_1) is 54 [mm].

The design proceedure for a simple air-core solenoid magnet with rectangular cross section (Fig. 1) was described by Agatsuma.[1] In this training course, we used a NbTi fine multi-filamentary wire of 0.7 [mm] diameter. The critical current of this wire is 100 [A] at 8.5 [T] and 383 [A] at 4 [T] in liquid helium. During winding the coil, the tension was controlled at about 4 [kgf]. It took three days to wind the coil, which was not resined.

Winding dimensions for the magnet are as follows. Inner winding radius (a_1): 27.1 [mm], outer winding radius (a_2): 51.27 [mm], winding length ($2b$): 67.5 [mm], average number of turns in a layer: 96, total number of turns: 3838.

From basic design theory,[1] the axial magnetic field at the center of the coil B_0 [T] can be written as follows:

$$B_0 = j\lambda F(\alpha, \beta) \times 10^{-6}. \tag{1}$$

where j is the current density, $\alpha = a_2/a_1$, $\beta = b/a_1$, λ is the packing factor, and the Fabry factor $F(\alpha, \beta)$ is written as follows:

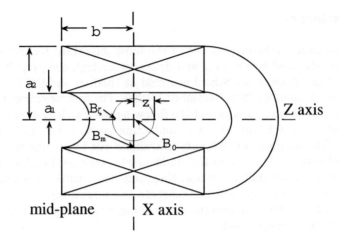

Figure 1. Cross section of air-core solenoid magnet. $\alpha = a_2/a_1$, $\beta = b/a_1$.

$$F(\alpha,\beta) = \frac{4\pi}{10}\beta\ln\frac{\alpha + \sqrt{\alpha^2 + \beta^2}}{1 + \sqrt{1 + \beta^2}}. \qquad (2)$$

On the other hand, the maximum magnetic field B_m is experienced at the center of the inner winding layer, and the ratio $\kappa = B_m/B_0$ is a function of α and β,as shown by Agatsuma.[1]

For our magnet, $\alpha = 1.89$, $\beta = 1.25$, $\lambda = 0.90$, $F(\alpha,\beta) = 0.74$ and $\kappa = 1.1$. Therefore, $B_m = 7.7$ [T] and $B_0 = 7.0$ [T] would be expected at the critical current of 150 [A].

Magnetic fields were calculated by using the program coded by Prof. K. Hosoyama on Mathematica.[7]Mathematica The inductance of this magnet was calculated to be 0.66 [H].

3 Field measurements

The magnet was cooled down to liguid helium temperature and its temperature was monitored by a carbon resistor attached to its surface. A dummy resistor for protection was connected in parallel with the magnet. The impedance of the dummy resistor was measured to be 0.156 [Ω] by Group A.[3] The central magnetic field was measured by a Hall sensor.

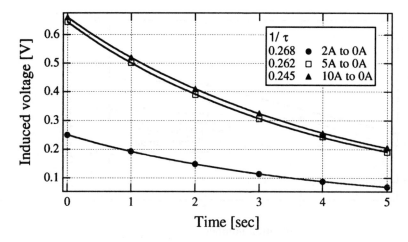

Figure 2. Decay constants measured by switching off the power supply. We made measurements it at currents of 2, 5 and 10 [A].

The decay constant can be measured by switching off the power supply at a small current. Figure 2 shows the voltage of the dump resistor vs time. The decay constant was measured as 0.268 at 2 [A], 0.262 at 5 [A] and 0.245 at 10 [A]. Thus, the inductance for these currents was 0.58, 0.60 and 0.64 [H], respectively. This is in good agreement with simulation results.

Figure 3 shows the quench training history. The quench current was gradually increased and saturated to the quench current of 149.5 [A]. The maximum central field B_0 was 7.17 [T] at 149.5 [A].

4 Conclusion

In this hands-on training, we constructed a superconducting solenoid magnet. The measured inductance was in good agreement with the simulation result. During the quench training, the quench current was gradually increased. Finally, we suceeded in obtaining a high quality magnet, with a central magnetic field of 7.17 [T] at the quench current of 149.5 [A]. And we also succeeded in international collaboration through the hands-on training at AAS.

530

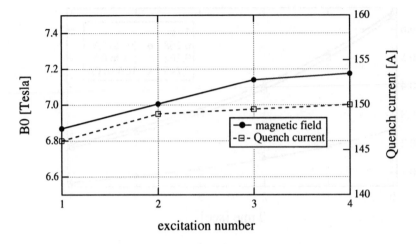

Figure 3. Quench training history.

Acknowledgments

The authors would like to thank Prof. S. Kurokawa for giving us a chance to join this school and Profs. K. Hosoyama, K. Agatsuma, H. Sato, Y. Hao, S. Han, and Drs. A. Kabe, Y. Kojima and K. Hara for giving a good hands-on training and for their encouragement.

References

1. K. Agatsuma, Basic Solenoid Superconducting Magnet Design, in these proceedings.
2. Mathematica, Wolfram Research, Inc.
3. Report from Group A, in these proceedings.

REPORT FROM GROUP D

J. CHENG, Q. PENG, W. CHEN, Y. YINGJIE

IHEP, China

K. ZHAO

Beijing University, China

K. ZHANG

IFPCAEP, China

A. PUNTAMBEKAR

CAT, India

L. BOONANAN

NSRC, Thailand

K. GOTO

Hiroshima University, Japan

M. MASUZAWA

KEK, Japan

A Superconducting magnet with a 40-layer (~4000-turn) solenoid winding configuration was fabricated over the course of three days at IHEP during the Asian Accelerator School. The magnet was excited up to the quench level four times, and a central magnetic field of ~7 Tesla was achieved in the end. Ten people from China, India, Thailand and Japan participated in the fabrication and measurements. This report summarizes the fabrication process and the magnet measurement results.

1 Coil Winding

A superconducting wire was used for fabricating the 40-layer solenoid coil. In order to prevent the wire from moving upon coil excitation and causing a quench, the following measures were adopted:

• Keep the tension on the wire constant (~4 kgf) throughout the winding process;

• Wind the wire neat and tight, though not too tightly as the wire may stretch;

• Pay special attention to the transition between two layers, filling the void with epoxy putty;

• Permit no kink in the wire.

The wire was wound evenly in each layer, 96 turns per layer (expect for one layer which had 95 turns). Figure 1(a) shows the number of turns as a function of layer number. It took three days to wind 3839 turns. Our progress can be seen in Figure 1(b), where the average time spent finishing one layer is plotted. The time includes the winding and the filling of the void. The treatment of the void was usually more time consuming than the winding itself.

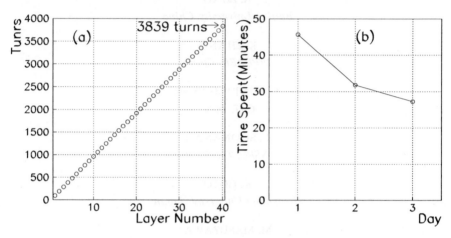

Figure 1: (a) Number of turns vs layer number and (b) average time consumed per layer.

The diameter of the coil was measured after completing one layer in order to make sure that the winding was done uniformly. The measuring points are indicated in Figure 2.

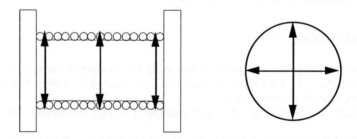

Figure 2: Measuring points. Both horizontal and vertical diameters were measured at three locations along the coil.

The thickness of each layer was obtained from the diameter measurements, and the results are plotted in Figure 3. The average thickness is 0.62 mm with a

standard deviation of 0.10 mm.   According to the specifications of the wire, the
diameter is 0.648 mm.   The thickness of the layers is consistent with the case of
close packing of 0.648 ϕ wire.

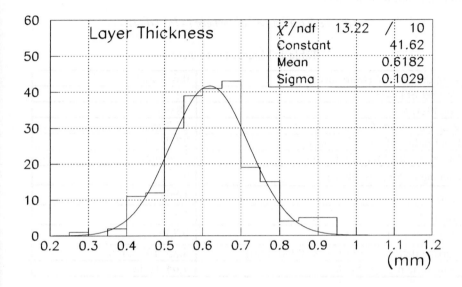

Figure 3: Layer thickness.

2 Magnet Parameters

Table 1 summarizes the parameters of the magnet. The central field is calculated from the following formula:

$$B_0 = j \lambda a_1 F(\alpha, \beta) \times 10^{-6}, \text{ where } j = I/(\pi d^2/4).$$

Table 1: Magnet Parameters.

Wire diameter d (from spec)	0.648 mm
Number of layers	40
Total turns	3839
Winding length ($=b$)	67.3 mm
Inner radius of the coil ($=a_1$)	27.0 mm
Outer radius of the coil ($=a_2$)	51.4 mm
Packing factor λ (with d)	0.771
$a_2/a_1 (= \alpha)$	1.904
$b/a_1 (= \beta)$	1.246
*Fabry factor F(α, β)	0.73
Current density j@150 A (d=0.648 mm)	4.55 10^8 A/mm^2
Central field B_0 @150 A	7.05 Tesla

*Fabry factor F(α, β) was estimated from a reference plot (not shown in this report), using α and β.

3 Measurements

3.1 Shutdown Test

The inductance of the coil was calculated by observing the decay of the current. It is modeled by the circuit shown in Figure 4. R and L represent the resistance and the inductance, respectively, of the coil.

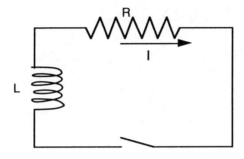

Figure 4: The RL circuit.

The current in the circuit is obtained by solving the following equation:

$$L \, dI/dt + RI = 0, \quad I = I_0 e^{-\alpha t}, \qquad \text{where } \alpha = R/L.$$

We applied a DC current I_0 to the coil and then turned it off (shutdown test). By recording the current as a function of time, the decay constant α in the equation above is obtained. Figure 5 shows the current decay for $I_0 = 2$, 5, and 10 A. The measurements of the decay constant α obtained by fitting the data are summarized in Table 2. Using $R = 0.165 \ \Omega$ (measured by Group C), the inductance is calculated to be 0.682 H, which is ~6.5 % larger than the design value.

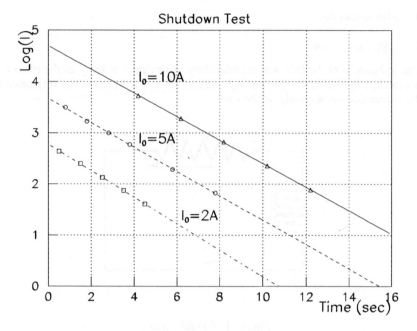

Figure 5: Current (log scale, arbitrary units) vs time for various initial currents.

Table 2: The decay constant.

I_0	2 A	5 A	10 A	Average
α (/sec)	0.259	0.238	0.230	0.242

3.2 Quench Test

The magnet was excited gradually until it quenched. The first quench took place when the current on the magnet reached 147 A. The ramp-up speed was kept constant at 150 A/500 seconds. The magnetic field at the center was measured to be 6.92 Tesla. We reached 149 A with the second trial. The central magnetic field exceeded 7 Tesla on the fourth trial. Figure 6 shows the training curve of the magnet.

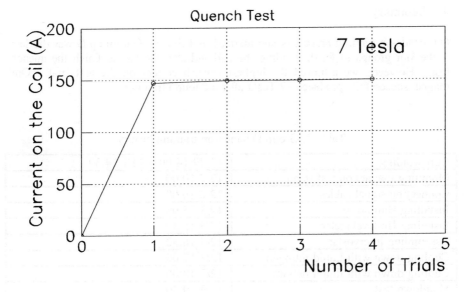

Figure 6: The training curve.

4   Summary

Our hands-on training process is summarized in Table 3. Our group D was the last of the four groups to finish winding the coil and also the last to finish the quench test. However, we achieved the highest quench current on the first trial. Our magnet successfully produced ~7 Tesla after the fourth training.

Table 3: Group D hands-on training history.

Coil winding	11/29 14:00 ~ 12/1 14:30
Mounting magnet on test stand	12/2 10:00
Connecting signal cables	12/2 16:00
Installing Hall probe	12/3 15:00
Installing He-level meter	12/3 15:30
Assembling in cryostat	12/3 16:30
Cooling down with liquid nitrogen	12/3 17:00
Cooling down with liquid helium	12/3 18:00
Shutdown test	12/3 19:30
System check with 75 A	12/3 20:30
Quench test	12/3 21:00 ~ 22:30

5   Acknowledgments

All the members of our group would like to express our thanks to the lecturers and to the staff members who helped us during the fabrication and measurements. It was a valuable experience for us to actually fabricate a superconducting magnet and test it.

REPORT FROM GROUP E

S. YOSHIMOTO AND T. TAKAHASHI

KEK, High Energy Accelerator Research Organization,
1-1 Oho, Tsukuba-shi, Ibaraki-ken, 305-0801, Japan

T. DATTA

Nuclear Science Center

J. HAO

Beijing University

H. KIM

Seoul National University

Z. LI, Y. QU, H. SUN, Y. WANG AND S. ZHAO

Institute of High Energy Physics

H. WEN

Institute of Electrical Engineering

W. ZHOU

Shanghai Synchrotron Radiation Center

This paper describes low power measurements of the L-band superconducting RF cavity. We obtained the unloaded Q value as 3.1×10^8 and the accelerating field as $E_{acc} = 2.1$ MV/m.

1 Introduction

For very high Q cavities such as superconducting cavities, it is difficult to measure the Q value. It that case we can make use of the long decay time for measurement of both β and Q. By observing the emitted power after the drive has been shut off, we can obtain the the loaded Q factor of the cavity by measuring the decay time.

2 Principles of the measurements

2.1 Q and β determination

After the drive power has been shut off, the stored energy in the cavity decreases exponentially. The decaying energy is given by

$$\frac{dU(t)}{dt} = -\frac{1}{\tau}U(t) = -(P_c + P_r + P_t) \tag{1}$$

where τ is the decay time of the cavity, P_c is the power dissipation in the cavity, and P_r and P_t are emitted power though the coupler ports. The relationships of the unloaded quality factor, Q_0, the external quality factor of the input coupler, Q_i, and the external quality factor of the transmitting coupler, Q_t, are as follows:

$$\omega U = P_c Q_0 = P_r Q_i = P_t Q_t \,. \tag{2}$$

The loaded Q factor, Q_L, is related by

$$\frac{1}{Q_L} = \frac{1}{Q_0} + \frac{1}{Q_i} + \frac{1}{Q_t} \,. \tag{3}$$

Using Eqs. (1), (2), and (3), the loaded Q becomes

$$Q_L = \omega\tau \,. \tag{4}$$

The emitted power from each coupler is proportional to $e^{-\omega t/Q_L}$. Then Q_L can be written as

$$Q_L = \omega\tau_{1/2}/\ln 2 \tag{5}$$

where $\tau_{1/2}$ is the half-time of the cavity energy. The coupling constant of the input and transmitting coupler is

$$\beta_i = \frac{Q_0}{Q_i} = \frac{P_i}{P_c} \quad , \quad \beta_t = \frac{Q_0}{Q_t} = \frac{P_t}{P_c} \,. \tag{6}$$

Then we get the unloaded Q from the loaded Q as follows:

$$Q_0 = (1 + \beta_i + \beta_t)Q_L \,. \tag{7}$$

By measuring τ, P_i, and P_t and P_r, we can determine Q_L, β, and Q_0 respectively.

2.2 Accelerating field of the cavity

Shunt impedance is defined by

$$R_s \equiv \frac{V_c^2}{P_c} = \frac{(E_{acc}L_{eff})^2}{P_c} \tag{8}$$

where V_c is the cavity voltage, E_{acc} is the accelerating field, and L_{eff} is the effective length of the cavity.

$$E_{acc} = \frac{1}{L_{eff}} \sqrt{R_s P_c} \tag{9}$$

$$= \frac{1}{L_{eff}} \sqrt{(R_s/Q_0)Q_0 P_c} \tag{10}$$

$$= A\sqrt{Q_0 P_c}. \tag{11}$$

R_s/Q is independent of the cavity size and material. Then the parameter A can be calculated by computer codes and we can also obtain the accelerating field by using measured value of Q_0 and P_c.

3 Calculation of cavity parameters

Fig. 1 shows a photograph and a cross-sectional view of the superconducting cavity used in this experiment. In order to calculate a cavity shape and obtain the parameters for the measurements, we used the computer code SUPERFISH. Fig. 2 shows the electrical field lines and the electrical field distribution along the z-axis. Table 1 shows the input parameters and the results of the calculation.

Table 1. The input parameters and the results of the calculation.

Resonant frequency	f_0	1296.93	MHz
Particle rest mass energy	E_0	0.511	MeV
Coupling factor	β	1.0	
Stored energy	U_0	0.0258	J
Superconductor surface resistance	R_{sur}	664.53	$n\Omega$
Operating temperature	T_0	4.2	K
Power dissipation	P_0	5.192	mW
Unloaded quality factor	Q_0	4.1×10^8	
Shut impedance	R_s	5.39×10^5	$M\Omega$
	R/Q	122.13	Ω

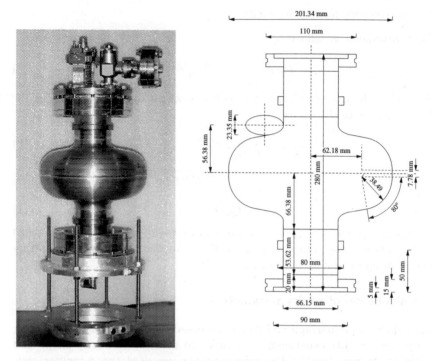

Figure 1. Shape and the cross-sectional view of the cavity.

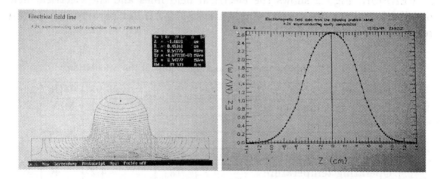

Figure 2. Electrical field lines and electrical field distribution along the z-axis of the L-band superconducting cavity.

4 Experimental Setup

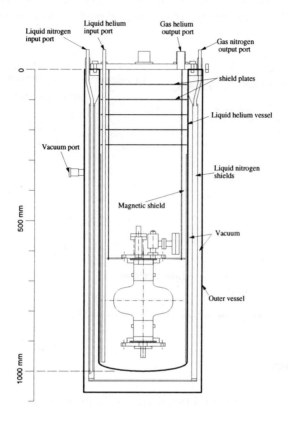

Figure 3. Cross-sectional view of the cryostat.

4.1 Cryogenics

Fig. 3 shows a cross-sectional view of the cryostat for this experiment. Liquid helium is usually stored in metal vacuum-insulated vessels, as shown here.

4.2 RF system

Fig. 4 shows a block diagram of the RF system used in this measurement. The standard signal generated at the signal generator (SG) is amplified and fed to the cavity. The phase deviation between the input signals and the cavity pickup signals is detected by the DBM (double-balanced mixer). The output signal of the DBM is fed back to the DC-FM port of the SG. The phase shift is set to control the frequency at SG, which should be the resonance frequency of the cavity.

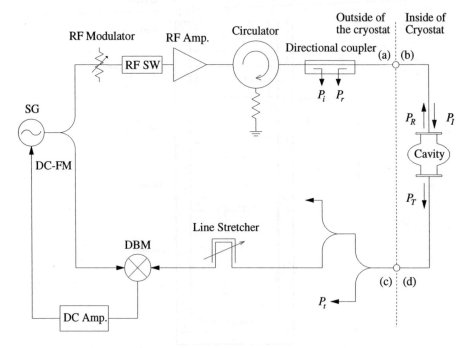

Figure 4. Block diagram of the RF system.

5 Measurement procedure

The measurement procedure is as follows:

1. precooling with LN$_2$

2. cooling with LHe to 4.2 K

3. cable correction

4. searching for the resonant frequency and turning on the PLL

5. measuring the decay constant of transmitted power with CW mode

6. measuring the input power P_i, the reflected power P_r, and the transmitted power P_t

5.1 Cooling the cavity

Fig. 5 shows the LHe level and temperature as a function of time.

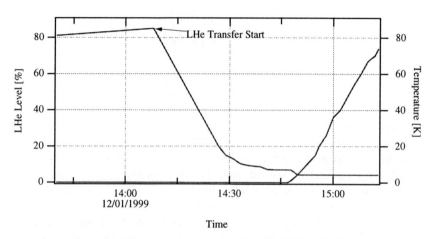

Figure 5. LHe level and temperature as a function of time.

5.2 Cable correction

We need to know the cable correction factors, which include cable loss, coupling of the directional coupler, loss by insertion devices, and so on, because we need to base our calculation on the measured powers. The correction factors are defined as

$$P_I = C_i P_i, \quad P_R = C_r P_r, \quad C_r P_T = C_t P_t \tag{12}$$

where P_I is the input power of the cavity, P_i is the power from the forward port of the directional coupler, P_R is the reflected power of the cavity, P_r is the power from the reflection port of the directional coupler, P_T is the transmitted power of the cavity, and P_t is the power from the transmitting port. The procedure for determining these factors is shown below.

1. Calibration of the directional coupler with the input cable
First, measure the power, P_1, at point (a). Second, terminate point (a) using the matched load and measure the power, P_2, at the forward port

of the directional coupler. Third, set the short terminal at point (a) and measure the power, P_3, at the reflection port of the directional coupler. Finally, connect (a) and (b), and measure the power, P_4, at the reflection port of the directional coupler. Then the correction factors C_i and C_r can be written as follows:

$$C_i = \frac{P_1}{P_2}\sqrt{\frac{P_3}{P_4}}, \tag{13}$$

$$C_r = \frac{P_1}{P_3}\bigg/\sqrt{\frac{P_3}{P_4}}. \tag{14}$$

2. Calibration of the transmitting cable:

First, the RF power is fed to point (d) with the circulator. Set the short terminal at port #2 of the circulator and measure the power, P_5, at port #3 of the circulator. Second, feed the RF power to port (d) and measure the power, P_6, at port #3 of the circulator. Third, set the matched load at port #3 of the circulator and measure the power, P_7, at port # 2 of the circulator. Finally, feed the RF power to point (c), and measure the power, P_8, at the transmitting monitor port. Then the correction factor C_t can be written as follows:

$$C_t = \sqrt{\frac{P_5}{P_6}\frac{P_7}{P_8}}. \tag{15}$$

5.3 Searching for the resonant frequency

We search for the resonant frequency by sweeping the frequency of the signal generator. It is very difficult to find the frequency because the Q value of the superconducting cavity is very high at LHe temperature. When the resonant frequency is found, we need to turn on the PLL to keep the measurement stable.

5.4 Measuring the decay constant

The rectangular pulse power is fed to the cavity using the RF switch and pulse generator. The decay constant is measured by observing the time response of the transmitted the power, which is monitored by an oscilloscope. Fig. 6 shows the decay curve of the transmitted power.

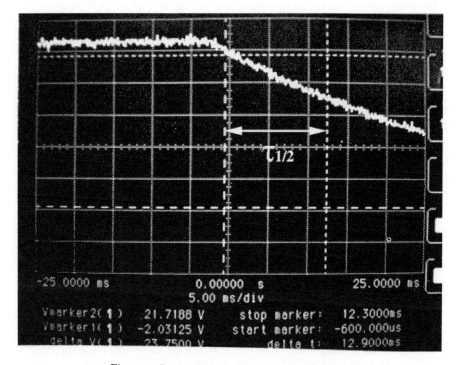

Figure 6. Decay curve of the transmitted power.

5.5 Measuring the input power P_i, the reflected power P_r, and the transmitted power P_t

Set the input power for CW operation and adjust the antenna length of the input coupler to the coupling $\beta = 1$. Then the reflected power has its minimum value. Measure the input power P_i, the reflected power P_r, and the transmitted power P_t using the power meter.

6 Results of measurement

Table 2 shows the results of this measurement. The calculated cavity parameters are shown in Table 3.

Table 2. The results of the measurement.

Resonant frequency	f_0	1297.91160	MHz
Decay constant	$\tau_{1/2}$	12.9	ms
Coupling factor	β	1.0	
Input power	P_i	24.2	mW
Reflected power	P_r	5.54	μW
Transmitted power	P_t	4.17	mW

Table 3. The results of calculations.

Input power	P_I	1.57	W
Reflected power	P_R	0.8	mW
Transmitted power	P_T	21.0	mW
Cavity dissipation	P_c	1.55	W
Loaded Q	Q_L	1.529×10^8	
unloaded Q	Q_0	3.1×10^8	
Stored energy	U_0	0.0588	J
Accelerating field	E_{acc}	2.10	MV/m
Cavity voltage	V_c	0.242	MV
	R/Q	144.3	Ω

Acknowledgments

We would like to thank Professor S. Kurokawa for giving us the opportunity to attend this school. Our thanks are also due to the ASS staff for good support.

PROGRAM FOR THE ASIAN ACCELERATOR SCHOOL AT IHEP

29 November - 4 December, 1999.

Time	Sunday 28 Nov.	Monday 29 Nov.	Tuesday 30 Nov.	Wednesday 1 Dec.	Thursday 2 Dec.	Friday 3 Dec.	Saturday 4 Dec.
9:00	Excursion to Forbidden City & Summer Palace	Meeting at the lecture hall building					Presentation of the Results
9:30		Hands-on training of superconducting technology					
10:00							
10:30							
11:00							
11:30							Closing
12:00		***L u n c h***					
12:30							
13:00			Student Presentation				
13:30							
14:00		Hands-on training of superconducting technology					
14:30							
15:00							
15:30							
16:00							
16:30							
17:00							
17:30							Farewell Party
18:00							
18:30							
19:00							

PROGRAM FOR THE ASIAN ACCELERATOR SCHOOL

22 - 27 November, 1999

Time	Monday 22 Nov.	Tuesday 23 Nov.	Wednesday 24 Nov.	Thursday 25 Nov.	Friday 26 Nov.	Saturday 27 Nov.
8:30	Registration	Fundamental of Electron	Accelerator for Synchrotron Light Source I	Beam Instability I	Beam Instability II	Beam Instability III
9:00	Opening	Accelerators IV Guo Zhiyuan	Zhao Zhentang	Alex Chao	Alex Chao	Alex Chao
9:30	Photo Session	Injectors	Accelerator for Synchrotron	RF System I	RF System II	RF System III
10:00	Fundamental of Electron	Wang Shuhong	Light Source II Zhao Zhentang	Kazunori Akai	Kazunori Akai	Kazunori Akai
10:30	Accelerators I Qin Qing	Injection	Fundamental of Superconductivity III	Vacuum System I	Cryogenic System I	Superconducting Magnet III
11:00	Fundamental of Electron	Luo Xiaoan	Maury Tigner	Masanori Kobayashi	Kenji Hosoyama	Martin Wilson
11:30	Accelerators II Qin Qing	Electron Positron Collider I	Fundamental of Superconductivity IV	Vacuum System II	Cryogenic System II	Superconducting Cavity III
12:00	Fundamental of Electron	Zhang Chuang	Maury Tigner	Masanori Kobayashi	Kenji Hosoyama	Takaaki Furuya
12:30	Accelerators III Guo Zhiyuan				Brief Introduction on Hands-on Training	Closing
13:00			*** Lunch ***			
13:30	*** Lunch ***					
14:00		Fundamental of Superconductivity I		Superconducting Magnet I	Superconducting Magnet II	
14:30	Beam Instrumentation &	Maury Tigner		Martin Wilson	Martin Wilson	
15:00	Feedback I John Fox	Fundamental of Superconductivity II	Excursion	Superconducting Cavity I	Superconducting Cavity II	Move
15:30	Beam Instrumentation &	Maury Tigner	to	Takaaki Furuya	Takaaki Furuya	to
16:00	Feedback II John Fox	Electron Positron Collider II	Great Wall		**Seminar** History of	IHEP
16:30	**Seminar** History of	Zhang Chuang		Student Presentation	Chinese Accelerator Science Development Chen Jiaer	
17:00	High Energy Accelerators	Student Presentation				
17:30	in Japan Ken Kikuchi					
18:00	Reception		*** Dinner ***			
18:30						

List of Participants

Name	Institution	Country
Agatsuma, Koh	Electrotechnical Laboratory	Japan
Akai, Kazunori	KEK	Japan
Bae, Young-soon	POSTECH	Korea
Bhattarai, Sitaram	Suryodaya Secondary School	Nepal
Boonanan, Lek	National Synchrotron Research Center	Thailand
Chao, Alexander W.	SLAC	U.S.A.
Chen, Jiaer	Peking Univ.	China
Chen, Wan	IHEP	China
Cheng, Jian	IHEP	China
Datta, Tripti Sekhar	Nuclear Science Center	India
Fox, John D.	SLAC	U.S.A.
Furuya, Takaaki	KEK	Japan
Garcia, Alipio T.	Univ. of the Philippines	Philippines
Ge, Jun	Tsinghua Univ.	China
Goto, Kiminori	Hiroshima Univ.	Japan
Gou, Weiping	IHEP	China
Guo, Xiaohong	Inst. of Modern Physics	China
Guo, Zhiyuan	IHEP	China
Hajra, Debaprasad	Variable Energy Cyclotron Center	India
Han, Shiwen	IHEP	China
Han, Yi	IHEP	China
Hao, Jiankui	Peking Univ.	China
Hao, Yaodou	KEK / IHEP	China
Hara, Kazufumi	KEK	Japan
Harada, Kentaro	Univ. of Tokyo	Japan
Hosoyama, Kenji	KEK	Japan
Hu, Yanle	Peking Univ.	China
Kabe, Atsushi	KEK	Japan
Kikuchi, Ken	JSPS	Japan
Kim, Han-sung	Seoul National Univ.	Korea
Kobayashi, Masanori	KEK	Japan
Koikegami, Hajime	Ishikawajima-Harima Heavy Industries Co.Ltd	Japan
Kojima, Yuuji	KEK	Japan
Kumita, Tetsuro	Tokyo Metropolitan Univ.	Japan
Kuriki, Masao	KEK	Japan

Li, Dazhi	IHEP	China
Li, Shaopeng	IHEP	China
Li, Yingjie	IHEP	China
Li, Zhongquan	IHEP	China
Luo, Xiaoan	IHEP	China
Masuzawa, Mika	KEK	Japan
Meng, Tiejun	Peking Univ.	China
Morita, Yoshiyuki	KEK	Japan
Nakai, Hirotaka	KEK	Japan
Nakanishi, Kota	Mitsubishi Heavy Industries, Ltd.	Japan
Natsir, Muhamad	National Nuclear Energy Agency	Indonesia
Ni, Ganlin	IHEP	China
Peng, Quanling	IHEP	China
Puntambekar, Avinash M.	Center for Advanced Technology	India
Qin, Qing	IHEP	China
Qu, Ying	IHEP	China
Rahman, Md. Mashiur	Bangladesh Atomic Energy Commission	Bangladesh
Ruwali, Kailash	Center for Advanced Technology	India
Sanami, Toshiya	KEK	Japan
Sato, Akio	National Research Inst. for Metals	Japan
Satoh, Masanori	Univ. of Tokyo	Japan
Sugiyama, Harue	Nagoya Univ.	Japan
Sun, Hong	IHEP	China
Sun, Yi	IHEP	China
Takahashi, Takeshi	KEK	Japan
Takasu, Yuko	Univ. of Tokyo	Japan
Tang, Yuxing	Peking Univ.	China
Tawada, Masafumi	KEK	Japan
Tigner, Maury	Cornell Univ.	U.S.A.
Tran, Dinh Phu	Inst. of Physics	Vietnam
Wang, Lanfa	IHEP	China
Wang, Shuhong	KEK / IHEP	China
Wang, Yanshan	IHEP	China
Wen, Huaming	Inst. of Electrical Engineering	China
Wilson, Martin N.	Oxford Instrument	England
Win, Su S.	National Synchrotron Research Center	Thailand

Xu, Jingwei	IHEP	China
Yang Wenli	IHEP	China
Yin, Lixin	SSRC	China
Yoshimoto, Shin-ichi	KEK	Japan
Zeng, Ziqiang	Atomic Energy Research Academy	China
Zhang, Chuang	IHEP	China
Zhang, Fengxia	IHEP	China
Zhang, Kaizhi	Academy of Engineering Physics	China
Zhao, Gang	IHEP	China
Zhao, Hongwei	Inst. of Modern Physics	China
Zhao, Kun	Peking Univ.	China
Zhao, Shengchu	IHEP	China
Zhao, Zhentang	SSRC	China
Zhou, Anqi	Univ. of Science and Technology of China	China
Zhou, Qiaogen	SSRC	China
Zhou, Weimin	SSRC	China

Organizers

Hayashi, Yoko	KEK	Japan
Hou, Rucheng	IHEP	China
Ikeda, Takashi	KEK	Japan
Kurokawa, Shin-ichi	KEK	Japan
Liu, Jie	IHEP	China
Motohashi, Setsuko	KEK	Japan
Nagashio, Yoshiko	KEK	Japan
Oba, Takashi	KEK	Japan
Shi, Jingyan	IHEP	China
Shibahara, Terunori	KEK	Japan
Zhang, Jinjun	IHEP	China
Zhou, Xiaobing	IHEP	China